PRAISE FOR *HUMANITARIAN LOGISTICS*

'Humanitarian logist... ...it has evolved into a mature discipline. Organizations have developed systems, people and expertise. Progress has been amazing, but the world has also substantially changed (think COVID-19, climate change, AI/data analytics, technology, environmental footprint, etc.). This book should be required reading for anyone interested in seeing humanitarian logistics as a young and evolving science rather than just a skill. It wonderfully combines the past, present and future of this discipline.'
Professor Luk Van Wassenhove, Emeritus Professor of Technology and Operations Management, INSEAD University

'We have not ended humanitarian crises yet. Fortunately *Humanitarian Logistics* continues to document the ever-changing challenges and learning to keep us all on the front foot. A must-read for all who strive to improve humanitarian supply chains.'
Martijn Blansjaar, Head of International Supply and Logistics, Oxfam GB

'The need for responsiveness to humanitarian emergencies has never been greater. This excellent book provides valuable insights into how logistics capabilities can be developed to better cope with crises, before, during and after they happen.'
Martin Christopher, Emeritus Professor of Marketing and Logistics, Cranfield University

'The book *Humanitarian Logistics* could not be more timely. With all the challenges facing those involved in the management of the logistics of disaster relief, offering potential solutions to the problems is truly welcome. This book has been, and will even more so be, a must-read for students, academics and practitioners who want to understand how to tackle the complexity of the networks involved in humanitarian logistics and the world we live in today.'
Professor Karen Spens, President of BI Norwegian Business School

Humanitarian Logistics

Meeting the challenge of preparing for and responding to disasters and complex emergencies

FOURTH EDITION

Graham Heaslip
and Peter Tatham

Publisher's note

Every possible effort has been made to ensure that the information contained in this book is accurate at the time of going to press, and the publishers and authors cannot accept responsibility for any errors or omissions, however caused. No responsibility for loss or damage occasioned to any person acting, or refraining from action, as a result of the material in this publication can be accepted by the editor, the publisher or the author.

First published in Great Britain and the United States in 2011 by Kogan Page Limited
Fourth edition published in 2023

Apart from any fair dealing for the purposes of research or private study, or criticism or review, as permitted under the Copyright, Designs and Patents Act 1988, this publication may only be reproduced, stored or transmitted, in any form or by any means, with the prior permission in writing of the publishers, or in the case of reprographic reproduction in accordance with the terms and licences issued by the CLA. Enquiries concerning reproduction outside these terms should be sent to the publishers at the undermentioned addresses:

2nd Floor, 45 Gee Street	8 W 38th Street, Suite 902	4737/23 Ansari Road
London	New York, NY 10018	Daryaganj
EC1V 3RS	USA	New Delhi 110002
United Kingdom		India

www.koganpage.com

Kogan Page books are printed on paper from sustainable forests.

© Graham Heaslip, Peter Tatham and Martin Christopher, 2023

The right of Graham Heaslip, Peter Tatham and Martin Christopher to be identified as the authors of this work has been asserted by them in accordance with the Copyright, Designs and Patents Act 1988.

ISBNs

Hardback	978 1 3986 0716 3
Paperback	978 1 3986 0714 9
Ebook	978 1 3986 0715 6

British Library Cataloguing-in-Publication Data

A CIP record for this book is available from the British Library.

Library of Congress Cataloging-in-Publication Data

Names: Heaslip, Graham, editor. | Tatham, Peter, editor.
Title: Humanitarian logistics: meeting the challenge of preparing for and responding to disasters and complex emergencies / [edited by] Graham Heaslip and Peter Tatham.
Description: Fourth edition. | London; New York, NY: Kogan Page Inc, 2023. | Includes bibliographical references and index.
Identifiers: LCCN 2022042971 (print) | LCCN 2022042972 (ebook) | ISBN 9781398607149 (paperback) | ISBN 9781398607163 (hardback) | ISBN 9781398607156 (ebook)
Subjects: LCSH: Humanitarian assistance. | Emergency management. | Business logistics.
Classification: LCC HV553 .H855 2023 (print) | LCC HV553 (ebook) | DDC 361.2/6–dc23/eng/20221021
LC record available at https://lccn.loc.gov/2022042971
LC ebook record available at https://lccn.loc.gov/2022042972

Typeset by Integra Software Services, Pondicherry
Print production managed by Jellyfish
Printed and bound by CPI Group (UK) Ltd, Croydon CR0 4YY

CONTENTS

List of figures and tables xi
List of abbreviations xv
List of contributors xxi

Introduction 1
Graham Heaslip and Peter Tatham

Managing supply networks under conditions of uncertainty 6
Lessons from best practice 11
The way forward 14
Endpiece 18
Notes 19

01 Impacts of funding systems on humanitarian operations 21
Tina Wakolbinger and Fuminori Toyasaki

Abstract 21
Introduction 21
Structure of funding systems 23
Impacts of financial flows on disaster response 24
Incentives provided by donors 30
Fundraising–humanitarian operations interface in literature 33
Summary and conclusion 35
Acknowledgement 36
References 37

02 Supplier relationships in humanitarian organizations 43
Jihee Kim, Stephen Pettit and Anthony Beresford

Abstract 43
Introduction 44
Humanitarian supply chain management and partnership 45

Framing the key issues – power, trust and commitment 52
Situational factors 55
Organizational factors 59
Summary and conclusion 60
Acknowledgement 62
References 62

03 Providing logistics services for humanitarian relief 67
Diego Vega and Christine Roussat

Abstract 67
Introduction: Logistics services 67
Commercial providers 69
Humanitarian providers 72
A continuum of actors 75
Summary and conclusion 78
Note 79
References 79

04 Risky business revisited: Disasters within disasters 83
Paul D Larson

Abstract 83
Review and classification of the literature 85
Security evaluation and management 91
Research agenda 95
References 96

05 The journey from a patchy to a comprehensive supply chain in UNHCR (2005–2015) 101
Svein J Håpnes

Abstract 101
Introduction to the United Nations High Commissioner for Refugees (UNHCR) 101
Building the understanding of supply chain management in UNHCR (2005–2006) 103
Establishing the need to improve supply chain management in UNHCR (late 2006) 105

The internal 'battle' for change (2007) 107
Establishing in Budapest, building the new supply chain organization (2008–2015) 109
The development in forcible displaced in the world and UNHCR budgets (2005–2015) 114
Improvements made to the UNHCR supply chain management function (2005–2015) 117
Key lessons identified 123
Summary and conclusion, UNHCR SCM 2015 onwards 126
References 128

06 Humanitarian supply chain service performance 131
Ruth Banomyong, Puthipong Julagasigorn, Paitoon Varadejsatitwong and Thomas E Fernandez

Abstract 131
Introduction 132
Service quality performance measurement used in commercial supply chains 133
Service quality performance measurement in relief operations 136
Developing HUMSERVPERF questionnaire 140
Summary and conclusion 150
Acknowledgement 150
References 151
Appendix 6.A: HUMSERVPERF questionnaire 155
Appendix 6.B: Result of HUMSERVPERF for the flood context 157

07 Network design for pre-positioning emergency relief items 159
Gerard de Villiers

Abstract 159
Introduction 160
Setting the scene 160
Covid-19 context for pre-positioning 162
Humanitarian logistics 163
Planning hierarchy 164

Network design 166
Location of disaster events 170
Typical emergency relief items to be pre-positioned 174
UNHRD network 175
ESUPS project 177
Centre-of-gravity analysis technique 178
Centre-of-gravity analysis application 181
Channel strategy 185
Summary and conclusion 188
Notes 188
References 189

08 Competing for scarce resources during emergencies: A system dynamics perspective 191
Paulo Gonçalves

Abstract 191
Introduction 192
Challenging aspects of humanitarian response to emergencies 195
Characteristics of complex systems 197
Behaviour in complex systems 200
Components of complex systems 203
Summary and conclusion 211
References 212

09 Preparing for cash and voucher assistance: Developing capabilities and building capacities 217
Russell Harpring

Abstract 217
Introduction 217
A brief introduction to cash and voucher assistance 220
What does preparedness mean for cash and voucher assistance? 224
Preparedness throughout the cash and voucher assistance operations cycle 226

Summary and conclusion 242
Acknowledgements 243
References 243

10 Pandemic response and humanitarian logistics 247
Gyöngyi Kovács, Tina Comes and Ioanna Falagara Sigala

Abstract 247
Introduction 248
Humanitarian healthcare logistics 249
Insights from past pandemic response 251
Lessons learnt from the Covid-19 pandemic: The case of personal protective equipment 253
Supply chain tools to accelerate vaccination programmes 256
Summary and conclusion 262
Acknowledgements 262
Notes 262
References 263

11 Helping people and planet: Making the humanitarian supply chain more sustainable 265
Maria Besiou, Sarah Joseph, Sophie t'Serstevens and Jonas Stumpf

Abstract 265
Introduction 266
Sustainability in the context of uncertainty 267
The environmental impact of HSCS and current initiatives 270
Shifting perspectives: Opportunities for HSCS 276
Summary and conclusion 281
References 282

12 What next for humanitarian logistics? 285
George Fenton and Tikhwi Jane Muyundo

Abstract 285
Introduction 286
The right product 287

The right cost 293
The right place 296
The right time 302
The right quantity 305
References 314

13 The way forward: Current trends in humanitarian logistics 317
Gyöngyi Kovács

Abstract 317
Introduction 317
Don't throw the baby out with the bathwater 319
What is new under the sun? 323
Famous last words 325
Acknowledgements 326
References 326

Annex 1: Operational assessment 329
Index 331

LIST OF FIGURES AND TABLES

FIGURES

Figure 0.1	There are two generic categories of risk	12
Figure 2.1	Categories of HO suppliers	51
Figure 3.1	Roles and activities in different humanitarian relief phases	71
Figure 3.2	Roles and activities of HOs as service providers	73
Figure 3.3	A continuum of logistics service providers for humanitarian relief	78
Figure 4.1	Attacks on aid workers, 1997–2021	84
Figure 4.2	Articles on humanitarian aid worker security, 2000–2021	86
Figure 4.3	From security triangle to security pentagon	90
Figure 5.1	UNHCR displacement and funding statistics 2005–2015	116
Figure 6.1	Developing the HUMSERVPERF	140
Figure 6.2	Organizational structure	142
Figure 7.1	Humanitarian logistics and supply chain management	164
Figure 7.2	Planning hierarchy	165
Figure 7.3	Decentralized distribution network	167
Figure 7.4	Centralized distribution network – Consolidated warehousing at the origin	167
Figure 7.5	Centralized distribution network – Consolidated warehousing at the destination	168
Figure 7.6	Location of natural disasters – 1986 to 2015	171
Figure 7.7	Number of disaster by continent and top 12 countries in 2021	172
Figure 7.8	Crisis country clusters	173
Figure 7.9	Location of UNHRD facilities (UNHRD, 2017)	176

Figure 7.10   Centre-of-gravity technique 179
Figure 7.11   Centre-of-gravity analysis of all natural disaster events in Africa (January 2014 to June 2017) 183
Figure 7.12   Potential regional clusters of natural disasters in Africa (January 2014 to June 2017) 184
Figure 7.13   Centre-of-gravity analysis for potential regional clusters of natural disasters in Africa 184
Figure 8.1   Stock-and-flow/causal loop diagram for beneficiaries served 207
Figure 8.2   Complete stock-and-flow/causal loop diagram for response 208
Figure 8.3   HO's mental model: Its competitor HO's relief capacity is exogenous 210
Figure 9.1   CVA conditionality and restrictions matrix 221
Figure 9.2   Changes typical humanitarian supply chain flows undergo through CVA programming 222
Figure 9.3   CVA operational cycle 227
Figure 9.4   Hierarchy of capacities, capabilities, and competencies in an organization 229
Figure 9.5   Complex task competency framework with learning feedback loops 232

TABLES

Table 1.1   Main objectives and challenges in fund management 30
Table 2.1   Main objectives and challenges in fund management 48
Table 2.2   Context of humanitarian organizations (HOs) 58
Table 5.1   Summary of important developments and documents 2005–2007 109
Table 5.2   Summary of important developments and documents 2008–2015 115
Table 5.3   Review of UNHCR's supply chain organization, 2013 figures 119
Table 6.1   Ten SERVQUAL's dimensions 134
Table 6.2   Five of SERVQUAL's dimensions 135

Table 6.3	SERVQUAL vs. SERVPERF 135
Table 6.4	The changes made by the respondents with supporting reasons 144
Table 7.1	Advantages and disadvantages of relief procurement 169
Table 7.2	CRED Disaster Centre-of-Gravity Analysis – Africa (Extract) 182
Table 7.3	Matching humanitarian situations with generic supply chains 187
Table 7.4	Supply chain strategies and network design 188
Table 8.1	Challenging factors in humanitarian response 196
Table 8.2	A disaster management perspective of complexity 199
Table 8.3	A dynamic perspective of complexity 200
Table 9.1	Types of CVA 220
Table 9.2	Individual-level CVA capacity and capability assessment 234
Table 9.3	Team-level CVA capacity and capability assessment 234
Table 9.4	Cash Working Group members and interviews conducted 236

LIST OF ABBREVIATIONS

3DP	3D printing
3PL	third party logistics
4PL	fourth party logistics
ABS	acrylonitrile butadiene styrene
ALNAP	Active Learning Network on Accountability and Performance
API	active pharmaceutical ingredients
ART	antiretroviral therapy
ARV	antiretroviral
AY	African Union
BINGOs	big international non-governmental organizations
BLOS	'beyond line of sight'
BPM	'big picture map'
BRAT	basic rapid assessment tool
CCIC	Canadian Council for International Cooperation
CDEM	civil defence and emergency management
CERF	Central Emergency Response Fund
CFW	cash for work
CHF	Common Humanitarian Funds
CHL	Certificate in Humanitarian Logistics
CHP	combined heat and power plants
CHS	Core Humanitarian Standard
CIDA	Canadian International Development Agency
CILT	Chartered Institute of Logistics and Transport
CIPD	Chartered Institute of Personnel and Development
CRED	Centre for Research on the Epidemiology of Disasters
CSIR	Council for Scientific and Industrial Research
CSR	corporate social responsibility
CRIs	core relief items

CTPs	cash transfer programmes
DAC	Development Assistance Committee
DCs	distribution centres
DCA	Danish Church Aid
DESC	Cabinet Committee on Domestic and External Security Coordination
DFID	Department for International Development
DMISA	Disaster Management Institute of Southern Africa
DPMC	Department of the Prime Minister and Cabinet
DRM	disaster relief management
DRVCA	disaster response value chain analysis
ECB	Emergency Capacity Building
ECC	Emergency Coordination Centre
ECHO	European Civil Protection and Humanitarian Aid Operations
EFQM	European Foundation for Quality Management
EFSVL	Emergency Food Security and Vulnerable Livelihoods
EM	Emergency Management
EMIS	Emergency Management Information System
EM-DAT	Emergency Events Database
EOC	Emergency Operations Centre
ERF	Emergency Response Funds
ERP	emergency response preparedness
ERUs	emergency response units
ESUPS	Emergency Supplies Prepositioning Strategy
EU	European Union
FAA	Federal Aviation Administration
FDM	fused deposition modelling
FEMA	Federal Emergency Management Agency
FFA	Forest Fire Association
FMCG	fast moving consider goods
FSM	Swiss Mine Action
GAP	Global Accountability Partnership
GDP	gross domestic product

GIMPA	Ghana Institute of Management and Public Administration
GOs	governmental organizations
GRI	Global Reporting Initiative
Groupe URD	Urgence, Réhabilitation Développement
H-CLOP	humanitarian common logistic operations picture
HAP	Humanitarian Accountability Project
HAP-I	Humanitarian Accountability Partnership International
HCAs	hybrid cargo ships
HDR-SCM	humanitarian and disaster relief supply chain management
HL	Humanitarian logistics
HLA	Humanitarian Logistics Association
HLCF	humanitarian logistics competency framework
HQ	headquarters
HSCM	humanitarian supply chain management
IATA	International Air Transport Association
ICAO	International Civil Aviation Organization
ICTs	information communication technologies
IDP	integrated development plan
IDPs	internally displaced persons
IFRC	International Federation of Red Cross and Red Crescent Societies
IHOs	international humanitarian organizations
INGO	international non-governmental organization
IPs	implementing partners
IS/IT	information systems/information technology
ISCRAM	Information Systems for Crisis Response and Management
ISO	International Organization for Standardization
ITAR	International Traffic in Arms Regulations
ITU	International Telecommunications Union
JIT	just-in-time
KPIs	key performance indicators
KYC	'know your customer'

LALE	'low altitude, long endurance'
LMMS	Last Mile Mobile Solutions
LRT	logistics response team
LSPs	logistics service providers
MCDEM	Minster's Office
MSF	Médicins Sans Frontières
MSG	Meteosat Second Generation
MREs	meals ready to eat
NDMC	South African National Disaster Management Centre
NDMIS	National Disaster Management Information System
NDMO	National Disaster Management Organization
NEPAD	New Partnership for Africa's Development
NESA	National Emergency Supply Agency
NESO	National Emergency Supply Operations
NFIs	non-food items
NGOs	non-governmental organizations
NSS	National Security System
NVQ	National Council for Vocational Qualifications
NZTA	New Zealand Transport Agency
OCHA	Office for the Coordination of Humanitarian Affairs
ODA	official development assistance
OECD	Organization for Economic Co-operation and Development
OR	Operations research
PoCs	persons of concern
PDMCs	The Provincial Disaster Management Centres
PEPFAR	US President's Emergency Plan for AIDS Relief
PMTCT	prevent mother-to-child transmission
PTD	People that Deliver
PVO	Private Voluntary Organization
RPAS	remotely piloted aircraft systems
SAC	Satellite Application Centre
SATs	standard attainment tests
SCHR	Steering Committee for Humanitarian Response
SCI	supply chain integration

LIST OF ABBREVIATIONS

SCM	supply chain management
SCOR	supply chain operations reference
SCRM	supply chain risk management
SCs	supply chains
SEAM	security evaluation and management
SKUs	stock keeping units
SRTs	ShelterBox response teams
UAS	unmanned aerial systems
UAV	unmanned aerial vehicles
UN	United Nations
UNCERF	United Nations Central Emergency Response Fund
UNCTAD	United Nations Conference on Trade and Development
UNDESA	United Nations Department of Economic and Social Affairs
UNDP	United Nations Development Programme
UNGM	United Nations Global Marketplace
UNHCR	United Nations High Commissioner for Refugees
UNHRD	United Nations Humanitarian Response Depot
UNICEF	United Nations Children's Fund
UNISDR	UN Office for Disaster Risk Reduction
UNOPS	United Nations Officer for Project Services
VfM	value for money
VMI	vendor managed inventory
VTOL	vertical take-off and landing
WoG	whole of Government
WASH	water, sanitation and hygiene
WATSAN	water and sanitation
WFP	World Food Programme
WHO	World Health Organization
WHS	World Humanitarian Summit
WIM	warehousing and inventory management
WV	World Vision
VCA	value chain analysis

LIST OF CONTRIBUTORS

Ruth Banomyong is Dean of Thammasat Business School, Thammasat University in Thailand. His research interests include multimodal transport, international logistics, logistics policy development, and supply chain performance measurement.

Anthony Beresford is a Professor of Logistics and Transport at Cardiff Business School, Cardiff University, UK. He was awarded his PhD in Environmental Sciences at the University of East Anglia in 1982 for research focusing on climate change in East Africa. He has travelled widely in an advisory capacity within the ports, transport and humanitarian fields in Europe, Africa, Australasia and North America. He has been involved in a broad range of transport-related research and consultancy projects including: transport rehabilitation, aid distribution and trade facilitation for the United Nations Conference on Trade and Development and for, for example, the Rwandan Government. He has also advised both the United Kingdom and Welsh Governments on road transport and port policy options. More recently, he has conducted research on humanitarian supply chain operations in the context of both man-made emergencies and natural disasters.

Maria Besiou is Dean of Research and Professor of Humanitarian Logistics at Kühne Logistics University. She is also the Academic Director of the Center for Humanitarian Logistics and Regional Development (CHORD). Her research focuses on humanitarian and sustainable operations. Her research appears in Production and Operations Management (POM), Journal of Operations Management (JOM), Manufacturing and Service Operations Management (MSOM), California Management Review (CMR), and the European Journal of Operational Research (EJOR).

Tina Comes is a Professor in Decision Theory & ICT for Resilience at the Faculty Technology, Policy and Management at the TU Delft, Netherlands. Dr Comes is a Visiting Professor at the Université Dauphine, France, a member of the Norwegian Academy for Technological Sciences and a member of the Academia Europaea. She serves as the Scientific Director of the 4TU Centre for Resilience Engineering, as the Director of the TPM Resilience Lab, and as the theme lead for Disaster Resilience for the Delft Global Initiative. Dr Comes' research focuses on decision-making and information technology for resilience and disaster management, which is reflected in more than 100 publications.

Gerard de Villiers has been a professional transportation engineer for more than 42 years who specialises in transport economics, freight logistics, humanitarian logistics and supply chain management. His career includes Global Supply Chain Director of World Vision International for four years where he established a global office for supply chain management. Gerard has various local and international post-graduate qualifications and he lectured part time at universities in South Africa, Switzerland and Liverpool. He has contributed chapters in various text books on topics such as Transport, Network Design and Channel Strategy, Humanitarian Logistics, Inland Intermodal Terminals and City Logistics. He was Honorary Professor in the Department of Business Management, Faculty of Economic and Management Sciences at the University of Pretoria from 2009 to 2011, and he started Supply Chain Advancement (Pty) Ltd specialist consulting company in September 2020.

George Fenton is an experienced consultant and evaluator, working with both the aid and private sectors, on emergency operations, procurement, logistics and standards, digital cash transfers and market-based interventions. He is an expert and thought leader in humanitarian supply chain management with over 35 years of experience. George is also co-founder and Chief Executive of the Humanitarian Logistics Association, a membership organization that supports research and knowledge management in the sector.

Thomas E Fernandez is a lecturer and researcher at the International School of Management of the University of the Thai Chamber of Commerce. His research interests include supply chain management, humanitarian logistics, and cross-cultural management.

Paulo Gonçalves is a Professor of Management and the Director of the Humanitarian Operations Group at the Università della Svizzera italiana (USI), Switzerland, and Research Fellow at the University of Cambridge Judge Business School (CJBS). His research combines system dynamics, behavioral experiments, optimization, and econometrics, to understand how managers make strategic, tactical, and operational decisions in humanitarian settings. Paulo holds a PhD in Management Science and System Dynamics from the MIT Sloan School of Management and an MSc from the Massachusetts Institute of Technology (MIT).

Svein J Håpnes is a professional logistician who has specialized in Military and Humanitarian Logistics and Supply Chain Management. He holds two Master's degrees in Logistics from the University of Westminster and Molde University College. He has served a 20-year career in the Norwegian Army with the Army Transport Corps, 15 years with the United Nations (UN DPKO, UNHCR and UNICEF), and is currently working as a Senior Project Manager in WilNor Governmental Services at Wilh. Wilhelmsen, developing new Military Logistics Support Solutions. During his UN career he gained experiences from UN Headquarters, managing global support functions and logistics change processes, and has field experience from Europe (Balkans), the Middle East and Africa. Svein has worked together with the academia since 2008, being a co-author on an article (M. Jahre et al, 2016) in the *Journal of Operations Management*, and has been, and is continuing, to give guest lectures in (Humanitarian) Logistics and Crises Management at BI Oslo, Lund University, National University of Ireland, and University of Westminster. He has also been invited to present at academic conferences as the HOCM/POMS, IAME and HHL.

Russell Harpring is a PhD student at Hanken School of Economics in the department of Supply Chain Management and Social Responsibility. His research focuses on humanitarian logistics with a special interest in the use of cash and voucher assistance in complex emergencies, including how services are procured and implemented. Russell has worked with UN agencies and humanitarian NGOs, both in field locations and headquarter offices, including North and Central America, the Caribbean, East Africa, and Central Europe. He received his Master of Engineering in Industrial Engineering from the University of Louisville, with a thesis on the use of simulation techniques in the public sector to improve efficiency.

Graham Heaslip is Professor of Logistics and Head of the School of Engineering at the Atlantic Technological University In Galway, Ireland. Prior to joining ATU Graham was Associate Professor of Logistics at UNSW, Australia. Graham completed his PhD studies in the area of Civil Military Cooperation/Coordination at the Logistics Institute, University of Hull, for which he was awarded the James Cooper Memorial Cup for best PhD in Logistics and Supply Chain Management by the Chartered Institute of Logistics and Transport. Prior to entering academia Graham spent fourteen years working in the Irish Defence Force. Graham's research interests are broadly in the intersections between global logistics/supply chain management, humanitarian logistics and organizational management development.

Sarah Joseph is a Postdoctoral researcher at the Kühne Logistics University (KLU), based at the Center for Humanitarian Logistics and Regional Development (CHORD). Her research is focused on increasing transparency and sustainability in supply chains, specifically concentrated on food supply chains, sustainable food production, regional supply, humanitarian logistics, Life Cycle Analysis (LCA), and spatial econometrics. As part of CHORD, she is also working on projects with humanitarian organizations to improve sustainability in supply chains, including analysing and identifying sustainability-based metrics to measure impact of operations.

Puthipong Julagasigorn is a research associate and consultant at Centre for Logistics Research (CLR) at Thammasat University in Thailand. His research interests include consumer behaviour and psychology, marketing, and humanitarian reliefs.

Jihee Kim is a lecturer in supply chain management at Anglia Ruskin University, Cambridge, UK. Before moving into academia, she worked in a number of organizations including the Korean National Assembly, Samsung, LG and Korean Air. She was awarded an MA in Management from Durham University in 2010 and a PhD from Cardiff University in 2021. Funding for her PhD research was provided by the Economic and Social Research Council. Her research interests involve cooperation, coordination, collaboration, supply chain integration and partnerships in humanitarian and disaster relief supply chain management. In particular, she seeks to understand the suitability of the key assumptions regarding business supply chain management in the humanitarian space.

Gyöngyi Kovács is the Erkko Professor in Humanitarian Logistics, and the Subject Head of Supply Chain Management and Social Responsibility at the Hanken School of Economics, in Helsinki, Finland. She is a founding Editor-in-Chief of the Journal of Humanitarian Logistics and Supply Chain Management (JHLSCM) and is on the editorial board of several other journals. She was the first Director of the Humanitarian Logistics and Supply Chain Research Institute (HUMLOG Institute) and has published extensively in the areas of humanitarian logistics and sustainable supply chain management. She was awarded humanitarian logistics researcher of the year 2020 by the American Logistics Aid Network ALAN. Currently, she is leading a Horizon 2020 (EU) Covid-19 project called 'HERoS' (Health Emergency Response in Interconnected Systems).

Paul D Larson is the CN Professor of SCM at the University of Manitoba, and a former Head of SCM Department and Director of the Transport Institute. He is also a HUMLOG International Research Fellow at Hanken School of Economics in Helsinki, Finland. During 2006–2007, Paul was the principal investigator on a $200,000 pandemic planning project, led by the late Dr. Allister Hickson. That project resulted in the 2008 report: *Manitoba Nutrition Supply in Event of a Pandemic* – and inspired Larson's Supply Chain Risk Evaluation and Management (SCREAM) framework. In 2016, he was lead researcher and author of a report titled *Supplier Diversity in Canada*, published by the Canadian Centre for Diversity and Inclusion (CCDI). Professor Larson is a senior editor for the *Journal of Business Logistics* and serves on the editorial review boards of the *International Journal of Physical Distribution and Logistics Management* and the *Journal of Humanitarian Logistics and Supply Chain Management*. His current research interests include sustainability and security/risk management in humanitarian logistics. Recent funded research projects focus on active transportation and transit ridership in Canadian CMAs; impact of the pandemic on Canadian food banks, with special reference to nutrition; and household food waste in North America. From 1979 to 1981, Paul worked with the Ministry of Cooperatives in Fiji, as a United States Peace Corps Volunteer. More recently, Larson has been on short-term aid missions to Colombia, Haiti, Tanzania, and Trinidad & Tobago. In February 2012, as a member of the CARE Canada Kilimanjaro expedition, and again in February 2017, he stood at Uhuru peak, Tanzania, the highest point in Africa. He has also topped out on several of California's fourteeners.

Tikhwi Muyundo is a supply chain professional, with extensive experience in Procurement, Warehousing, Distribution and Operational Logistics. She has worked in the private, public and humanitarian sectors. The roles in the humanitarian sector, enabled her to work in countries across the world, leading logistics teams in complex emergency operations, providing supply chain and operational support. She continues to provide technical support as an independent resource in supply chain to organizations globally. The complexity of these roles enhanced her skills in global supply chain and she continues to serve as an independent resource to organizations globally. She was a member of the founding groups for the East Africa Inter-Agency working Group, the Global Fleet Forum, and Women's Initiative for Supply Chain Excellence (WISE). Currently volunteering as the Africa Regional Representative for the Humanitarian Logistics Association (HLA) and recently assigned as the Coordinator for the Health & Humanitarian Supply Chain Advocacy Group, an initiative by International Association of Public Health Logisticians (IAPHL), People that Deliver (PtD) and the HLA.

Stephen Pettit is a Professor in Logistics and Operations Management at Cardiff Business School, Cardiff University, UK. He was awarded a PhD from the University of Wales in 1993 and he has worked at Cardiff Business School since 2000. He has been involved in a wide range of research with his most recent work focusing on humanitarian aid logistics and supply chain management. Stephen has contributed to many journal papers, conference papers and reports primarily on port development, port policy and humanitarian aid logistics. His humanitarian supply chain research has focused on operational and strategic aspects of humanitarian and emergency response, and his published output has been widely cited.

Christine Roussat is an Assistant Professor of Business Science at the Clermont-Auvergne University (France) and researcher at the CRET-LOG research centre (Aix-Marseille University). Her major fields of interest question the future of logistics: crowd logistics, role and strategy of logistics service providers, innovation in the field of services, sustainable supply chain management and interactions between supply chain management and environmental scanning approaches. She has published several papers on these topics and works more specifically nowadays on supply chain management and the Anthropocene.

Sophie t'Serstevens is a research analyst at the Kühne Logistics University (KLU). After her master's studies in supply chain management, Sophie worked for five years in consulting. She focused on supply chain planning and optimization and helped global organizations set in place planning tools and processes. Sophie joined the KLU in 2021 with the wish to leverage her experience from the commercial sector to help build efficient and sustainable humanitarian supply chains. She is part of CHORD (Center for Humanitarian Logistics and Regional Development) and is working on research and consulting projects with different humanitarian organizations in sustainable supply chain management.

Ioanna Falagara Sigala is a Project Researcher at Hanken School of Economics. Ioanna holds a PhD in Economics and Social Science from Vienna University of Economics and Business. Her primary research interests include the digitalization of humanitarian supply chain, the medical supply chains and COVID-19 response, the outsourcing of humanitarian logistics and the role public-private partnerships in building supply chain resilience.

Jonas Stumpf obtained his Masters in Logistics and International Management from the University of Mannheim in 2009. Before he joined the Kühne Foundation, he worked at the United Nations World Food Programme (WFP) where he conducted supply chain analyses and provided training courses at headquarters and various field offices in the Central and Eastern Africa region. From 2013 to 2016 he began HELP Logistics operations in Asia where he established a broad network of humanitarian actors. In his current role as HELP's Director Global Programs and CHORD's Operations Manager, he focuses on the design, development and coordination of education, research and consulting services for global partners.

Peter Tatham is an Adjunct Professor of Humanitarian Logistics at Griffith Business School, Queensland, Australia. Prior to entering academia, he served for 34 years as a logistician in the (UK) Royal Navy reaching the rank of Commodore (1*). His doctoral thesis investigated the role of shared values within UK military supply networks, and this work received the 2010 Emerald/EFMD prize for the year's best logistics/supply network management-related PhD. In addition to investigating the challenges of achieving agile defence and business logistic systems, his main research field is that of humanitarian logistics and, in particular, the use of emerging technologies in the preparation and response to natural disasters and complex emergencies. He is a member of the editorial boards of the Journal of Humanitarian Logistics and Supply Chain Management, the International Journal of Physical Distribution and Logistics Management, and the Journal of Defense Analytics and Logistics.

Fuminori Toyasaki is Associate Professor of School of Administrative Studies, Faculty of Liberal Arts & Professional Studies at York University in Toronto, Canada. He explores topics related to supply chain management with a specific focus on humanitarian logistics, health system management, and circular economy. He also conducts patent data analysis for pharma/biotech companies' R&D partnerships. His work was published in journals such as Decisions Sciences, Production and Operations Management, European Journal of Operational Research, International Journal of Production Economics, and International Journal of Production Research

Paitoon Varadejsatitwong is a research associate and consultant at Centre for Logistics Research (CLR) at Thammasat University in Thailand. His research interests include logistics and supply chain management, performance measurement/assessment, and humanitarian logistics.

Diego Vega is an Assistant Professor of Supply Chain Management and Social Responsibility with a particualr fous on Humanitarian Logistics at Hanken School of Economics and also the Director of the HUMLOG Institute. He is an Associate Editor of the Journal of Humanitarian Logistics and Supply Chain Management, and sits on the Editorial Advisory board of the International Journal of Logistics Management, and the SCM programming board of the World Humanitarian Forum. Diego's research interests include logistics services in humanitarian operations, emergency relief logistics, and competence-based strategic management for humanitarian organizations.

Tina Wakolbinger is Professor of Supply Chain Services and Networks and serves as the head of the Research Institute for Supply Chain Management at WU (Vienna University of Economics and Business). In her research, she explores topics related to supply chain management with a specific focus on humanitarian logistics and sustainability. Her work was published in journals such as Decisions Sciences, European Journal of Operations Research and International Journal of Production Economics. She serves as senior editor of the Disaster Management Department of Production and Operations Management.

Introduction

GRAHAM HEASLIP AND PETER TATHAM

Why a fourth edition of *Humanitarian Logistics*? The simple answer is that, sadly and perhaps inevitably, since the 3rd edition, humanitarian crises have continued to emerge and challenge the international community in how best to respond. As prime examples, the ongoing Syrian crisis, the Covid-19 pandemic and events emerging in Ukraine cause significant disruptions to humanitarian supply chains, putting those in need in the most vulnerable of positions.

The Covid-19 pandemic in particular highlighted the need to understand and orchestrate the many direct and indirect feedbacks that impact medical supply chains. This required the mapping of operational gaps and bottlenecks in these complex networks in order to support an efficient and effective response. This challenge was exacerbated by the fact that the Covid-19 pandemic was managed differed vastly across countries. Some countries imposed vast quarantines, others closed their borders and introduced travel bans. An EU-wide export ban for some medical protective equipment in a bid to keep sufficient supplies within the bloc was put in place. Countries were outbidding one another for badly needed PPE and manufacturing plants closed down. The collective effect of these policies, quarantines, bans, and shutdowns was to decrease the availability of essential medical items on the market, impacting the delivery of medical aid, as well as leading to secondary effects through cascades in the supply chain.

While in the commercial context Covid-19 brought to light unseen vulnerabilities, in the humanitarian world it magnified problems that already existed in humanitarian supply chains. First and foremost is the massive uncertainty surrounding, in particular, rapid onset events – where, when, what intensity – underlining the '6W Problem' of 'who wants what where when and why'. Thus, while we know that such events will unquestionably occur, their timing and location is hugely difficult to predict with any significant degree of certainty. This leads to the emergence of unexpected demand for products and short lead times for supplies.

Secondly, the humanitarian field faces the challenge of a de-coupling of financial and material flows. As a result, aid agencies are placed in the difficult position of having to second guess the needs of the beneficiaries who are frequently solely focused on the business of staying alive – and yet, at the same time, the agencies must satisfy the increasingly demanding governance requirements of the donor community. The high risks involved with deliveries, and the shortage of human, physical and financial resources are all characteristics that not only add to the difficulties of managing humanitarian logistics, but they also compound them. Therefore, while many management gurus would argue strongly that the voice of the customer should always be paramount in an organization's thinking, the absence of clarity over the identity of the humanitarian logistician's customer remains unhelpful and can lead to behaviour that, inadvertently, works against the best interests of those impacted by the disaster event.

Thirdly, almost by definition, the infrastructure surrounding the disaster will be devastated to a greater or lesser extent. Thus, generic prescriptions such as the substitution of information for inventory face a particular challenge in this environment. Finally, of course, the price of failure in terms of unnecessary loss of life or prolonged hardship is significantly greater than that of reduced profits.

The need for increased visibility across a construct that typically incorporates hundreds of suppliers sees a shift from linear supply chains to more integrated networks connecting many players. This shift is enabled by technologies, such as blockchain, that provide valuable data on the physical location of goods within the chain as well as their condition. Importantly, however, the pandemic has highlighted and reinforced the strategic importance of supply chains, causing humanitarian organizations to re-evaluate their supply chains and ensure they are fit for the future. Covid-19 has also helped to underline the need for the humanitarian supply chain of the future to be agile, flexible, efficient, resilient and digitally networked for improved visibility. Furthermore, Covid-19 has also served to stress the vital role of the humanitarian logistician in overseeing and managing these super-complex networks.

With this introduction in mind, this volume seeks to understand the nature of the challenges facing those who are involved in the management of the logistics of disaster relief, and to offer some potential solutions that can be developed in the near and longer term. Many of those contributing have spent considerable periods thinking about such issues, be this in a commercial, a humanitarian or a military context. We aim, therefore, to try to bring these perspectives together as a means of offering ways in which particular aspects of this complex and evolving problem might be tackled. In doing so, we have retained some 50 per cent of the chapters published in the third edition, but invited these authors to update their thoughts in line with developments over the last three years.

At the same time, we have introduced a number of new chapters, which are designed both to increase the level of input from the practitioner community and also to focus on a number of subjects that have received only limited consideration previously. Examples of the latter include a chapter on sustainability, insights from the migrant crises and the Covid-19 pandemic, emerging technologies such as blockchain, cash-based humanitarian logistics systems, updates from humanitarian service performance and a new chapter focusing on ways in which humanitarian logistics can overcome the challenge of obtaining scarce resources. Sadly, this does not mean that the issues that no longer feature in the book have been solved – rather that their relative importance has, to our editorial eyes, diminished slightly, albeit we would wish to stress the 'relative' qualifier that has guided our thinking.

But why focus on logisticians? The answer is simple. Be it in the context of a rapid or slow onset disaster or emergency, the imperative is to procure and move the required material – water, food, shelter, clothing, medicines, etc. – from point A to point B in the most efficient and effective way possible. But although simply stated, the reality is hugely complicated, and indeed costly, not least because of the difficulty of forecasting when and where the next crisis will occur. It is unsurprising, therefore, that recent estimates would suggest that some 60–80 per cent of the expenditure of aid agencies is on logistics[1] within which we include the procurement, transport into the affected country, warehousing and internal and 'last mile' distribution processes. Given that the overall annual expenditure of humanitarian agencies is of the order of $20 billion, the resultant logistic spend of some $15 billion provides a huge potential area for improvement, and consequential benefit to those affected by such disasters or emergencies.

However, in the period since the first edition of the book was produced, there have been a number of important advances in thinking. Firstly, the recognition that from a logistics perspective, rapid onset disasters go through a number of phases, each of which potentially requires a different logistic response. Thus, in the initial (emergency) period of one to two weeks, there will only be limited availability of demand information from the affected area and, as a result, a 'push' approach may be entirely appropriate. This also reflects the relatively limited range of material, equipment, etc. that is needed to provide the necessary life-saving response. For example, the Oxfam equipment catalogue is of the order of 300 stock keeping units (SKUs), which is two orders of magnitude less than the equivalent for a large supermarket chain that may run to 30,000–40,000 SKUs[2]. In light of the potential mass movements away from an affected area, the issue in this phase is, therefore, more about how many people/family units are to be found in a particular geographic location and the associated 'last mile' delivery challenge in the face of a severely disrupted logistics and communications infrastructure.

After the first two weeks, it is likely that needs assessments will have been completed and that a clearer picture of the requirements is available, leading to the ability to employ a significantly more targeted 'pull' approach. At this point, in order to reduce the potential of inefficient overlaps and/or life-threatening gaps in the response, the benefits of improved inter-agency coordination are becoming clearer. However, while the World Food Programme (WFP) Logistics Cluster is making an increasingly effective impact in driving coordination across the UN family and more broadly, there remains a strong case for greater effort in this regard. In a commercial context, major players have recognized the benefits of the development of a multisource overview of their supply network, which enables them to intervene proactively in order to meet the emerging demand picture. It is strongly argued, therefore, that the development of such a 'humanitarian common logistic operations picture' (H-CLOP) is increasingly becoming a necessary next step on the path from operational humanitarian logistics to strategic supply network management.

The development of the H-CLOP concept does, however, imply the need for a far greater level of pan-agency process integration than is currently the case. Historically, there has been a marked reluctance to share data with other agencies for a number of reasons, not least because, in part, they can be perceived as competitors for donor funding. This is slowly changing, especially across the larger agencies, as demonstrated by the use of the same software supply chain management systems by multiple organizations. But, in doing so, it raises the larger question of whether a system of certification of agencies (and, by implication, their logistics teams) should be introduced – not least because, at the operational level, such a certification process could incorporate more standardized policy/process/procedural requirements which would, in turn, lead to a greater efficiency and/or effectiveness in the response process.

By extension, the development of such common processes would become the basis on which education and professional curricula could be developed and which would, in turn, not only enhance the professional status of the humanitarian logistician, but also help to reduce the burden of training for new staff as well as for implementing partners within a given country or region. Importantly, such an approach already exists in the related context of the international search and rescue community, and is being developed by the UN's health cluster that is overseen by the World Health Organization (WHO). In parallel, the Humanitarian Logistics Association (HLA) is actively developing a 'body of knowledge' that is designed both to assist those carrying out the key logistic functions (especially for those who are new to this role), while at the same time assisting in the development of a commonality of practice.

With these opening remarks in mind, we propose to begin by exploring the challenge of how to manage supply networks when future requirements are manifestly uncertain, which is, of course, one of the most challenging aspects of humanitarian relief programmes. The following section summarizes this challenge and the editors of this edition of *Humanitarian Logistics* would wish to express their enormous appreciation to the former editor, Professor Martin Christopher, who was instrumental in developing these ideas, concepts and approaches.

Managing supply networks under conditions of uncertainty

One of the distinguishing features of modern supply networks – both in the world of business as well as in the humanitarian arena – is that they are characterized by uncertainty and, hence, unpredictability. For some time now, commercial supply network managers have become accustomed to the idea that they can no longer rely on the traditional rules and techniques that have allowed them to plan ahead with a degree of confidence.

Thus, although conventional supply network management typically assumes a degree of stability with planning horizons that extend some months into the future, the last few decades have seen a considerable increase in turbulence in the wider business environment. Demand can no longer be easily forecast and supply conditions have become more volatile in almost every industry. As a result of this uncertainty, new business models have emerged to enable organizations to make the transition from the classic forecast-driven approach to a much more event- or demand-driven capability.

Organizations doing business in turbulent markets have learnt that one of the key elements to ensure survival is agility. This can be defined as the ability to respond rapidly to unexpected changes in demand or supply conditions and, indeed, to changes in the wider business environment.

It can thus be argued that the logistic capabilities required by aid agencies and others to deal successfully with large scale, sudden onset disasters are not dissimilar to those required in commercial organizations faced with rapidly changing conditions. There is, therefore, an excellent opportunity to learn from the experiences of companies who have become adept in responding rapidly to unpredictable events.

Because all organizations are part of a wider network of suppliers, intermediaries and customers, it is important to recognize that agility is not just about achieving internal responsiveness, but rather about how the end-to-end supply network can become more agile. Thus, the concept of agility has significant implications for how organizations within the supply/demand network relate to each other, and how they can best work together to maximize the efficiency and effectiveness of the network as a whole. It has been suggested[3] that there are a number of key prerequisites to the design and management of such agile supply networks. Specifically, the concept of agility implies that they are demand or event driven, they are network based, they are process oriented and they are virtually integrated through shared information.

Demand and event driven

Traditional management practice has been based upon the principle of planning ahead, usually based upon a forecast. In conditions of turbulence and unpredictability, however, the challenge is to create a capability to facilitate a rapid response to events as they happen. A fundamental enabler of demand/event driven responsiveness is time compression. Much of the time that is consumed in supply networks could be termed 'non-value adding time'. In other words, it is time when nothing is happening to achieve the goal of the 'right product in the right place at the right time'. Sometimes this non-value adding time is incurred because of cumbersome planning and decision-making processes. At other times it may arise because of queues at bottlenecks, or because of inadequate coordination across the different stages in the supply network. As a result, many commercial organizations have transformed their responsiveness by a strong focus on what has been called 'business process re-engineering'[4] whereby every underpinning process in the supply network is put under the spotlight with the intention of squeezing out as much non-value adding time as possible.

Demand and event driven supply networks are also often characterized by their strategic use of inventory and capacity. Conventional wisdom is typically driven by the desire to follow 'lean' principles of reducing inventory and eliminating idle capacity. Agile supply networks on the other hand recognize that in conditions of uncertainty – both on the demand side and the supply side – a certain level of 'slack' is essential. Ideally, such strategic inventory is held as far upstream as possible and in a generic form to enable 'risk pooling' – in other words, rather than disperse the inventory in its final form and run the risk of having the wrong product in the wrong place, it is held centrally, shipped and configured on a just-in-time (JIT) basis. Clearly this approach will incur a cost penalty compared to the 'leaner' alternative, but that is the price of responsiveness.

Network based

One way that organizations can enhance their agility is by making use of the capacity, capabilities and resources of other entities within the network. It could be financially crippling for one organization to have to carry enough capacity and inventory to, for example, cope with any demand eventuality. However, if close working relationships can be established with other organizations that can provide access to their own resources, then a real opportunity exists for creating high levels of flexibility in the supply network.

A good example of how network partners can enable a more agile capability in the commercial world is provided by the Spanish clothing manufacturer and retailer, Zara. Because Zara competes in a market characterized by unpredictability and short product life cycles, the need for agility is high. One way that Zara achieves this agility is by making use of a network of small, independent workshops that do the final sewing of many of their products. Zara has established strong working relationships with these suppliers and regards them as part of their 'extended enterprise'. These external workshops reserve capacity for Zara even though they will not know the precise requirements until a few days before the garment is to be manufactured.

In other cases, organizations can benefit by sharing resources across a network even with competitors. Thus, for example, petrol companies such as Shell, BP and Total will often share refinery capacity, while in the airline industry different airlines will pool their inventory of service parts and position these strategically around the world. In a similar way, the armed forces of NATO countries use a common parts identification system that facilitates an equivalent approach.

Indeed, in the world of humanitarian logistics, such a resource sharing model has recently been created to enable access to a common inventory, with the United Nations Humanitarian Response Depot (UNHRD) network, which is coordinated by the World Food Programme (WFP) in Italy, being a case in point. In a similar way, the water and sanitation (WATSAN) cluster and associated NGOs who operate in this space have recently agreed to both standardize and pool their equipment resources.

Process oriented

One feature of organizations that can respond rapidly to unpredictable events is that they have achieved a high level of cross-functional working. Most conventional businesses tend to be organized around functions, e.g. the production function, the distribution function, etc. This type of organizational structure may be administratively convenient, but it often leads to an inwardly-focused 'silo' mentality. It also means that there are usually multiple hand-offs from one department to another. The end result is that the decision-making process is lengthy, and that lead times are extended.

The alternative is to break down the silos by adopting a cross-functional team-based approach that reflects the key business processes, particularly the supply network processes. Processes are the horizontal, market-facing sequences of activities that create value for customers. In the context of supply networks, they include such key underpinning processes as order-to-delivery, capacity and demand management and supplier management. For each of these processes a 'process owner' should be appointed whose task is to bring together a cross-functional team and to seek to create a seamless and more rapid achievement of the process goals. Thus, for example, the order-to-delivery process will consider how a customer's requirement can be met in shorter time frames with greater reliability by 'project managing' the order from the moment it is captured until it is delivered. Usually when processes are managed in this way, opportunities for process simplification and improvement quickly become apparent.

Furthermore, if the supply network is to work effectively across multiple independent entities, it is critical that processes are aligned across organizational boundaries. A good example of such process alignment is provided by the concept of vendor managed inventory (VMI). Under a VMI arrangement, the sales outlet (say a supermarket) does not formally place an order to the supplier; rather they provide the supplier with regularly updated information (easily extracted from the point of sale systems) on the rate at which the customer's inventory of the product in question is being depleted. The supplier then automatically replenishes the inventory. It is akin to a closed-loop supply network process.

Virtually integrated

By definition, for global networked supply networks to achieve high levels of agility there must be a corresponding level of *connectivity*. Historically, such connectivity may have been achieved through ownership and control – a state often described as 'vertical' integration. Today, the likelihood is that the supply network will be fragmented and dispersed with each entity independent from the other. However, the need for integration is still as vital as ever, but now the essential integration is not achieved through ownership and control but rather through shared information and collaborative working. This type of connectivity is often called 'virtual integration'.

The underpinning idea of virtual integration is that an agile capability can be enabled through enhanced visibility. Ideally, all parties in the network should share information in as close to real time as possible. This information will include the actual requirement from the field (demand), current inventory dispositions, the supply schedule and event management alerts.

Many traditional supply networks have poor upstream and downstream visibility with little shared information. Hence, they are prone to mismatches of supply and demand at every interface – a situation made worse by the so-called 'bullwhip' effect, which amplifies disturbances in the demand signal as orders are passed up the supply network. Bullwhips can be dramatically reduced or even eliminated if the different echelons in the supply network can be linked through shared information.

The barrier to improved visibility is, however, no longer technological. The tools exist to enable the highest levels of connectivity in even the most fragmented global network. The real challenge is the reluctance that still exists within some organizations to share information across boundaries – be those internal or external. The most agile supply networks are typified by a mindset of collaborative working with other partners in the network based upon a spirit of trust and shared goals.

Lessons from best practice

It may sometimes seem banal or inappropriate to ask the question 'what can humanitarian logistics learn from best practice in the commercial sector?' While there can be no question that the challenge of saving lives is significantly more important than improving on-the-shelf availability of consumer products in a retail outlet, we would argue that there *are* lessons that can be learnt and through which humanitarian logistics practice can be improved.

We have suggested that the key connection between the worlds of commerce and humanitarian logistics is that of uncertainty. We have highlighted how, to a certain extent, such uncertainty can be conquered through agility, but one of the biggest remaining barriers to supply network agility is complexity. In a global supply network this complexity comes in many forms, but one of its most potent manifestations is in the multitude of nodes and links that constitute the network.

As Figure 0.1 suggests, what are often referred to as 'supply chains' are not really chains; rather they are networks or webs of interconnected and interdependent entities. The resulting complexity can be considerable and, unless a means is found of managing across these nodes and links, the system will be prone to disturbance and disruption. The challenge is to synchronize activities across the network so that a more agile response to changes in demand can be achieved. One idea that is attracting attention is the concept of supply network 'orchestration' and a good example of such orchestration is provided by the Hong Kong-based company Li & Fung.

Li & Fung work on behalf of clients, mainly retailers, who are seeking to source products made to their own specification. Thus, for example, the global retailer Walmart might decide that for the next winter season in the United States they want to introduce a range of

FIGURE 0.1 There are two generic categories of risk

- Supply chains comprise nodes and links

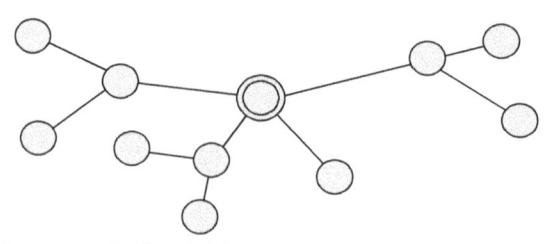

- Nodes – organizational risk
- Links – connectivity risk

low-priced skiwear. Acting on their behalf, Li & Fung will identify the appropriate designers, they will source the different fabrics, fasteners and zips, they will contract with appropriate manufacturers and manage the whole supply network from raw materials through to Walmart's stores. Li & Fung's capability as an orchestrator comes from their specialist knowledge of the industry, their long-standing relationships with suppliers, and their information systems that enable them to coordinate and synchronize the flows of material and product across a complex network.

Sometimes the supply network orchestrator is termed a 'lead logistics provider' or a fourth party logistics (4PL) provider, and companies like DHL, UPS and FedEx are increasingly taking on this role on behalf of global corporations. For example, Cisco, one of the world's leading suppliers of communication network equipment, use UPS to coordinate a large part of their global network of contract manufacturers, distribution service providers and component suppliers to enable a high level of synchronization in what has become a very volatile marketplace. Again, this synchronization is greatly enabled by real time information that is shared across the partners in Cisco's global network.

One important feature of humanitarian emergencies is that each one is different. Hence it has to be recognized that the conventional 'one-size-fits-all' approach to supply chain management clearly does not apply. Even within a specific emergency, the likelihood is that, as the focus moves from the immediate relief effort through to eventual reconstruction, the nature of the supply chain response will need to change. Thus, in the early stage where time is critical the emphasis will be on agility, but later when cost-effectiveness is the objective then 'lean' becomes the watchword.

The implication is that humanitarian relief supply networks must be configured to meet the specific contingencies of the prevailing situation. Meeting this challenge requires a high degree of what has been termed 'structural flexibility'[5]. Structural flexibility refers to the capability of a supply network to adapt rapidly to changing conditions and environments. It is a step beyond the idea of agility previously referred to – rather it suggests an ability to reconfigure the entire end-to-end supply network as and when required. To facilitate this capability, it is important that access to appropriate capacity and assets be easily obtained. In a commercial setting this access is often achieved through collaborative arrangements with other organizations, even, sometimes, with competitors. It is also increasingly commonplace to find assets being shared between supply chain partners rather than being reserved for exclusive use. Equally, it is often seen to be preferable to rent those assets rather than to own them.

However, structural flexibility is as much about 'mindset' as it is about tangible assets. It implies a willingness to work closely with other players in the supply network, to share information and to invest in systems with a high degree of interoperability. This is an idea that is slowly gaining ground in the commercial world as uncertainty and turbulence continue to be the hallmarks of the 21st-century business environment. We would argue that a similar transformation in the way we think about supply networks could be of great benefit in the humanitarian logistics arena.

The way forward

From the above discussion, a key thread running through the development and operation of agile and flexible supply networks is a focus on synchronization enabled by shared information. Clearly, there are other enablers of agility and flexibility, such as process alignment and collaboration across inter- and intra-organizational boundaries, but 360 degree visibility appears to be the critical element. Given that meeting the challenges facing the humanitarian logistic community would seem to demand the ultimate in agile response, it is heartening to recognize that, as reflected in the contributions contained within this book, this message is now gaining traction and that there is a growing commitment to breaking down the barriers to much closer collaboration across organizational boundaries.

In approaching the development of this fourth volume discussing the humanitarian logistics challenge, as editors we were keen to ensure that a broad range of perspectives was presented for consideration by the reader. Furthermore, we were also keen to follow the approach adopted in previous editions and to focus on areas that had not previously been discussed in any level of detail – an obvious example being the impact of the Covid-19 pandemic on both high- and low-income countries and their populations. In particular, we believed that an improved understanding of the challenges and solutions could be gained by garnering contributions from both a geographically diverse community and, critically, from practitioners (as distinct from academics).

In the opening chapter of this edition, Professor Tina Wakolbinger from WU Vienna, Austria, and Dr Fuminori Toyasaki of York University, Canada, reflect on one of the key challenges facing the humanitarian logistic system as a whole. This emanates from the basic structure and complexity of the system for funding the preparation and response mechanisms. Wakolbinger and Toyasaki go on to analyse the potential impact of emerging trends, particularly in relation to fundraising, and also to suggest a number of key areas for further research.

In Chapter 2, one particular aspect of the overall humanitarian logistics supply chain is investigated in depth by Dr Jihee Lim of Anglia Ruskin University, UK, and Professors Stephen Pettit and Anthony Beresford of Cardiff University, UK. These authors focus on the relationship between humanitarian organizations and their suppliers and how this can be improved moving forward.

Continuing this focus on aspects of the supply chains supporting the humanitarian logistician in Chapter 3, Dr Diego Vega, Assistant Professor at the Hanken School of Economics, Finland and Director of the HUMLOG Institute, considers the breadth of providers within such supply chains and how they are able to contribute to the improvement of the response to those impacted by disasters.

Self-evidently, the humanitarian logistician as an individual is an absolutely core component of the overall supply chain, but in the next chapter Professor Paul Larson from the University of Manitoba, Canada, shines a spotlight on one of the more complex and unfortunate aspects of responding to disasters, namely the increasing level of attacks on humanitarian aid workers. Sadly, all of the statistics demonstrate that these are growing, reflecting at least in part, the rise in inter- and intrastate conflicts. Paul offers a focused agenda for further research to understand and ameliorate this emerging challenge.

However, the delivery of humanitarian logistics support also depends on the ability of the providing organization to provide the necessary support, and this element of the overall picture is considered in depth by Svein Håpnes, who was a key member of the supply chain management team in UNHCR for over a decade. In Chapter 5, he explores the challenges within this key humanitarian organization and the associated lessons identified and key success criteria.

Continuing this focus on the ways in which the disaster response supply chain might be improved, Chapter 6, authored by Professor Ruth Banomyong, Puthipong Julagasigorn and Paitoon Varadejsatitwong from Thammasat University in Thailand, and Thomas Fernandez of the University of the Thai Chamber of Commerce, propose the use of a performance measurement tool (HUMSERVPERF) that will enable the monitoring of, and hence the potential for improvement to, humanitarian supply chains.

Following the same generic theme of applying academic models to the humanitarian logistic challenge, in Chapter 7 Gerard de Villiers describes how the centre of gravity analysis that is widely used by commercial companies can be used to improve the efficiency and effectiveness of post-disaster supply chains, with a particular focus on the importance of pre-positioning key items at appropriate locations across the globe.

Chapter 8, authored by Professor Paulo Gonçalves from Università della Svizzera Italiana (USI), Switzerland, considers another way in which the complex challenges associated with the competition for scarce resources during humanitarian response can be mitigated through the adoption of a systems dynamics methodology. This approach sees systems as evolving, dynamic, adaptive, self-organizing and governed by feedback, and with this in mind alternative policy options and approaches can be developed and brought to bear on the humanitarian response effort.

One of the potential ways in which the logistic response to a disaster might be improved is through the use of cash and/or vouchers to provide financial assistance to those affected by the event. In Chapter 9,

Russell Harpring, a researcher at the Hanken School of Economics, Finland, demonstrates the importance of the associated information-sharing networks as well as the continual development of the skills and knowledge that are necessary for such programmes to work.

In Chapter 10 Professor Gyöngyi Kovács from the Hanken School of Economics, Finland, Professor Tina Comes of Delft University of Technology, Netherlands, and Dr Ioanna Falagara Sigala, also of the Hanken School of Economics, Finland, review the impact that the Covid-19 pandemic has had on supply chains that are designed to support those in need. These authors analyse the concrete lessons learnt across multiple elements of the humanitarian supply chain and make recommendations for tools and approaches that can be applied to this specific health logistics challenge, with the aim of helping to ensure that any such future pandemics can be more efficiently and effectively managed from a logistics perspective.

In a similar vein, Professor Maria Besou, Dr Sarah Joseph, Sophie t'Serstevens and Jonas Stumpf of the Kühne Logistics University, Germany, consider another key emerging challenge that is increasingly impacting the humanitarian logistics response – namely that of climate change. In Chapter 11, these authors consider the impacts of short-term tactical responses that, inevitably, are the initial focus of attention in the aftermath of a disaster and the longer-term implications for both local communities and more broadly.

Written from the personal perspective of a number of practitioners, Chapter 12 developed by George Fenton, who is the Executive Director and a co-founder of the Humanitarian Logistics Association, and Tikhwi Jane Muyundo, a supply chain professional based in East Africa, consider a broad range of developments and challenges faced by those working in the humanitarian logistic sector over the last decade as well as the impact of Covid-19 in the recent period.

The final chapter has been written by Professor Gyöngyi Kovács from the HUMLOG Institute at Hanken University, Finland. Universally acknowledged as one of the thought leaders in the humanitarian logistic field, she has risen to the editors' challenge of peering into the mythical crystal ball in order to discern and capture the emerging trends. In doing so, she has first reflected on the accuracy of the predictions that she made when penning a similar concluding chapter to the earlier editions, before considering what other developments have taken place during the last four years with, inevitably, particular reference to the impact of the Covid-19 pandemic. Finally, she offers a new set of thoughts over where the theory and practice of humanitarian logistics is heading. In doing so, not only has she touched on many of the key themes that have been considered in the chapters of this book, but she also highlights a number of emerging challenges.

Endpiece

As editors of this fourth edition of *Humanitarian Logistics*, we wish to take this opportunity to acknowledge the enormous contribution to the field by the outgoing editor, Professor Martin Christopher. Widely recognized as a global leader in developing our understanding of how commercial logistics and supply chain management could be improved, he was able to bring this knowledge and understanding to the humanitarian sector. This has, unquestionably, led to improvements in the efficiency and effectiveness of the preparation and response to disasters, and we hope that we have been able to continue to use these insights in overcoming emerging challenges.

Finally, we would like to reiterate the observation that we made at the end of the previous editions of this book in which we noted that, in inviting contributions, we were keen to stress that authors were at liberty, if not positively encouraged, to offer unusual or controversial viewpoints. Inevitably, therefore, the book will continue to contain perspectives that can be contrasted or may even be thought to be in downright opposition. We have not sought to ameliorate these conflicting viewpoints – indeed, to do so would seem to imply a degree of arrogance as it would suggest that we are aware of the correct and proper approach to a particular challenge. Certainly, this is not the case; rather we hope that this volume will provide some tangible assistance to those tasked with prosecuting the enormously complex business of humanitarian logistics. However, what we would claim is that we have yet to meet a 'wicked problem' in this field that cannot be at least partially tamed through the application of prescriptions drawn from the commercial environment.

Notes

1 Tatham, P H, and Pettit, S J (2010) Transforming humanitarian logistics: The journey to supply network management, *International Journal of Physical Distribution and Logistics Management*, **40**(8/9), pp. 609–22
2 Fernie, J and Sparks, L (2004) *Logistics and Retail Management*, Kogan Page: London
3 Christopher, M G (2016) *Logistics and Supply Chain Management* (5th ed), Pearson Education: London
4 Hammer, M and Champy, J (1993) *Re-engineering the Corporation*, Nicholas Brenley Publishing: London
5 Christopher, M G and Holweg, M (2011) Supply Chain 2.0: Managing supply chains in the era of turbulence, *International Journal of Physical Distribution and Logistics Management*, **41**(1), pp. 63–82

01

Impacts of funding systems on humanitarian operations

TINA WAKOLBINGER AND FUMINORI TOYASAKI

ABSTRACT

Funding systems and financial flows play an important role in humanitarian operations. They, directly and indirectly, affect the scope, speed, effectiveness and efficiency of disaster response. Constraints imposed by funding systems are increasingly considered in models of humanitarian supply chains, however, models that simultaneously optimize fundraising and operational decisions are still rare. This chapter explores the interdependence of financial flows and material flows in humanitarian relief operations. Specifically, this chapter demonstrates how the structure of funding systems and the characteristics of financial flows impact humanitarian operations, how this relationship is captured in current research and how it could be further explored in future work.

Introduction

The year 2021 was challenging for the humanitarian community and 2022 is expected to be difficult as well.

> In 2022, humanitarian action will need to adapt to new and challenging realities. The COVID-19 pandemic is taking a heavy toll in developing countries, civilians continue to be the most affected by conflict and extreme poverty is rising. Climate change effects are devastating, forced displacement is at record levels and 161 million people face acute food insecurity.
>
> (UN OCHA 2021: 19)

Tackling these issues requires reducing inefficiencies in humanitarian operations. Inefficiencies in humanitarian operations have many causes. Current funding systems are one of the causes that have been cited in the literature (e.g. Thomas and Kopzcak, 2005; Jahre and Heigh, 2008; Gupta et al. 2016, Development Initiatives 2016). Funding systems limit the scope of humanitarian response and they, directly and indirectly, affect the speed, effectiveness and efficiency of disaster response. The impact of funding systems is increasing as aid agencies are facing multiple changes and challenges in their environment: increasing demand for disaster relief amid the Covid-19 crisis, increasing numbers of aid agencies leading to more intense competition for donations, earmarking of donations, increasing use of cash transfer programs, new funding mechanisms such as the Central Emergency Response Fund (CERF) and the Country-Based Pooled Funds (CBPFs), the emergence of joint fundraising initiatives and donors that are more demanding in terms of performance, accountability, quality and impact (Thomas and Kopczak, 2005; Beamon and Balcik, 2008; Street, 2009; Development Initiative, 2021).

In this chapter, we explore the interaction between funding systems and humanitarian relief operations. The chapter is organized as follows: first, we provide an overview of the structure of humanitarian funding systems, then we describe how characteristics of financial flows impact the efficiency and effectiveness of humanitarian operations and we show how incentives provided by donors can lead to misallocation of resources. Based on insights concerning how literature has analysed the link between funding systems and humanitarian operations, we offer our perspective on the expected impacts of recent trends in fundraising. Lastly, we provide recommendations concerning future research projects.

Structure of funding systems

Funding systems typically involve multiple stakeholders with diverse objectives. The structure of the humanitarian funding system and the number of stakeholders involved impact the characteristics of funding flows and the power of the stakeholders. Furthermore, they affect the percentage of donations that reach beneficiaries, since intermediaries typically keep a percentage of the money as transaction costs and, thereby, reduce the amount of money that can be used for beneficiaries (Walker and Pepper, 2007).

Funds to deal with the effects of disasters come from different sources. Traditionally, governments provided a significant portion of funds; therefore, they strongly influenced the sector (Thomas and Kopzcak, 2005). Over the years, however, contributions from foundations, individual donors, and the private sector have increased in importance. (Thomas and Kopzcak, 2005; Kovács and Spens, 2007).

Donors need to decide how to allocate their money. Funds are either provided directly to aid providers or channelled through intermediaries (Macrae, 2002). Brokers can help overcome problems of matching donors and humanitarian organizations (Stapleton et al., 2010). Providers of aid include international aid agencies, local NGOs and community-based organizations (Oloruntoba and Gray, 2006). International aid agencies can be divided into three categories: entities operating under the United Nations umbrella; international organizations such as the International Federation of the Red Cross and Red Crescent Movement (IFRC); and global non-governmental organizations (NGOs) (Thomas and Kopczak, 2005). Intermediaries include the World Bank, international organizations and NGOs (Macrae 2002).

Cash transfer programmes (CTPs) are increasingly employed in humanitarian response as a substitute or complement to in-kind aid. CTPs transfer purchasing power directly to beneficiaries in the form of currency for them to obtain goods and/or services directly from the local market (Falagara Sigala and Fuminori, 2018). For a successful implementation of CTPs, humanitarian organizations' collaboration with donors, the private sector (especially financial institutions), local authorities and national governments is indispensable. Literature on CTPs is growing (see Maghsoudi et al. 2021 and literature within).

Administrative costs involved in fund allocation play an increasingly important role in donors' allocation decisions. Furthermore, networks of international aid agencies are being established that collaborate in fundraising (Toyasaki and Wakolbinger, 2019). The advantages of joint fundraising include a reduction in excessive competition for funds, reduced information costs (Rose-Ackerman, 1982), and reduced solicitation costs due to economies of scale in fundraising (Weinblatt, 1992). Disadvantages include reduced discretion in the allocation of funds, concerns over a possible loss of market share (Westhead and Chung, 2007) or independence (Chua and Wong, 2003; Nunnenkamp and Öhler, 2012). Since financial considerations are very important in encouraging aid agencies to participate in a joint fundraising organization (Chan, 1998 in Chua and Wong, 2003), allocation rules applied by the intermediary strongly determine its desirability for aid agencies.

Besides deciding how to allocate money, donors and intermediaries also need to decide on restrictions that they impose on the use of financial funds and reporting requirements. Accountability is of increasing concern to many donors, and this has led to increases in reporting requirements for aid agencies. Donors who want to ensure that their resources are used for their intended purpose need to consider the impact of their restrictions with respect to the use of their resources. While more restrictions allow for greater control, they also potentially reduce the effectiveness and efficiency of aid agencies' operations, as the next section shows.

Impacts of financial flows on disaster response

Financial flows are an important input in humanitarian supply chain operations and, therefore, humanitarian operations are strongly affected by the characteristics of funding flows (Development Initiatives, 2009a). When analysing the impact of financial flows, traditionally, significant emphasis has been put on the total amount of donations received.

However, speed and timing, fluctuation and predictability and flexibility of funds also influence the efficiency and effectiveness of disaster response. These characteristics strongly impact the value of donations from the perspective of an aid agency.

Volume

While need for humanitarian assistance is growing, international humanitarian assistance plateaued in 2020, hence the funding gap is growing (Development Initiatives, 2021). Given the increasing need for disaster response, this situation is not likely to change. Aid agencies need to compete for the resources that are available. The amount of funds that an aid agency receives determines the scope of the relief operations that it can conduct. Large amounts of donations provide aid agencies with the opportunity to take advantage of economies of scale and to gain influence and negotiation power with suppliers. Furthermore, large amounts of resources provide aid agencies with visibility to donors. Due to the importance of financial funds, a large literature on fundraising strategies and issues exists.

When analysing the impact of increased donation amounts, it is essential to distinguish between donations that are earmarked and donations that are not. Earmarking/restricting donations means that donors put conditions on their gifts and select what projects or activities to fund within the recipient organization (Barman, 2008). In the case of earmarked funds, increasing donation amounts is not always desirable for aid agencies, as too much money is allocated to specific emergencies. This is especially true for disasters/emergencies with considerable media attention. Aid agencies that conduct relief operations for these disasters/emergencies tend to receive large amounts of earmarked donations that cannot be sensibly spent within the allocated timeframe. In these situations, aid agencies sometimes discourage donors from donating money, e.g., Doctors Without Borders discouraged donors from donating money for the 2004 Indian Ocean earthquake and tsunami (Eggerston, 2006).

Fluctuation and predictability

Fluctuations in funding levels can be observed in funding from private and government donors. Private donation levels are strongly impacted by the amount and type of news coverage (Bennett and Kottasz, 2000; Tomasini and Van Wassenhove, 2009). The 2004 Indian Ocean earthquake and tsunami, for example, led to a massive response from the donor community, while other emergencies that were neglected by the media received few resources. In the case of government donations, money that has been pledged is not always delivered. In the refugee crises in Darfur, Western Sudan and after Hurricane Mitch, for example, aid agencies only received a third of promised funds (Oloruntoba, 2005).

Wild fluctuations with respect to donations make it difficult for aid agencies to use their resources efficiently. A sudden inflow of financial resources might overload an aid agency's capacity to handle these resources, while minimal donations might force an aid agency to reduce valuable resources and capabilities. Kovács and Spens (2007) divide disaster relief operations into three phases: preparation, immediate response and reconstruction. Adequate funding for the preparation stage strongly determines how quickly and efficiently an aid agency can respond to a disaster (Jahre and Heigh, 2008). Of special importance is multi-year funding, since it allows for long-term planning (Development Initiatives, 2021).

Speed and timing

Once a disaster occurs, an immediate response is critical. Time delays can lead to loss of lives (Kovács and Spens, 2007; Beamon and Balcik, 2008).

> The response to the Covid-19 pandemic has illustrated weaknesses in current approaches to crisis financing, with finance often found only after a disaster strikes, and where disbursements can be slow, poorly coordinated and therefore inequitable. (Development Initiatives, 2021)

How quickly an aid agency can respond to a disaster depends on its preparedness. However, it also depends on how fast the aid agency is able to receive money to set up operations. Some aid agencies have resources that they can use to pre-finance their operations before they get aid from outside sources (Development Initiatives, 2009a). The IFRC, for example, has a Disaster Relief Emergency Fund. The Fund is used immediately after a disaster and allows the IFRC to respond quickly in many emergencies, for example, in the case of the Gujarat earthquake (Chomilier et al., 2003). Aid agencies that do not have the financial resources to pre-finance their operations need to wait until they receive the aid before they can respond, resulting in costly time delays of their response.

While many donors are aware of the importance of quick aid, they are also increasingly concerned about financial accountability. The increasing desire for financial accountability slows down releasing funds from official sources, which can take up to 40 days (Walker and Pepper, 2007).

Flexibility

Aid agencies receive earmarked and non-earmarked donations. Government aid is often earmarked concerning regions and use. Private donations are also often earmarked for a particular disaster; however, they are typically not earmarked concerning how an aid agency needs to spend the money. Private donors are typically willing to reallocate donations if the need arises (Development Initiatives, 2009a). Earmarked donations reduce aid agencies' flexibility in their allocation decisions. Aid agencies may be forced to allocate money and resources to emergencies, activities and projects that provide little benefit to them. Interest in the OR community on the topic of earmarking and its implications on efficiency, service levels and operational performance has been increasing (Burkart et al., 2016 and references therein).

ALLOCATION TO EMERGENCIES

Private as well as government donors frequently earmark donations with respect to the emergency that it should be used for. Resources are often not allocated according to need but according to donors' preferences, which are frequently driven by media attention in the case of private donors and political and strategic considerations in the case of government donors.

Too many resources allocated to one area not only take away from resources being used in other areas, but they can also lead to increased competition, increased prices and wasted resources (Van Wassenhove, 2006; Beamon and Balcik, 2008). Furthermore, donations and equipment that are earmarked for certain areas also restrict aid agencies' flexibility concerning their allocation and reallocation of resources (Besiou et al., 2012). Postponement strategies have been shown to potentially lead to large cost savings in humanitarian supply chains (Oloruntoba and Gray, 2006; Jahre and Heigh, 2008), but earmarked donations severely restrict postponement strategies.

ALLOCATION TO RESOURCES AND ACTIVITIES

Donor funding tends to focus on direct programme and project inputs and does not provide enough funding for disaster preparedness, infrastructure, information and logistics systems (Gustavsson, 2003; Thomas, 2007; Beamon and Balcik, 2008). IFRC, for example, found it challenging to obtain funds for disaster preparedness and capacity building (Chomilier et al., 2003). Besides too little funding for the preparation phase, there is also often too little funding for the long-term phase of reconstruction (Kovács and Spens, 2007).

Resources need to be available in the right amounts in each disaster relief phase: preparation, immediate response and reconstruction. An extreme focus on short-term relief leads to a lack of planning, capacity building and investment in infrastructure and employee training as well as a lack of long-term reconstruction (Kovács and Spens, 2007; Oloruntoba, 2007; Perry, 2007; Beamon and Balcik, 2008; Goncalves, 2008; Jahre and Heigh, 2008). Lack of planning leads to high competition for available resources, overuse of expensive and unsafe transportation modes and increased supply chain costs (Oloruntoba, 2007; Jahre and Heigh, 2008). Not enough investment into areas such as computer systems and employee training leads to wasted time, reduced efficiency and increased costs (Perry, 2007; Beamon and Balcik, 2008).

Aid agencies also benefit from flexibility concerning changing the allocation of resources at any time. The situation in the emergency areas frequently changes. Hence, it is vital to be able to change the allocation of resources accordingly. Still, donors' increasing desire for financial accountability has led to a large amount of funding being allocated against requests for proposals which has, in turn, strongly limited aid agencies' flexibility in their allocation decisions (Walker and Pepper, 2007).

Volume, speed, fluctuation, predictability and flexibility of funds from different donors and intermediaries determine how desirable they are for aid agencies. Private funds, for example, are typically preferable to official funds with respect to the speed with which they are available as well as the flexibility with which they can be used (Development Initiatives, 2009a). However, not every aid agency values these characteristics equally. The speed of financial flows, for example, is typically more important for small organizations that do not have enough money to pre-fund activities than for large organizations with funding reserves. Also, the importance of these characteristics differs with respect to the phase of disaster relief. Speed of aid provision, for example, is crucial for immediate disaster relief and private funds are therefore often used in this phase.

Jahre and Heigh (2008) provide a table of preferred funding models for each disaster relief phase differentiating between earmarked and non-earmarked as well as long- and short-term funds, and in doing so, highlight the importance of flexibility and speed in the response phase of disaster relief. Table 1.1 provides an overview of the main objectives of fund management and their associated challenges with respect to the characteristics discussed in this section.

Aid agencies need to determine the optimal portfolio of funding sources based on the characteristics of funds from different sources, their specific needs, the type of disaster they respond to, as well as the phase of the disaster response. Some previous papers analyse the importance and trade-offs between some of these characteristics. For example, Toyasaki and Wakolbinger (2014) analyse the trade-off between size and flexibility, while papers such as Jahre and Heigh (2008), Balcik and Beamon (2008), Martinez et al. (2011), and Besiou et al. (2012) highlight the importance and value of flexibility.

TABLE 1.1 Main objectives and challenges in fund management

Disaster response Phase	Main Objectives	Challenges
Preparation Phase	Forecast costs and needs for various disaster scenarios	Uncertainty concerning future needs
	Allocate resources to prepare for future disasters and disaster response	Volume: Shortage of funds because of donors' lack of interest
Immediate Response Phase	Quickly secure adequate amount of funds for disaster response under limited information concerning need	Speed: Slow disbursement of government funds
		Predictability: Strong role of media and strong fluctuation in private funds
		Flexibility: Increasing earmarking of private and government funds
		Volume: Too little or too many funds depending on media response
Reconstruction Phase	Allocate funds to long-term projects considering need, impact and cost effectiveness	Volume: Shortage of funds because of donors' lack of interest
	Consider interaction between reconstruction phase and immediate response phase, especially for areas where emergencies occur frequently	Predictability: Reallocation of funds if new emergencies arise
		Coordination: Different organizations and funding sources for disaster relief and long-term development

SOURCE Jahre and Heigh, 2008

Incentives provided by donors

As the previous section highlights, donors can directly influence how their donations are used. Besides this direct impact, donors also indirectly influence how aid agencies use their resources. When aid agencies make decisions concerning resource allocations, they need to consider how these decisions impact donors' perception of their work and future donation streams.

Donors are interested in making sure that their donations are used in the best possible way; however, they cannot directly measure the impact of their donations (Tatham and Hughes, 2011). Although current efforts exist to measure the performance of aid agencies, this is not an easy task due to the intangibility of services offered and characteristics of humanitarian operations (Beamon and Balcik, 2008). Since donors cannot directly observe the quality of an aid agency's work, they need to rely on indicators of the quality of aid agencies' operations. Examples of indicators that are currently used are visibility, fundraising cost percentage and overhead costs. These indicators have the potential to provide aid agencies with incentives for inefficient resource allocations.

Visibility

Aid agencies that want to make sure that their efforts are noted by donors, especially private donors, need to focus their activities on emergencies and activities with high donor visibility. Aid agencies' desire for visibility leads aid agencies to provide aid in areas with considerable media attention, which are typically areas that receive a large volume of aid (Oloruntoba, 2007). Furthermore, it leads aid agencies to focus their activities on the response phase, since it provides more possibilities for activities with high visibility than the preparation phase (Jahre and Heigh, 2008). It also encourages aid agencies to participate in projects and activities with high donor visibility, for example, provision of water as opposed to infrastructure development. In addition, the need to be visible to donors can also reduce collaboration and coordination between aid agencies (Tomasini and Van Wassenhove, 2009) since aid agencies want to emphasize their own contribution.

Financial indicators

Government and private donors increasingly focus on financial accountability. In the case of government donors, this leads to slower disbursement of donations and reduced flexibility for aid agencies as discussed in the previous section. Private donors who want to make informed decisions when donating money can base their decisions on information provided by organizations such as Charity Navigator, Give Well and BBB Wise Giving Alliance. These organizations provide rankings and/or evaluations of aid agencies based on a multitude of financial and non-financial factors. Financial indicators that are used include overhead costs and fundraising cost percentages. Problems of both indicators are that they are currently not consistently defined and measured and that data quality is often low. Furthermore, these indicators are influenced by many different factors, e.g. NGOs that receive official funds typically have lower fundraising costs but higher administrative costs than comparable aid agencies without official funds (Nunnenkamp and Öhler, 2012). Hence, a detailed and often costly analysis is necessary in order to receive valid information about an organization's financial efficiency. In addition to these general problems, each indicator also has its unique associated problems.

Overhead costs include support activities (Burkart et al., 2018). By supporting aid agencies that have low overhead costs, donors encourage aid agencies to focus on the response phase, and they discourage them from allocating sufficient amounts of money to the preparation phase with all the negative consequences discussed in the previous sections. Furthermore, competition to lower costs might lead to underreporting of administrative expenses (Krishnan et al., 2006).

The fundraising cost percentage is defined as total fundraising costs divided by total funds raised. This indicator is typically reported by aid agencies. It represents the percentage of donations that cannot be directly used for disaster relief activities. Donors and policymakers generally prefer aid agencies with a low fundraising cost percentage (Sargeant and Kaehler, 1999; Hopkins, 2002). However, when looking at fundraising cost percentages, one must consider that dollars raised for 'unpopular' emergencies typically require more fundraising activities than fundraising for 'popular' emergencies. Trying to reach a low fundraising cost percentage might encourage aid agencies to focus their fundraising activities on popular emergencies. Hence, this might further emphasize emergencies that already receive a disproportionate amount of aid. This problem is further increased due to the increasing amount of private funds earmarked for certain disasters and donors' increasing willingness to enforce these allocation decisions.

A strong focus on visibility and financial accountability can provide aid agencies with incentives to use their resources inefficiently. It can potentially encourage aid agencies to provide aid for emergencies that are already crowded with relief groups. It can also encourage them to focus on short-term objectives instead of long-term goals. That said, donors are starting to realize that visibility and financial indicators do not always reflect the quality of aid agencies' work. However, given that better indicators are still largely missing, the development of improved measurement for aid agencies' performance, need in different regions, the impact of improvements in logistics systems and time value of money are of utmost importance. Furthermore, a stronger focus on programme effectiveness instead of financial efficiency is necessary (Lowell et al., 2005).

Fundraising–humanitarian operations interface in literature

Burkart et al. (2016) provide a literature review of papers addressing the funding–humanitarian supply chain interface and a framework for categorizing the articles. The authors show that the number of publications has been increasing over the years but that more research is necessary to fully capture the impact of funding on humanitarian supply chains.

In terms of the literature in the field of humanitarian supply chain management, we see an evolution of research. The first papers focused strongly on either fundraising issues without any operational considerations or literature that considers funding as a constraint in operational models. More recently, we see literature that combines both areas. We provide examples of papers from each research stream.

Fundraising issues have been analysed by researchers in the areas of economics, operations research and marketing, among others. This literature highlighted, for example, implications of auditing (e.g. Privett and Erhun, 2011), earmarking (e.g. Toyasaki and Wakolbinger, 2014 Ülkü et al., 2015; Aflaki and Pedraza Martinez, 2020; Fuchs et al., 2020; Özer et al., 2021), financial and performance indicators (e.g. Burkart et al., 2018; Kotsi et al., 2020) and information disclosure (e.g. Zhuang et al.

2014; Fajardo et al. 2018; Yang and Hsee, 2022) as well as fundraising coordination (Toyasaki and Wakolbinger, 2019; Aflaki and Pedraza Martinez, 2020) on donation amounts. Also, the issue of crowdfunding platforms is increasing in attention (see e.g. Mejia et al., 2019; Behl and Dutta, 2020a; Behl and Dutta, 2020b). These models, however, often ignore the interplay between aid agencies' decisions on operations and donors' reaction to aid agencies' fundraising.

Literature that highlights the impact of funding constraints on operational decisions is increasing. Several researchers shed light on the impact of budgetary constraints on aid agencies' procurement and inventory decisions (e.g. Balcik and Beamon 2008; Natarajan and Swaminathan, 2014; Natarajan and Swaminathan, 2014; Park et al., 2018; Fard et al., 2019; Fard et al., 2021). Recently, several researchers applying the newsvendor approach considered aid agencies' operational challenges stemming from budget limitations and how this applies in the context of prepositioning of inventory (see Chen et al., 2018; Chakravarty, 2014; Acar and Kaya, 2021; Eftekhar et al., 2021). Also, the issue of fleet coordination and management was addressed considering budget constraints and earmarking (Besiou et al., 2014; Aflaki and Pedraza Martinez, 2020). The aforementioned literature treats donations or budgets as exogenously given parameters or exogenous random shocks. Exploring aid agencies' decisions on their budget allocation in the context of the interplay between aid agencies' operational decisions and donors' reactions is an important future research avenue.

Literature that actively manages fundraising and operational decisions jointly considering their interactions is rare. Most recently, Arikan et al. (2022) endogenously treated the interaction between aid agencies' fundraising and resulting donors' reactions in the context of prepositioning decisions. They compared a budget-constrained case with a budget-unconstrained counterpart from the perspective of relief operation effectiveness and efficiency. Using the case of the International Federation of Red Cross and Red Crescent Societies (IFRC), Turrini et al. (2020) empirically explored the impact of operational expenditures on the donation amount that the IFRC receives.

Summary and conclusion

Due to the increasing demand for disaster relief and limited resources, aid agencies must use the available resources in the best possible way. Currently, misallocation of resources reduces the efficiency and effectiveness of humanitarian operations. Resource misallocations are partly caused by aid agencies' difficulty in determining the optimal allocation of resources. Aid agency workers are often not aware of the value of logistics and information systems and, therefore, do not invest enough in these areas. Researchers in the area of OR/MS are working on projects that highlight the value of logistics operations and information systems. Furthermore, while models have been developed that help aid agencies in allocating appropriate resources to different phases and activities in humanitarian operations, very often these models do not consider financial constraints. The inclusion of financial constraints into such models of logistics operations and the analysis of their impact is of key importance.

Resource misallocations are also partly caused by funding systems. They would not disappear until funding systems are improved and donors are educated about the consequences of their decisions. Donors provide resources for aid agencies and, hence, have a strong influence on aid agencies' allocation decisions. On the one hand, donors can directly impact allocation decisions by earmarking donations for specific emergencies or activities. On the other hand, they can provide incentives that guide aid agencies towards a particular decision. Currently, donors frequently explicitly and implicitly provide incentives for aid agencies' behaviour that leads to too many resources being allocated to direct response instead of preparedness and reconstruction, and also to excess resources being provided for those emergencies that have gained media attention while others are largely neglected.

Aid agencies, UN agencies and donors start being aware of the shortcomings of current funding systems. They are re-evaluating their previous approach to humanitarian funding and are trying to improve the system through many programmes such as the Good Humanitarian Donorship Initiative (2003). Other new initiatives such as joint fundraising can provide sound fundraising systems as they can contribute to lowering competition and excessive fundraising efforts. However, even in the case of joint fundraising initiatives, they can worsen the situation if they are not carefully implemented (Toyasaki and Wakolbinger, 2019).

Operations researchers and operations research tools could significantly contribute to establishing sound humanitarian funding systems. The number of publications in this research field has increased over previous years (Burkart 2016). Further research is necessary to capture key elements and stakeholders of funding systems, their characteristics and impacts. System dynamics models can highlight the interaction between money and product flows as well as trade-offs between short-term and long-term goals (Goncalves, 2008; Besiou et al., 2011). Principal-agent models (Seabright, 2001) can provide insights that indicate how to improve the alignment of the interests of the different stakeholders in humanitarian supply chains. Development of indicators of need and quality of response (Beamon and Balcik, 2008) can reduce information asymmetries between donors and aid agencies, which can contribute to improving the quality of aid agency evaluations and rankings. Optimization models can highlight the trade-offs between different characteristics of funding flows, allowing aid agencies to create appropriate funding portfolios. The effects of competition and collaboration in fundraising activities have been described in the economics literature (Rose-Ackerman, 1982; Bilodeau and Slivinski, 1997) and they are currently analysed in the operations research literature (Toyasaki and Wakolbinger, 2019; Aflaki et al., 2020). Further research is necessary concerning the benefits and drawbacks of joint fundraising modes and the effects of allocation rules. We expect to see more research papers that address these issues in the future.

Acknowledgement

This work was partially funded by a Summer Research Grant from the Fogelman College of Business and Economics, University of

Memphis, a start-up fund at Faculty of Liberal and Professional Studies, York University, CIHR Canadian 2019 Novel Coronavirus (COVID-19) Rapid Research Funding and YUFA Leave Fellowship Fund. This support is gratefully acknowledged.

References

Acar, M and Kaya, O (2021) Inventory decisions for humanitarian aid materials considering budget constraints, *European Journal of Operational Research*, https://doi.org/10.1016/j.ejor.2021.07.029 (archived at https://perma.cc/F3PN-P9ZE)

Aflaki, A and Pedraza Martinez, A (2020) Competition and collaboration on fundraising for short-term disaster response: The impact on earmarking and performance http://dx.doi.org/10.2139/ssrn.3705595 (archived at https://perma.cc/RT3R-G65X)

Arikan, E, Silbermayr, L and Toyasaki, F (2022) Interplay between humanitarian procurement operations and fundraising. York University Working Paper.

Balcik, B and Beamon, B M (2008) Facility location in humanitarian relief, *International Journal of Logistics: Research and Applications*, **11**(2), pp 101–21

Barman, E (2008) With strings attached, *Nonprofit and Voluntary Sector Quarterly*, **37**(1), pp 39–56

Beamon, B M and Balcik, B (2008) Performance measurement in humanitarian relief chains, *International Journal of Public Sector Management*, **21**(1), pp 4–25

Behl, A and Dutta, P (2020a), Social and financial aid for disaster relief operations using CSR and crowdfunding: Moderating effect of information quality, *Benchmarking: An International Journal* **27**(2), pp 732–759 https://doi.org/10.1108/BIJ-08-2019-0372 (archived at https://perma.cc/UT9Y-PTGA)

Behl, A and Dutta, P (2020b) Engaging donors on crowdfunding platform in Disaster Relief Operations (DRO) using gamification: A Civic Voluntary Model (CVM) approach, *International Journal of Information Management* **54**, 10.1016/j.ijinfomgt.2020.102140 (archived at https://perma.cc/X2FD-7G26)

Bennett, R and Kottasz, R (2000) Emergency fundraising for disaster relief, *Disaster Prevention and Management*, **9** (5), pp 352–60

Besiou, M, Stapleton, O and Van Wassenhove, L N (2011) System dynamics for humanitarian operations, *Journal of Humanitarian Logistics and Supply Chain Management*, **1**(1), pp 78–103

Besiou, M, Pedraza-Martinez, A J and Van Wassenhove, L N (2012) The effect of earmarked funding on fleet management for relief and development, INSEAD Working Paper No. 2012/10/TOM/ISIC, https://papers.ssrn.com/sol3/papers.cfm?abstract_id=1991068 (archived at https://perma.cc/2BTB-BN4F)

Besiou, M, Pedraza-Martinez, A J and Van Wassenhove, L N (2014) Vehicle supply chains in humanitarian operations: Decentralization, operational mix and earmarked funding, *Production and Operations Management* **23**(11), pp 1950–1965

Bilodeau, M and Slivinski, A (1997) Rival charities, *Journal of Public Economics*, **66**, pp 449–67

Burkart, C, Besiou, M and Wakolbinger, T (2016) The funding–humanitarian supply chain interface, *Surveys in Operations Research and Management Science* **21**(2), pp 31–45

Burkart, C, Wakolbinger, T and Toyasaki, F (2018) Funds allocation in NPOs: the role of administrative cost ratios, Central European Journal of Operations Research 26, pp 307–330

Chakravarty, A K (2014) The humanitarian relief chain: rapid response under uncertainty, *International Journal of Production Economics* **151**(C), pp 146–157

Chen, J, Liang, L and Yao, D (2018) Pre-positioning of relief inventories: a multi-product newsvendor approach, *International Journal of Production Research* **56**(18), pp 6294–6313

Chomilier, B, Samii, R and Van Wassenhove, L N (2003) The central role of supply chain management at IFRC, *Forced Migration Review* **18**, pp 15–18

Chua, V C H and Wong, C M (2003) The role of united charities in fund-raising: the case of Singapore, *Annals of Public and Cooperative Economics* **74**(3), pp 433–464

Development Initiatives (2009a) Public support for humanitarian crises through aid agencies, United Kingdom www.globalhumanitarianassistance.org/Projects.htm (archived at https://perma.cc/AB34-BFMZ)

Development Initiatives (2009b) Global Humanitarian Assistance Report 2009, United Kingdom www.globalhumanitarianassistance.org (archived at https://perma.cc/A5JJ-AEZF)

Development Initiatives (2016) Global Humanitarian Assistance Report 2016, United Kingdom www.globalhumanitarianassistance.org (archived at https://perma.cc/A5JJ-AEZF)

Development Initiatives (2021) Global Humanitarian Assistance Report 2021, United Kingdom www.globalhumanitarianassistance.org (archived at https://perma.cc/A5JJ-AEZF)

Eftekhar, M, Song, J-S J and Webster, S (2021) Pre-positioning and local-purchasing for emergency operations under budget and supply uncertainty, *Manufacturing & Service Operations Management* 24(1). https://doi.org/10.1287/msom.2020.0956 (archived at https://perma.cc/3AJZ-ZA43)

Eggerston, L (2006) Tsunami donations help worldwide, *Canadian Medical Association Journal* **274**(3), p 299

Falagara Sigala, I and Fuminori, T (2018) Prospects and bottlenecks of reciprocal partnerships between the private and humanitarian sectors in cash transfer programming for humanitarian response. In: Kotsireas, I, Nagurney, A and Pardalos, P (eds) *Dynamics of Disasters*. DOD 2017. Springer Optimization and Its Applications, vol 140. Springer, Cham. https://doi.org/10.1007/978-3-319-97442-2_3 (archived at https://perma.cc/P6YS-3B7A)

Fajardo TM, Townsend, C and Bolander, W (2018) Towards an optimal donation solicitation: Evidence from the field of the differential influence of donor-related and organization-related information on donation choice and amount, *Journal of Marketing* **82**(2), pp 142–152

Fard, M K, Eftekhar, M and Papier, F, An approach for managing operating assets for humanitarian development programs, *Production and Operations Management* **28** (8), pp 2132–2151

Fard, M K, Ljubić, I and Papier, F (2021) Budgeting in international humanitarian organizations. *Manufacturing & Service Operations Management*, forthcoming. https://doi.org/10.1287/msom.2021.1016 (archived at https://perma.cc/DU63-HJ9T)

Fuchs, C, de Jong, M G, and Schreier, M (2020) Earmarking donations to charity: Cross-cultural evidence on its appeal to donors across 25 countries, *Management Science* **66**(10), pp 4820–4842

Goncalves, P (2008) System dynamics modeling of humanitarian relief operations, working paper, MIT Sloan Research Paper No 4704–08

Good Humanitarian Donorship (2003) 23 principles and practices of good humanitarian donorship www.ghdinitiative.org/ghd/gns/home-page.html (archived at https://perma.cc/J726-LDSU)

Gupta, S, Starr, M K, Farahani, R Z and Matinrad, N (2016) Disaster management from a POM perspective: mapping a new domain, *Production and Operations Management* **25**(10), pp 1611–1637

Gustavsson, L (2003) Humanitarian logistics: Context and challenges, *Forced Migration Review,* **18**, pp 6–8

Hopkins, B R (2002) *The law of fund-raising*, Wiley: New York

Jahre, M and Heigh, I (2008) Does failure to fund preparedness mean donors must prepare to fund failure in humanitarian supply chains?, in Autere, V, Bask, A H, Kovács, G, Spens, K and Tanskanen, K (eds), *Beyond Business Logistics*, NOFOMA Conference Proceedings, Helsinki, Finland, 265–282

Kotsi, T, Aflaki, A, Goker, A and Pedraza Martinez, A (2020) Allocation of Nonprofit Funds Among Program, Fundraising, and Administration, Available at SSRN: https://ssrn.com/abstract=3460795 (archived at https://perma.cc/6VXU-ZYJM) or http:// dx.doi.org/10.2139/ssrn.3460795 (archived at https://perma.cc/Y76P-BCUU)

Kovács, G and Spens, K M (2007) Humanitarian logistics in disaster relief operations, *International Journal of Physical Distribution & Logistics Management*, **37**(2), pp 99–114

Krishnan, R, Yetman, M and Yetman, R (2006) Expense misreporting in nonprofit organizations, *The Accounting Review*, **81**(2), pp 399–420

Lee, H W and Zbinden, M (2003) Marrying logistics and technology for effective relief, *Forced Migration Review*, **18**, pp 34–45

Lowell, S, Trelstad, B and Meehan, B (2005) The ratings game, *Stanford Social Innovation Review*, **3** (2), pp 38–45

Macrae, J (2002) The bilateralisation of humanitarian response: trends in the financial, contractual and managerial environment of official humanitarian aid, A background paper for UNHCR https://reliefweb.int/report/world/bilaterisation-humanitarian-response-trends-financial-contractual-and-managerial (archived at https://perma.cc/93QC-5CVN)

Martinez, A P, Stapleton, O and Van Wassenhove, L N (2011) Field vehicle fleet management in humanitarian operations: A case-based approach, *Journal of Operations Management*, **29**(5), pp 404–21.

Maghsoudi, A, Harpring, R, Piotrowicz, WD and Heaslip, G (2021) Cash and voucher assistance along humanitarian supply chains: A literature review and directions for future research, *Disasters,* https://doi.org/10.1111/disa.12520 (archived at https://perma.cc/B8PD-2K3B)

Mejia, J, Urrea, G and Pedraza-Martinez, A J (2019) Operational transparency on crowdfunding platforms: effect on donations for emergency response, *Production and Operations Management* **28**(7), pp 1773–1791

Natarajan, V K, Swaminathan, J M (2014) Inventory management in humanitarian operations: Impact of amount, schedule, and uncertainty in funding, *Manufacturing & Service Operations Management* **16**(4), pp 595–603

Nunnenkamp, P and Öhler, N (2012) Funding competition and efficiency of NGOs: An empirical analysis of non-charitable expenditure of US NGOs engaged in foreign aid, *KYKLOS*, **65**(1), pp 81–110

Oloruntoba, R (2005) A wave of destruction and the waves of relief: Issues, challenges and strategies, *Disaster Prevention and Management* **14**(4), pp 506–21

Oloruntoba, R (2007) Bringing order out of disorder: Exploring complexity in relief supply chains, in Laptaned, U (ed) *Proceedings 2nd international conference on operations and supply chain management: regional and global logistics and supply chain management*, Bangkok, Thailand

Oloruntoba, R and Gray, R (2006) Humanitarian aid: An agile supply chain? *Supply Chain Management*, **11**(2), pp 115–20

Özer, Ö, Urrea, G and Villa, S, To earmark or to non-earmark? The role of control, transparency, salience and warm-glow (August 16, 2021). Available at SSRN: https://ssrn.com/abstract=3907401 (archived at https://perma.cc/B2SJ-3MAW) or http:// dx.doi.org/10.2139/ssrn.3907401 (archived at https://perma.cc/8VDZ-VYQN)

Park, H J, Kazaz, B, Webster, S (2018) Surface vs. air shipment of humanitarian goods under demand uncertainty, *Production and Operations Management* 27(5), 929–948

Perry, M (2007) Natural disaster management planning. A study of logistics managers responding to the tsunami, *International Journal of Physical Distribution and Logistics Management*, 37(5), pp 409–33

Privett, N and Erhun, F (2011) Efficient funding: Auditing in the nonprofit sector, *Manufacturing and Service Operations Management*, 13(4), pp 471–88

Rose-Ackerman, S (1982) Charitable giving and 'excessive' fund-raising, *The Quarterly Journal of Economics*, 97(2), pp 193–212

Sargeant, A and Kaehler, J (1999) Returns on fund-raising expenditures in the voluntary sector, *Nonprofit Management and Leadership*, 10(1), pp 5–19

Seabright, P (2001) Conflicts and objectives and task allocation in aid agencies: general issues and application to the European Union, in Martens (ed), *The Institutional Economics of Foreign Aid*, Cambridge University Press, Cambridge

Stapleton, O, Van Wassenhove, L N and Tomasini, R (2010) The challenges of matching private donations to humanitarian needs and the role of brokers, *Supply Chain Forum: An International Journal*, 11(3), pp 42–53

Street, A (2009) Review of the engagement of NGOs with the humanitarian reform process, synthesis report commissioned by the NGOS and Humanitarian Reform Project

Tatham, P H and Hughes, K (2011) Humanitarian logistics metrics: where we are, and how we might improve, in Christopher, M G and Tatham, P H (eds) *Humanitarian Logistics: Meeting the challenge of preparing for and responding to disasters*, Kogan Page, London

The European Commission (ECHO) (2017) Guidance to partners funded by ECHO to deliver medium to large-scale cash transfers in the framework of 2017 HIPs and ESOP. Ref. res (2017)516771 - 31/01/2017

The UN High-level panel (2016) High-level panel on humanitarian financing report to the United Nations secretary-general: Too important to fail—addressing the humanitarian financing gap

Thomas, A S (2007) Humanitarian logistics: enabling disaster response, Fritz Institute, San Francisco, CA

Thomas, A S and Kopczak, L R (2005) From logistics to supply chain management: The path forward in the humanitarian sector, white paper, Fritz Institute, San Francisco, CA

Tomasini, R and Van Wassenhove, L N (2009) *Humanitarian logistics*, Palgrave Macmillan, Hampshire, UK

Toyasaki, F and Wakolbinger, T (2014) An analysis of impacts associated with earmarked private donations for disaster relief, *Annals of Operations Research* 221(1), pp 427–447

Toyasaki, F and Wakolbinger, T (2019) Joint Fundraising Appeals: Allocation Rules and Conditions that Encourage Aid Agencies' Collaboration, *Decision Sciences* 50(3), 612–648

Turrini, L, Besiou, M, Papies, D, Meissner, J (2020) The role of operational expenditure and misalignments in fundraising for international humanitarian aid, *Journal of Operations Management* 66(4), pp 379–417

UN OCHA (2021) Global Humanitarian Overview 2022. UN Office for the Coordination of Humanitarian Affairs, Geneva

Ülkü, M A, Bell, K M, Wilson, S G (2015) Modeling the impact of donor behavior on humanitarian aid operations, *Annals of Operations Research* 230(1), pp 153–168

Van Wassenhove, L N (2006) Humanitarian aid logistics: Supply chain management in high gear, *The Journal of the Operational Research Society*, 57(5), pp 475–89

Walker, P and Pepper, K (2007) Follow the money: A review and analysis of the state of humanitarian funding, background paper for the meeting of the Good Humanitarian Donorship and Inter Agency Standing Committee, 20th July 2007, Geneva, https://fic.tufts.edu/assets/Walker-Follow+the+Money-A+Review+and+Analysis+of+the+State+of+Humanitarian+Funding.pdf (archived at https://perma.cc/58Y7-88YD)

Weinblatt, J (1992) Do government transfers crowd out private transfers to non-profit organizations? The Israeli experience, *International Journal of Social Economics*, 19(2), pp 60–66

Westhead, R and Chung, M (2007) Aid coalition formed; Care Canada, Oxfam, Save the Children team up to cut down on donor fatigue, but other agencies fear loss of market share, Toronto Star, January 1, A4. Toronto, Ontario,

Yang AX, and Hsee, CK (2022) Obligatory Publicity Increases Charitable Acts, *Journal of Consumer Research* 48(5), pp 839–857

Zhuang, J, Saxton, G D and Wu, H (2014) Publicity vs. impact in nonprofit disclosures and donor preferences: a sequential game with one nonprofit organization and N donors, *Annals of Operations Research*, 221(1), pp 469–491

02

Supplier relationships in humanitarian organizations

JIHEE KIM, STEPHEN PETTIT AND ANTHONY BERESFORD

ABSTRACT

Few studies have investigated the relationships between humanitarian organizations and their suppliers. This leads to difficulties in specifying the scope, framework, theories and factors on this topic. This chapter discusses the relationship dynamics between humanitarian organizations (HOs) and their suppliers and the diverse factors that should be considered in the humanitarian context. An exploratory study based on expert interviews was conducted to validate the significance of such research and explore HO–supplier relationships and their contexts. This chapter identifies the characteristics of suppliers in the humanitarian context and frames key elements and factors in HO–supplier relationships and the situational and contextual boundaries that should be considered in the humanitarian sector.

Introduction

In the humanitarian context, supplier relationships have been under-researched, and there is a limited literature addressing this topic. To date, researchers have paid little empirical attention to humanitarian organizations' relationships with their suppliers and, in particular, to the implications of these relationships for humanitarian supply chain management (HSCM). The humanitarian context is clearly different from business contexts; hence there is a need to consider different forms of supply chain management (SCM) implementation in the humanitarian context. Nonetheless, very little research has been undertaken in this area. This leads to difficulties in defining the scope, framework, factors and dimensions of research into supplier relationships. Thus, the elements that are valid in the business sector are not necessarily valid in the humanitarian sector. Similarly, the theories that underpin conventional logistics and supply chain management practices cannot necessarily be transferred directly into the humanitarian space.

This chapter takes a broad focus on supplier relationships in the humanitarian sector and discusses in detail the issues and factors that need to be taken into consideration. It then presents an exploratory study that used semi-structured interviews as the primary source of data. These interviews had characteristics of elite interviews, that is, interviews with experts in the field. The study required interviewees with a rich experience and expertise in the humanitarian sector. Expert interviewees can also provide a big-picture perspective and in-depth information for early-stage research (Marshall and Rossman, 1989). At the exploratory stage of the study, there was a need for flexibility, especially during the interview process, which created a conversational atmosphere. However, at the same time, the interviews needed to be controlled to achieve the main objectives. Semi-structured interviews meet these requirements by providing a structured approach, as well as leeway for the interviewer to expand the discussion if productive to do so (Rubin and Rubin, 2012).

The study utilized non-probability sampling, which is often used in the exploratory stages of research projects. There are no rules for non-probability sampling techniques, and samples can be selected based on the researcher's subjective judgement (Saunders et al., 2009). To mitigate the potential drawbacks associated with relying on such subjective judgement when selecting candidates, experts for the exploratory interviews were carefully selected with consideration given to the length of their career, their job position, the reputation of the affiliated organization and their dedication to the humanitarian sector.

Semi-structured interviews were conducted with respondents who were academics or practitioners, so that perspectives from both academia and the humanitarian sector could be combined. The humanitarian experts were divided into two groups: international humanitarian organizations and consultants. The respondents had different backgrounds and between 6.5 and 27 years (mean 15.5 years) of work experience. The discussions and interviews with these experts facilitated the development of a more precise definition of the research area. The interviews provided a broad perspective about the area from that of the interviewees and identified issues around the HO–supplier relationship.

Humanitarian supply chain management and partnership

HSCM research is a relatively recent academic area of interest, with the greater proportion of research taking place since the early 2000s (Kovács and Spens, 2007). Therefore, a general agreement on its definition and boundaries has not been precisely agreed by scholars or practitioners (Tomasini, 2012). In humanitarian research, logistics perspectives have received greater consideration and have been discussed more than those of SCM by around three times as much in papers addressing such issues (Day et al., 2012). In 2005, Thomas and Mizushima suggested humanitarian logistics is 'the process of planning, implementing and controlling the efficient, cost-effective flow of and storage of goods and materials as well as related information, from point of origin to point of consumption for the purpose of meeting the end beneficiary's requirements' (Fritz Institute, 2021). This approach seems to be based primarily on the definition of SCM rather than that of logistics (Tatham and Spens, 2011), although it does try to define logistics. As such, the boundary between the two areas tends to be blurred. However, there are differences between these two concepts, and these differences are also applicable in the humanitarian sector (Day et al., 2012).

The difference between humanitarian logistics (HL) and HSCM that Day et al. (2012, p. 28) proposes is primarily associated with the 'unionist perspective', defining humanitarian and disaster relief SCM as follows:

> 'The system that is responsible for designing, deploying and managing the processes necessary for dealing with not only current but also future humanitarian/disaster events and for managing the coordination and interaction of its processes with those of supply chains that may be competitive/complementary. It is also responsible for identifying, implementing and monitoring the achievement of the desired outcomes that its processes are intended to achieve. Finally, it is responsible for evaluating, integrating and coordinating the activities of the various parties that emerge to deal with these events'

As shown in this definition, the authors attempt to cover the complete cycle of HSCM from the supply chain planning processes in the preparatory period through to evaluating aid performance in readiness for future events. Makepeace et al. (2017, pp. 46–47) point out that Day et al. (2012) do not discuss 'the internal cross-functional implications of the adoption of such a unionist perspective'. They also emphasize that 'a definition of SCM which adequately serves this sector must encompass both humanitarian and development modes' and focused on beneficiaries. Based on this view, they further develop the definition of Day et al. (2012) as follows:

> The system that supports the delivery of both humanitarian relief and development programmes through the identification and strategic management of all interfaces involved in the provision of goods and services, in order to optimise service quality to beneficiaries, and by extension, to donors. It includes the strategic development of global supply chain capacity and the identification of competing and complementary supply chains and the organisation's strategic response to them.

Makepeace et al. (2017) thus reflect 'emerging commercial SCM trends', which is a transition from a narrow meaning to broader and more strategic roles. Again, however, the boundary between humanitarian logistics and disaster relief SCM is not clearly defined, and this lack of precision leaves ambiguity both in the understanding of the two concepts and in the forms of response and implementation on the ground.

The unique characteristics of the humanitarian or disaster relief context leads to distinct and different features of HSCM compared to commercial SCM. It was only from the turn of the millennium that humanitarian relief organizations began to realize the importance of HSCM to improve the success of their relief operations. The necessary skills and techniques of HSCM had fallen behind the commercial counterpart (Larson 2012). However, there are some similarities between the private sector and humanitarian sector in terms of managing and understanding supply chains (Van Wassenhove, 2006). Shifting advanced knowledge of SCM established in the business sector to the humanitarian sphere was both positive and significant (Maon et al., 2009). Swanson and Smith (2013) point out the fundamental similarity with commercial logistics in terms of the goals and objectives fulfilling demand for needed products and services. HSCM already contains strong elements of SCM (Day et al., 2012). Given these, it is possible and useful to transfer well-established knowledge of the commercial SCM to the area of the humanitarian SCM based on the fundamentally similar principle of SCM and the advanced techniques of the commercial one.

Nonetheless, it is necessary to consider the unique features of humanitarian chains which are often context-driven. Table 2.1 highlights the main differences between business and humanitarian SCs. This comparison does not mean that humanitarian SCs are always temporary and operate only in interrupted environments, or that commercial SCs are always permanent, stable and only in uninterrupted environments. Business logistics and SCM generally operate in uninterrupted environments and risk management deals with SC interruptions and risk strategies. In contrast, humanitarian SCs are normally organized in interrupted environments and risk is accepted as an unavoidable aspect of the SC operation and dealt with accordingly as needs arise (Beresford and Pettit, 2005; McLachlin et al., 2009; Choi et al., 2010).

Swanson and Smith (2013) also address several different features serving to define disaster response. Firstly, this has different characteristics from other forms of logistics. Also, its consumers are not traditional ones and its contexts show different attributes, for instance, where infrastructures are damaged or non-existent. Additionally, there are a variety of stakeholders and aid actors in disaster response such as donor organizations, government agencies and NGOs. Balcik and Beamon (2008, p. 102) contend that humanitarian supply chains differ from their commercial counterpart particularly from the perspectives of 'strategic goals, customers and demand characteristics and environmental

TABLE 2.1 Main objectives and challenges in fund management

Category	Commercial SCs	Humanitarian SCs
Motivation/ purpose	For-profit/economic profit	Not-for-profit/social impact
Source of funds	Paying customers	Donors
Context	Normally uninterrupted -Reasonably stable conditions in terms of political and economic conditions; infrastructure in place; and critical actors (e.g., customers, suppliers, service provides and employees) on stage	High levels of interruption -Unpredictability -Emergency conditions -Disruptions to normal activities -Issues in matching multiple resources with a surge of needs
Representative characteristics	Stable SCs -Regular and repetitive routine -Procedures and capital investment valued	Unstable SCs -Non-routine activities -Networks with diverse organizations from different countries, established in a short time -Actual time communications and transportations assets focused on

SOURCE Adapted from Kleindorfer and Saad (2005); Larson (2012); Long and Wood (1995); McLachlin et al. (2009); Tatham and Kovács (2010)

factors'. Additionally, for the environmental factors, the political environment in HSCM makes for differences from commercial SCM (Long and Wood, 1995).

Supply chain partnership

From the 1990s into the 2000s there was a steady increase in research and research outputs considering supply chain integration (SCI) (Flynn et al., 2010; Spiegel et al., 2014). Research in the manufacturing sector largely led the field, with Kamal and Irani (2014) calculating that over 50 per cent of published research in SCI between 2000 and 2013 (147 out of 293 SCI papers) was manufacturing-related. This increasing body of research highlights the growing significance of SCI going forward, both from an academic perspective and as an area of interest to business, at a relatively general level and at a sector or company-specific level. More recent work has focused on both deepening and widening research into specific sectors, for example, automotive, food, retail, construction, electronics, transport and logistics and the maritime environment (Teng and Tsinopoulos, 2021; Yu et al., 2021).

Several forms of partnership have evolved from open market negotiation-based relationships to full collaboration (Spekman et al. 1998). There are key intermediate positions on this continuum that can be identified as open market negotiation, cooperation, coordination and collaboration, although collaboration is sometimes considered as a subset of integration. On the other hand, Kahn and Mentzer (1998) suggested two tiers of integration, specifically: an interaction perspective involving communication behaviour and a collaboration perspective linked to resources and goal-sharing performance. In summary, integration represents one of the approaches to interconnect supply chain systems and it is a core element of the complex relationships in evidence in supply chain networks Furthermore, supply chain integration has multidimensional characteristics such as internal integration, customer integration and supplier integration (Flynn et al. 2010). This multidimensionality may aid in the reduction of ambiguity in partnership studies.

By examining SC relationships in the humanitarian sector, the three main terms cooperation, coordination and collaboration, sometimes referred to as the '3Cs', are regularly used (Heaslip et al., 2012). Further, Heaslip and Barber (2014) indicate that each organization has a different understanding of the true meaning of the 3Cs. This, in turn, leads to challenges in reaching a consensus concerning the interpretation of these terms, both at individual and organizational levels. In the business environment, the 3Cs have been 'often used more or less interchangeably for describing integrative efforts among partners to improve the overall efficiency of the SC' (Prajogo and Olhager, 2012, p. 514). However, Spekman et al. (1998) clearly show the use of these terms is often as transitional concepts between cooperation and collaboration. Cooperation is considered as 'the starting point for SCM' and 'the next level of intensity is coordination' (Spekman et al., 1998, p. 55), which shows a more active exchange of workflow and information. As a final stage, collaboration is viewed as the highest level of partnership by integrating supply chains and sharing a vision. Spekman et al. (1998) also argue that these three concepts should be carefully applied in practice according to the strategic importance of the partner. For example, when dealing with a very important SC partner, the attitude of collaboration can be adopted in managing the relationship.

CHARACTERISTICS OF SUPPLIERS

The interviewees clearly indicated the importance of the HO-supplier relationship for HOs, and several interviewees emphasized the role of supplier relationship management (SRM) and notably, its absence. SRM is 'the process of engaging in activities of setting up, developing, stabilising and dissolving relationships with existing suppliers as well as the observation of prospective suppliers' (Moeller et al., 2006, 73). This means that SRM involves the whole process of managing suppliers, including potential ones. Given that SRM is a key factor for successful humanitarian SCM, the HO–supplier relationship cannot be ignored. There are many challenges and difficulties associated with establishing and maintaining relationships with suppliers, and many HOs struggle with their relationships with suppliers. Often,

HOs do not have an SRM framework due to a lack of opportunities to amass experience and know-how in SRM. In addition, SRM is important for HOs, in terms of managing and monitoring suppliers' performance. Although many HOs consider this a critical issue, they do not have the experience or action plans to implement SRM. Therefore, HO–supplier relationships require further study.

There are many types of suppliers with which HOs are involved (see Figure 2.1). First, there are 'donor type' and 'seller type' suppliers. The former (donor type) offer in-kind support, including products and services. Their offered support sometimes does not meet the needs of beneficiaries and is useless in the affected area. The latter (seller type), also called the traditional type of supplier, is a major supplier for HOs. Second, the seller type of supplier can be further divided into two groups: specialized in humanitarian business or not. Finally, according to the contract span, the relationships can be divided into one-off and longer-term. It seems that one-off contracts with suppliers dominate the space rather than longer-term relationships.

The first contains suppliers that typically focus on the humanitarian sector. The second group contains suppliers that often originate in the commercial sector. Suppliers in the first group specialize in humanitarian aid supplies (e.g. Better Shelter and Nutriset Group). Those in the second group usually deal with the commercial sector and occasionally with the humanitarian sector. For instance, the Dantherm Group provides products to both the humanitarian and commercial sectors. There may be some differences between these two types of suppliers. As such, it is necessary to consider the donor type of supplier when dealing with humanitarian supply chains (SCs).

FIGURE 2.1 Categories of HO suppliers

- Donor type of suppliers
- Seller type of suppliers
  - Humanitarian business
  - Commercial business

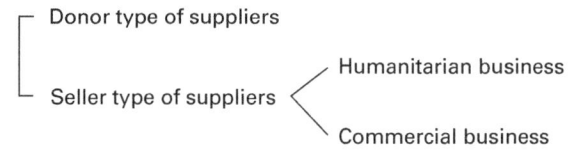

SOURCE Developed by the chapter authors

Framing the key issues – power, trust and commitment

It is evident that power dynamics exist between stakeholders in the humanitarian sector, such as HOs, their suppliers and donors. The most important issue is 'power' in the relationships between HOs and their suppliers. Power is considered by practitioners to be omnipresent in everyday business, particularly in supply chain relationships (Maloni and Benton, 2000). The interview findings of this study also show that the power issue is unavoidable in SC relationships in the humanitarian sector. In some cases, the interviewees reported seeing aid actors consciously recognized their power; in contrast, others did not realize that they held power and abused it. One of the respondents shared that no self-awareness about power abuse was the worst thing. This imbalance of power in relationships with suppliers influences the counterpart in setting up relationships.

First, financial resources can generate power asymmetry in HO–supplier relationships. One interviewee considered 'the money power' as a key resource of power, although this is not the only resource of power, described as follows: 'more often than not, it's about where the money is. It's the first thing about power. But it's not always the only thing'. This money power clearly affects the performance or decisions of the counterpart. For instance, an HO pressured to achieve a specific goal may exploit its financial power to achieve that goal.

Second, the strength of an organization's network can elevate its degree of power. When they do not possess financial power, many organizations fail to recognize that they have other power resources. Local or small organizations in particular do not realize their own non-financial power and strengths; they are more likely to have 'relational power', or networking power. For example, it can be easier for these organizations to access local authorities through the networks they have maintained in the local area. However, many organizations do not leverage such relationships due to a lack of self-awareness of their own power and abilities. This networking is also important in other ways. An organization with a well-integrated SC can quickly engage its network (contacts) to respond to disasters. The level of integration is determined by how rapidly appropriate programmes are evaluated and decisions are made about who needs to be contacted. This suggests that the ability to quickly establish a network is an asset for organizations in disaster relief situations.

Finally, the scarcity of aid commodities or services can lead to an imbalance of power between SC partners. It is quite common to see opportunistic behaviour exhibited by suppliers via pricing activities in emergency situations. One interviewee explained how the scarcity of products can lead to power dynamics by suppliers playing with prices in a monopolistic situation: 'there are suppliers who exploit the situation, where there is just one supplier for a certain commodity available'. In this situation, there are limited options for HOs, apart from following the supplier's suggestions.

This power imbalance has a greater impact on the SC relationship than their mutual interactions in the business sector (Maloni and Benton, 2000), and it can be assumed that power asymmetry has a great influence on HO–supplier relationships. The study findings indicate that this phenomenon is also present in the humanitarian sector.

In the humanitarian context, trust between HOs and their suppliers is also critical. When an HO–supplier relationship is contractual, the development of a trust-based relationship is challenging. Overall, HOs recognize that integrity is needed in their relationships with suppliers, however, they are not accustomed to working closely with their suppliers. Additionally, there is another aspect of developing such relationships: their development is based on the frequency of transactions and the quantity of supplies, and these can only be achieved by large organizations with global suppliers. It is not easy for small organizations and local suppliers to maintain sufficient frequency and quantity to develop trust-based relationships.

Many studies have concluded that trust is central to relational exchanges (Zaheer et al. 1998; Perrone et al. 2003). Morgan and Hunt (1994) suggested that 'cooperation is the only outcome posited to be influenced directly by both relationship commitment and trust'. Thus, it can be assumed that the degree of trust in the relationship can affect HO–supplier relationships and their activities with the affected communities in which they work.

When discussing interorganizational relationships, commitment also plays a critical role (Anderson and Weitz, 1992), and stability and sacrifice are considered the core of commitment (Wu et al., 2004). This means that SC members try to pursue longer-term and stable relationships with short-term sacrifices, which implies commitment to the relationship.

Many interviewees commonly described the phenomenon of there being few long-term relationships between HOs and their suppliers and many 'one-off procurement' or 'ad hoc one-off contracts'. This is known as 'emergency procurement' and generally occurs outside the normal public procurement approach. This often happens in the military realm and was most recently seen in procurement practices during the Covid-19 pandemic. Such activities can lead to cronyism and other unethical practices, which may not result in the most effective outcomes (Sian and Smyth, 2021). First, maintaining long-term relationships is not prioritized among the range of activities related to SRM. It appears that HOs tend to focus on practical activities, such as regular supplier meetings, supplier management and supplier performance management, rather than expending resources on maintaining long-term supplier relationships. Thus, as a result of short-term supplier relationships or one-off contracts, HOs seem to lose opportunities to improve SC efficiency by helping their suppliers build capacity: 'they do not help suppliers to become more capable. Just ask suppliers to be able to perform according to the term of reference, according to the amount of money that they have'.

These relationships with less commitment clearly have an impact on SCM efficiency, as they are usually short-term relationships. HOs want their suppliers to just follow the regulations; the HOs are not interested in helping the suppliers improve their performance. There is little interest in putting more effort into improving the HO–supplier relationships. One participant said, 'We don't spend a lot of energy on maintaining supplier relationships so long term.' Therefore, HOs are less flexible in tailoring specific relationships with their suppliers. Given this, long-term relationships with suppliers are not prioritized and HOs do not try to adopt suitable SRM practices for different types of suppliers. Thus, it can be understood that they do not want to spend much time or put effort into pursuing longer-term relationships with suppliers.

It is also worth noting that commitment is related to power and trust. A scarcity of aid commodities can lead to power imbalances between IHOs and their suppliers. This can also create ethical issues, since suppliers can demand high market prices in emergency situations. Many suppliers often exploit disaster circumstances, and in such situations, HOs have limited choices and are forced to accept suppliers' conditions. HOs are unwillingly driven to use the supplier that exploits the emergency situation when there is no alternative. These relationships are the result of emergency procurement or ad-hoc procurement and therefore cannot be developed into long-term relationships. Thus, it can be assumed that power asymmetry can influence relationship commitment.

In terms of trust, it is quite common to see 'swift-trust' formed between IHOs and their suppliers. One interviewee described swift-trust relationships with suppliers in the humanitarian context as follows: 'swift-trust like things have to happen so quick, so they might have their relationship but there might not be steady stable orders helping it'. This type of trust is probably fostered by the humanitarian context, in which regular demands cannot be guaranteed, and clearly influences the span of HO–supplier relationships.

Situational factors

HOs are not the traditional type of customer for their suppliers. One interviewee described a traditional customer as one who could request stable demands with consistent volume and make payments on time. However, in the humanitarian context, HOs have unpredictable, unforeseen and erratic requirements. This is why HOs may create problems in HO–supplier relationships. It is difficult for suppliers to deliver just-in-time because the demand is not constant; rather, there are many one-off demands in this sector. Additionally, many HOs do not maintain a certain level of frequency and quantity of transactions, and this is particularly the case for small organizations. It seems that this issue is related to the nature of the emergency context in which HOs operate. A participant described the emergency context and demands as follows: 'when you operate in the emergency context, it is hard to predict when and where the next disaster will actually happen and what's going to be needed [demand]'. In this difficult situation, it is very challenging to have supply chains ready to respond to a sudden influx of demands. Hence, SCs heavily influence the entire response. Unpredictable demand and incapable SCs can be regarded as problems. However, these features are inevitable in the emergency context; it is very challenging to forecast where and when a disaster will strike.

In contrast, the roles of local communities and community organizations are emphasized in disaster situations. In general, community organizations know the local conditions and structures in place. They are usually the first responders to disasters and are the key enablers of a speedy response and provision of relief. It can be assumed that HOs prefer to respond to disasters through these local communities or local organizations, which can access affected areas quickly and easily.

Furthermore, unstable and irregular demand for products and services makes it difficult for HOs to maintain longer-term relationships with their suppliers. Short-term relationships or one-off contracts may arise in the course of HOs responding immediately to unpredicted disasters and due to subsequent circumstances. As such, unstable circumstances are a key contextual factor impacting HO–supplier relationships.

One of the factors that distinguishes HOs from commercial sector organizations is that the 'funds flow' mostly rely on donations to fund their resources. Thus, donors are crucial for the successful operation of HOs, and they range from individuals to companies to government bodies. This means that donors can have a considerable influence on the decisions and performance of HOs. This, in turn, can lead to HOs expending significant effort to satisfying their donors and focusing on being accountable to them. Hence, donors provide financial resources to HOs, and at the same time, may create challenges for HOs, namely, meeting donors' requirements whatever the circumstances are during disasters. It can be assumed that donors can influence the activities and relationships of the HOs they support.

Indeed, as Bailey (2013) reported, on the ground, donors' geopolitical agendas are often prioritized over the humanitarian imperative and ultimate purpose of protecting the vulnerable from disasters.

The relationships between HOs and their suppliers can be defined as 'contractual relationships' and public procurement regulations indirectly influence the agreements made. One interviewee explained that public procurement regulations are contextually distinct in the humanitarian and private sectors. Individual HOs have similar procurement regulations, and they are like those of public healthcare and government agencies. In the humanitarian sector, public procurement regulations prohibit long-term relationships by regulating the frequency of the tender audit process, the number of bidders and the choice of the lowest bidder.

Additionally, there are basic contextual factors that should be considered as circumstances in which the HO–supplier relationships are situated. Table 2.2 outlines the humanitarian contexts suggested by several participants.

Given the procurement regulations to which HOs must abide, it seems that HOs and their suppliers must undergo lengthy procurement processes. In general cases, suppliers must participate in a bidding process, sometimes for each project or programme, which usually takes a lot of time. Difficulties can also arise when completing administrative work, such as providing documented financial evidence. This is particularly challenging when products are purchased via the centralized procurement system typically operated by an umbrella organization, and by departments or field offices that are located in different countries. Additionally, one interviewee described the tendering process as follows: 'you need to get to your tendering, then usually the rule in most NGOs is to have the three steps, you need to have three offers, three quotes and places'. This shows that procurement in the humanitarian sector has certain rules and regulations to control HOs. This could potentially be linked to the requirement that HOs be accountable to their donors. As a result, the procurement process can be delayed, leading to delays in aid delivery or urgent requirements. Hence, it can influence aid projects. The procurement process also leads to procurement of goods and services at higher prices, particularly in the case of

TABLE 2.2 Context of humanitarian organizations (HOs)

Categories	Coding for context
Two types of aid contexts	Differences between 1) disaster relief and 2) development aid
Diverse types of IHOs and stakeholders	The HO-supplier relationships vary depending on the organizations
	Due to more stakeholders, there are more challenges in integration compared to in the business sector
	Procurement strategies are highly dependent on the type of organization
Types of disasters	1. Development aid project: usually scheduled in advance, quite predictable and easier to be a supplier 2. Slow-onset disasters: often there is some advanced planning 3. Sudden-onset disasters: there is a need to quickly supply
Scale of disasters	Disaster category 1, 2 and 3 Category 1: disasters that the domestic national government can manage Category 2: disasters that require regional office involvement Category 3: disasters that require global aid support
Phases of disaster management	Length of aid period: 7 days and 90 days. The response and aid changes, depending on the length of time. (i.e. within 7 days: use a global warehouse of a global centre. Within 90 days: encourage the use of the local market)
Characteristics of regions	1. Asia: the most frequent natural disasters, a better established local market, improved logistics, and an improved airport 2. Middle East: military conflicts and less developed local market 3. Africa: more long-term ongoing crises due to conflicts rather than natural causes and a less developed local market

SOURCE Developed by the chapter authors

global procurement. Furthermore, tender audit processes that occur every two or three years can prohibit the formation of long-term HO–supplier relationships.

Organizational factors

Depending on the types of organizations among humanitarian relief providers, there will be different organizational objectives, structures, organizational cultures, different ways of operating (Larson, 2012; Long and Wood, 1995). 'Humanitarian relief environments engage various stakeholders like international relief organizations, host governments, the military, local and regional relief organizations and private sector companies, each of which may have different interests, mandates, capacity, and logistics expertise' (Yadav and Barve, 2015, p. 217). As a consequence, different disasters will require a different mix of aid providers, for example the International Federation of the Red Cross will be represented by either the Red Cross or Red Crescent depending on the location of the disaster. Also, each type of organization might have widely differing features in terms of agendas, religious beliefs, capabilities, fund availability, and the need for media attention (Chakravarty, 2014). In order to work together among these dissimilar organizations, great efforts are required to coordinate plans and share limited resources (Long and Wood, 1995). Hence, due to different perspectives about issues in humanitarian relief operations among them, the academic studies naturally reflect their differences.

HOs tend to have unique organizational structures, decision-making processes and financial systems. This makes it very difficult to develop one optimized model and apply one theory to all HOs in the academic research. Even among large HOs of the same type, each has its own organizational structure and work practices. For example, some HOs are more centrally structured, and others are decentralized, depending on the organization's decision-making process and financial system. Many HOs attempt to form a centralized finance or procurement system because it allows them to guard against potential corruption and achieve an economy of scale. Therefore, from a whole-of-organization perspective, it is cost-effective and more efficient to procure through mass contracts that lead to economies of scale.

In contrast, from the local office or programme perspective, producing financial reports for funders (donors) is time consuming, and centralized systems are often not flexible enough to allow localization and customization for each project and local office. Local offices may prefer a decentralized structure in which their activities are not (or only partly) controlled by a higher level within the organization. For instance, with decentralized structure, a local office of an HO may have more leeway in purchasing aid items or developing its own supplier relationships. Therefore, the organizational structure of an HO as well as other factors (e.g., procurement and finance systems) can affect its relationships with its suppliers. This indicates that each HO has different circumstances under which its relationships with its suppliers are formed and therefore these relationships may differ from one another depending on different circumstances. Thus, identifying and understanding the characteristics of organizations and their contexts is an important aspect of humanitarian SCM research.

Summary and conclusion

Set against theoretical frameworks, the discussion above provides new insights into examining aspects and situations in supplier relationships with international humanitarian organizations. The research highlights the elements that either directly or indirectly influence supplier partnerships in the humanitarian context. In particular, the structure of the organization and its context are shown to be important and complex. This confirms that there are a significant number of inputs that require consideration prior to implementing supplier relationship in the humanitarian sector.

This has specific and important implications for the suppliers themselves, and for HO–supplier relationships. This study also highlights the significant gap between the respective interpretations of supply chain management concepts in the commercial and humanitarian environments. Often these concepts are not compatible, or at most, they are only partly compatible. The need to conceptualize the term 'supplier' from the humanitarian perspective has been identified in this chapter. In particular, Figure 2.1 suggests the category of suppliers in the humanitarian sector; it shows the addition of donor type suppliers to the traditional concept of suppliers.

Through a series of exploratory interviews, the study found that there are many context-specific factors that influence HO–supplier relationships. For instance, donors can significantly affect HO performance because donations are the key financial resource of aid organizations, which is a different situation from that of private companies (Van Wassenhove 2006). An HO's organizational structure can also have a considerable impact on the power system and the standardization of regulations and norms.

The theoretical paradigms discovered throughout the interview process, such as power, trust and commitment aspects, are an intriguing component of this study. In supply chain research there are few structured frameworks that fully represent the broad scope of power asymmetry reality in interorganizational relationships (Cowan et al., 2015). Further, Belaya et al. (2009) show that there are a similarly very few scientific studies that adopt power aspects in the context of SC networks. There have been relatively few empirical studies examining the relationship between trust and commitment; but it is widely recognized that organizations are required to change continuously (Cowan et al., 2015). The implication of this is that relationships between organizations also change dynamically, largely in response to adjustment to the variations in operational conditions.

In this chapter, empirical evidence is presented of the specifics within the relationship dynamics in a humanitarian supply chain context. A prime example of research in this area is Tatham and Kovács (2010) who proposed a ground-breaking model of 'swift trust' in hastily formed networks during sudden onset disasters. Their study considered the necessity for developing trust as rapidly as possible within relationships between humanitarian organizations and their key partners, especially during sudden onset or protracted crises. From a theoretical perspective, the research presented here develops a more dynamic and comprehensive framework in order to capture the full diversity of organizational relationships in the humanitarian environment.

This chapter provides a better understanding of relationships with suppliers from the humanitarian supply chain management perspective. In addition, the findings provide insights into supplier relationship management for humanitarian practitioners and an opportunity to consider the diverse factors and elements that contribute to HO–supplier relationships as focal points for improving aid performance. These findings thus provide an opportunity for humanitarian practitioners to broaden their perspectives on managing supplier relationships, considering theoretical paradigms as well as humanitarian contexts. These could be highly fertile ground for future investigation.

Acknowledgement

We would like to thank the Economic and Social Research Council (ESRC) for their funding and support.

References

Anderson E, Weitz B (1992) The use of pledges to build and sustain commitment in distribution channels. *Journal of Marketing Research* **29**(1), pp 18–34

Bailey, R (2013) In Somalia, western donors made famine more, not less likely. *The Guardian*, 2 May 2013, www.theguardian.com/global-development/poverty-matters/2013/may/02/somalia-western-donors-famine-likely (archived at https://perma.cc/A3Z6-AFL4)

Balcik B, Beamon, BM (2008) Facility location in humanitarian relief. *International Journal of Logistics* **11**(2), pp 101–121

Belaya, V, Gagalyuk, T and Hanf, J (2009) Measuring asymmetrical power distribution in supply chain networks: What is the appropriate method? *Journal of Relationship Marketing* 8(2), pp 165–193

Beresford, A K C and Pettit, S J (2005) Emergency relief logistics: An evaluation of military, non-military and composite response models, *International Journal of Logistics: Research and Applications*, 8(4) pp 313–332

Blecken, A (2010) *Humanitarian logistics: modelling supply chain processes of humanitarian organisation*, Berne: Haupt

Chakravarty, A K, 2014. Humanitarian relief chain: rapid response under uncertainty. *International Journal of Production Economics,* **151**, pp 146–157

Choi, K-Y, Beresford, A K C, Pettit, S J and Bayusuf, F (2010) Humanitarian Aid Distribution in East Africa, A study in supply chain volatility and fragility, *Supply Chain Forum: An international journal,* **11**(3), pp 20–31

Cowan, K, Paswan, A K and Van Steenburg, E (2015) When inter-firm relationship benefits mitigate power asymmetry, *Industrial Marketing Management,* **48**, pp 140–148

Day, J M, Melnyk, S A, Larson, P D, Davis, E W and Whybark, D C (2012) Humanitarian and disaster relief supply chains: A matter of life and death, *Journal of Supply Chain Management,* **48**(2), pp 21–36

Flynn, B B, Huo, B and Zhao, X (2010) The impact of supply chain integration on performance: A contingency and configuration approach, *Journal of Operations Management,* **28**(1), pp 58–71.

Fritz Institute (2021) About Humanitarian Supply Chain Management, available at: https://fritzinstitute.org/humanitarian-scm/ (archived at https://perma.cc/82TV-KUF3)

Heaslip, G, Barber, E (2014) Using the military in disaster relief: Systemising challenges and opportunities, *Journal of Humanitarian Logistics and Supply Chain Management,* **4**(1), pp 60–81

Heaslip, G, Sharif, A M, Althonayan, A (2012) Employing a systems-based perspective to the identification of inter-relationships within humanitarian logistics, *International Journal of Production Economics,* **139**, pp 377–392

Jahre, M, Kembro, J, Rezvanian, T, Ergun, O, Håpnes, S and Berling, P (2016) Integrating supply chains for emergencies and ongoing operations in UNHCR, *Journal of Operations Management,* **45**, pp 57–72

Kahn, K B, Mentzer, J T (1998) Marketing's integration with other departments, *Journal of Business Research,* **42**(1), pp 53–62

Kamal, M M, Irani, Z (2014) Analysing supply chain integration through a systematic literature review: a normative perspective, *Supply Chain Management: An International Journal,* **19**(5/6), pp 523–557

Kleindorfer, P R, Saad, G H (2005) Managing disruption risks in supply chains, *Production and Operations Management,* **14**(1), pp 53–68

Kovács, G, Spens, K M (2007) Humanitarian logistics in disaster relief operations, *International Journal of Physical Distribution and Logistics Management,* **37**(2), pp 99–114

Larson, P D (2012) Strategic partners and strange bedfellows: Relationship building in the relief supply chain, in Kovács, G, and Spens, K (eds.) *Relief supply chain management for disasters.* Business Science Reference, PA, pp 1–15

Larson, P D, Halldórsson, Á (2004) Logistics versus supply chain management: An international survey, *International Journal of Logistics: Research and Applications*, 7(1), pp 17–31

Long, D C and Wood, D F (1995) The logistics of famine relief, *Journal of Business Logistics*, 16(1), pp 213–229

Makepeace, D, Tatham, P and Wu, Y (2017) Internal integration in humanitarian supply chain management: Perspectives at the logistics-programmes interface, *Journal of Humanitarian Logistics and Supply Chain Management*, 7(1), pp 25–26

Maloni, M and Benton, W C (2000) Power influences in the supply chain, *Journal of Business Logistics*, 21(1), pp 49–73

Maon, F, Lindgreen, A and Vanhamme, J (2009) Developing supply chains in disaster relief operations through cross-sector socially oriented collaborations: A theoretical model, *Supply Chain Management: An International Journal*, 14(2), pp 149–164

Marshall, C and Rossman, G B (1989) *Designing qualitative research*, Sage, London

McLachlin, R, Larson, P D and Khan, S (2009) Not-for-profit supply chains in interrupted environments: The case of a faith-based humanitarian relief organization, *Management Research News*, 32(11), pp 1050–1064

Moeller, S Fassnacht, M and Kiose, S (2006) A framework for supplier relationship management (SRM), *Journal of Business-to-Business Marketing*, 13(4), pp 69–94

Morgan, R M, Hunt, S D (1994) The commitment-trust theory of relationship marketing, *Journal of Marketing*, 58(3), pp 20–38

Perrone, V, Zaheer, A and McEvily, B (2003) Free to be trusted? Organizational constraints on trust in boundary spanners, *Organization Science* 14(4), pp 422–39

Prajogo, D and Olhager, J (2012) Supply chain integration and performance: The effects of long-term relationships, information technology and sharing, and logistics integration, *International Journal of Production Economics*, 135(1), pp 514–522

Rubin, H J, Rubin, I (2012) *Qualitative interviewing: The art of hearing data*, Sage, London

Saunders, M Lewis, P and Thornhill, A (2009) *Research methods for business students*, 5th edn, Pearson, Harlow

Sian, S and Smyth, S (2021) Supreme emergencies and public accountability: The case of procurement in the UK during the Covid-19 pandemic, *Accounting, Auditing and Accountability Journal*, 35(1), pp 146–157

Spekman, R E, Kamauff Jr, J W and Myhr, N (1998) An empirical investigation into supply chain management: A perspective on partnerships, *Supply Chain Management: An International Journal*, 3(2), pp 53–67

Spiegel, T, Vasconcelos, P E M, Porto, D L, Caulliraux, H M (2014) Supply chain integration research: An overview of the field. *International Journal of Supply Chain Management,* 3(1), pp 12–20

Start Network (2021) *About us: a new era of humanitarian action*, https://startnetwork.org/about-us (archived at https://perma.cc/Y9XY-YUV2)

Swanson, R D and Smith, R J (2013) A path to a public–private partnership: Commercial logistics concepts applied to disaster response, *Journal of Business Logistics,* 34(4), pp 335–346

Tatham, P, Kovács, G (2010) The application of 'swift trust' to humanitarian logistics, *International Journal of Production Economics,* 126(1), pp 35–45

Tatham, P, Spens, K (2011) Towards a humanitarian logistics knowledge management system, *Disaster Prevention and Management: An International Journal,* 20(1), pp 6–26

Teng, T and Tsinopoulos, C (2022), Understanding the link between IS capabilities and cost performance in services: the mediating role of supplier integration, *Journal of Enterprise Information Management,* 35(3) pp 669–700, doi.org/10.1108/JEIM-08-2020-0321 (archived at https://perma.cc/6F54-KUQU)

Tomasini, R M (2012) Humanitarian partnerships-drivers, facilitators, and components: The case of non-food item distribution in Sudan, in: Kovács, G, and Spens, K (eds.) *Relief Supply Chain Management for Disasters*, Business Science Reference, PA, pp 16–30

Wu, W Y, Chiag, C Y, Wu, Y J and Tu, H J (2004) The influencing factors of commitment and business integration on supply chain management, *Industrial Management and Data Systems,* 104(4), pp 322–333

Van Wassenhove, L N (2006) Blackett memorial lecture humanitarian aid logistics: Supply chain management in high gear, *Journal of the Operational Research Society,* 57(5), pp 475–489

Yadav, D K and Barve, A (2015) Analysis of critical success factors of humanitarian supply chain: An application of Interpretive Structural Modelling, *International Journal of Disaster Risk Reduction,* 12, pp 213–225

Yu, Y, Huo, B and Zhang, Z (2021) Impact of information technology on supply chain integration and company performance: Evidence from cross-border e-commerce companies in China, *Journal of enterprise information management,* 34(1), pp 460–448

Zaheer, A, McEvily, B and Perrone, V (1998) Does trust matter? Exploring the effects of interorganizational and interpersonal trust on performance, *Organization Science* 9(2), pp 141–59

03

Providing logistics services for humanitarian relief

DIEGO VEGA AND CHRISTINE ROUSSAT

ABSTRACT

Logistics services in humanitarian relief are provided by a myriad of actors. Academic literature has mainly focused on the involvement of commercial logistics service providers in the humanitarian sector, and the logistical activities performed by humanitarian organizations. However, little is said about other types of actors that are involved in the provision of humanitarian logistics activities. This chapter reviews the existent perspectives of service provision and presents a continuum of actors working with logistics in the humanitarian context, from commercial to humanitarian organizations, identifying different hybrid forms that have not been the focus of academic research so far.

Introduction: Logistics services

Logistics service provision in the humanitarian context is significant. Under the United Nations (UN) system, it represents the first category of services procured with over $2.7 billion for 2020, and the second largest procurement segment overall behind pharmaceuticals, contraceptives and vaccines in the same year[1]. The organization with the largest procurement volume, the World Food Program (WFP), is also the biggest customer of logistics services accounting for $1.3 billion. In the global humanitarian sphere, these services are provided by a myriad of actors (Jahre et al., 2009) that are either commercial logistics companies, or humanitarian organizations acting as logistics service providers. While the former include some of the industry leaders that have developed humanitarian specific divisions (see, for example, DHL's Aid & Relief Logistics, Kühne & Nagel's Emergency and Relief Logistics or UPS's Health & Humanitarian Relief), the former relates to humanitarian organizations' specialized logistics units providing services to their teams and other organizations (see, for example, MSF's MSF *Logistique* or MSF Supply, IFRC's Global Logistics Services or the UN's Global Logistics Cluster).

Service provision models have been extensively studied in various disciplines, with an important stream focusing on service operations management. In a previous edition of this book, Heaslip (2014) pointed out a number of service paradigms that have the potential of applying service management theories to humanitarian logistics research, including the service-dominant logic, value constellations, service profit chain and robust service frameworks. Among those, the service-process matrix (Schmenner, 1986) seems particularly relevant to the results from this research, as it would enable the categorization of the organizations involved in the provision of logistics services in supply chains with respect to their degree of customization and the degree of labour intensity. Contrary to Schmenner's (2004) argument, which focuses on productivity as the main power that leads service-based organizations to focus on low labour intensity, low customization strategies (thus becoming more productive), service providers in the humanitarian context seem to be taking the opposite direction, aiming at a more customized/labour intense strategy, favouring impact over productivity.

Over the past years, humanitarian logistics research on services has brought to light some important issues, such as the trend towards servitization (Heaslip, 2013; Heaslip et al., 2018), the roles that logistics companies can play in humanitarian supply chains (Vega and Roussat, 2015; Sigala and Wakolbinger, 2019) and the characteristics of humanitarian organizations acting as logistics companies (Abidi et al., 2015; Vega and Roussat, 2019). Other emerging themes, such as new organizational forms that provide logistics services or service provision and service modularity from a broader perspective, are yet to be further investigated. With this introduction in mind, this chapter considers logistics services in the humanitarian sector, from its roots in commercial logistics to its evolution, its current state and potential future avenues.

Commercial providers

During the early 1980s, a generalized outsourcing trend in the commercial sector caused an exponential growth of transportation and warehousing companies (Quinn, 1999), which resulted in a strong positioning of logistics service providers (LSPs) in the supply chain, both in the industry and the academic literature (Selviaridis and Spring, 2007). This interest in logistics service provision is continuously growing, as is the growth rate for logistics services (Langley and Infosys, 2020). Generally speaking, LSPs – or third-party logistics providers (TPL) – are external providers that 'manage, control, and deliver logistics activities on behalf of a shipper' (Hertz and Alfredsson, 2003, p. 140). Throughout the years, and in response to increasing market demands in terms of volume and scope of services, logistics companies have developed a series of value-added and informational activities (Anderson et al., 2011) and are able to provide a 'broad array of bundled services that also includes warehousing, inventory management, packaging, cross-docking and technology management' (Zacharia et al., 2011, p. 43). This evolution of LSPs in the commercial sector has led to numerous innovations and has not gone unnoticed by humanitarian logistics researchers.

Since the early 2010s, researchers have investigated the role of commercial logistics service providers in humanitarian logistics. Some of the most relevant works are allusive or incentivizing, pointing out the need for logistics specialized knowledge (Abidi et al.,

2015) or highlighting the role LSPs could play in the fulfilment of needs (Bealt et al., 2016) and the efficiency and effectiveness of humanitarian supply chains (Baharmand and Comes, 2019). A pioneering work from Vega and Roussat (2015) examined the diverse roles commercial firms could play throughout the different phases of humanitarian operations. Based on their analysis of 17 top worldwide service providers, the authors proposed three roles that LSPs could and do play in the humanitarian sector:

- As *members*, LSPs are likely to participate in the humanitarian field mostly through a corporate social responsibility strategy and in-kind donations, such as allocating free logistics capacity to humanitarian organizations for any of the different phases of a response.
- As *tools*, LSPs act as operators providing different services that can vary in their degree of specialization, from simple logistics activities (warehousing and transportation) to last-mile distribution and specialized logistics support, on a contract basis.
- Finally, as *actors*, LSPs support humanitarian organizations, sometimes as partners, offering all kinds of logistics activities with the ability to design, coordinate and implement the overall management of physical and informational flows through the preparedness/response/recovery phases, acting as an intermediary between different actors and filling the gaps as a last resort.

A list of activities to be potentially outsourced to LSPs was proposed by the authors (see Figure 3.1) and has since been validated and completed by Sigala and Wakolbinger (2019), mobilizing the humanitarian organizations' perspective.

Regarding the motivation of logistics outsourcing in the humanitarian field, the academic literature points out several arguments. Involving commercial LSPs in the humanitarian chain is considered to improve the delivery of goods (Cohen, 2016), to enhance the logistics capabilities of HOs (Bealt et al., 2016), while reducing risks (Baharmand et al., 2017).

FIGURE 3.1 Roles and activities in different humanitarian relief phases

	Member	Tool/operator	Actor
	The LSP participates on a CSR basis	The LSP as a 3PL provides the required services	The LSP as a 4PL coordinates (part of) the relief SC
Preparedness	In-kind donations	Transport of goods Warehousing of pre-positioned stocks	Procurement Inventory management of prepositioned supplies Transport coordination
Response	Allocation of free capacity (transportation/ warehousing) to humanitarian organizations Facilitation of employee volunteering	Fleet management Customs clearance Materials handling Logistics hub Last-mile distribution	Full-scale logistics Supply chain solutions Intermediary between NGOs/LSPs Overall coordination of the relief response
Recovery	In-kind donations	Transportation Warehousing Procurement	Supply chain management of a recovery project on behalf of an NGO/government
	Humanitarian logistics as a CSR commitment	***Humanitarian logistics as a source for ad hoc contracts***	***Humanitarian logistics as a strategic business unit***

Reactive ←─────────────────────────────────→ Proactive

SOURCE Reproduced with kind permission of Vega and Roussat (2015)

More concretely, LSPs can provide HOs with their specialized resources (technical knowledge, financial assets, access to data) and their capacity to scale and develop shorter ramp-up times (Sigala and Wakolbinger, 2019). Thus, outsourcing logistics services is considered as a strategic decision, and its success relies on time and budget frames (Akbari, 2018), best practices (Gossler et al., 2019) or providers' selection (Kim et al., 2019). Nevertheless, outsourcing logistics to LSPs in the humanitarian context can reveal differences that seriously challenge the collaboration between both actors. Obstacles can thus be linked to the lack of trust and transparency (Baharmand and Comes, 2019), the absence of contracts (McLachlin and Larson, 2011), conflicting goals or mandates (Heaslip et al., 2018), shaping cultural conflict (Kovács and Spens, 2007) or availability of resources (Nurmala et al., 2017). To solve these issues, and as a natural evolution of responding to the characteristics of the context, many humanitarian organizations have built their own logistics structures with specialized resources, reducing the need for specialized commercial providers and ensuring the level of responsiveness needed in case of emergencies.

Humanitarian providers

A growing body of literature in humanitarian logistics research has emerged focusing on the servitization of humanitarian aid (Heaslip, 2013; 2014; 2016; Heaslip et al., 2018). Particularly, this stream argues that there has been and is continuing to be a shift in humanitarian organizations from a product-based paradigm, i.e. tangible relief items, to a service view (Heaslip, 2013; Kovács, 2014) as many humanitarian organizations actually provide services to one another. Most of such services are found in the logistics arena, from basic transportation and warehousing, to procurement, consolidation and training and, most recently, cash-based assistance (Heaslip et al., 2018) but these are not exclusive areas of servitization. Privileged by their cultural compatibility, those organizations that provide services to their peers are defined as humanitarian logistics service providers (Dufour et al., 2018) or humanitarian service providers (Vega and Roussat, 2019). However, when compared to their commercial counterpart, these actors have not received the same amount of attention by the academic community.

When referring to logistics partnerships in the disaster chain, Kovács and Spens (2007) mention that one NGO can collaborate with another, while Heaslip (2014) points out that HOs have started to develop services for their peers. That 'general notion of HOs functioning as LSPs' (Heaslip, 2013, p. 44) was explored by Vega and Roussat (2019), where they define a humanitarian service provider (HSP) as 'a humanitarian organization (UN agency or NGO) that carries out activities on behalf of internal customers as well as other HOs or governmental structures' (p. 949). The authors emphasize that in their conceptualization of HSP the word 'logistics' is not included, as they demonstrate that the activities performed and services provided by HSPs go beyond

FIGURE 3.2 Roles and activities of HOs as service providers

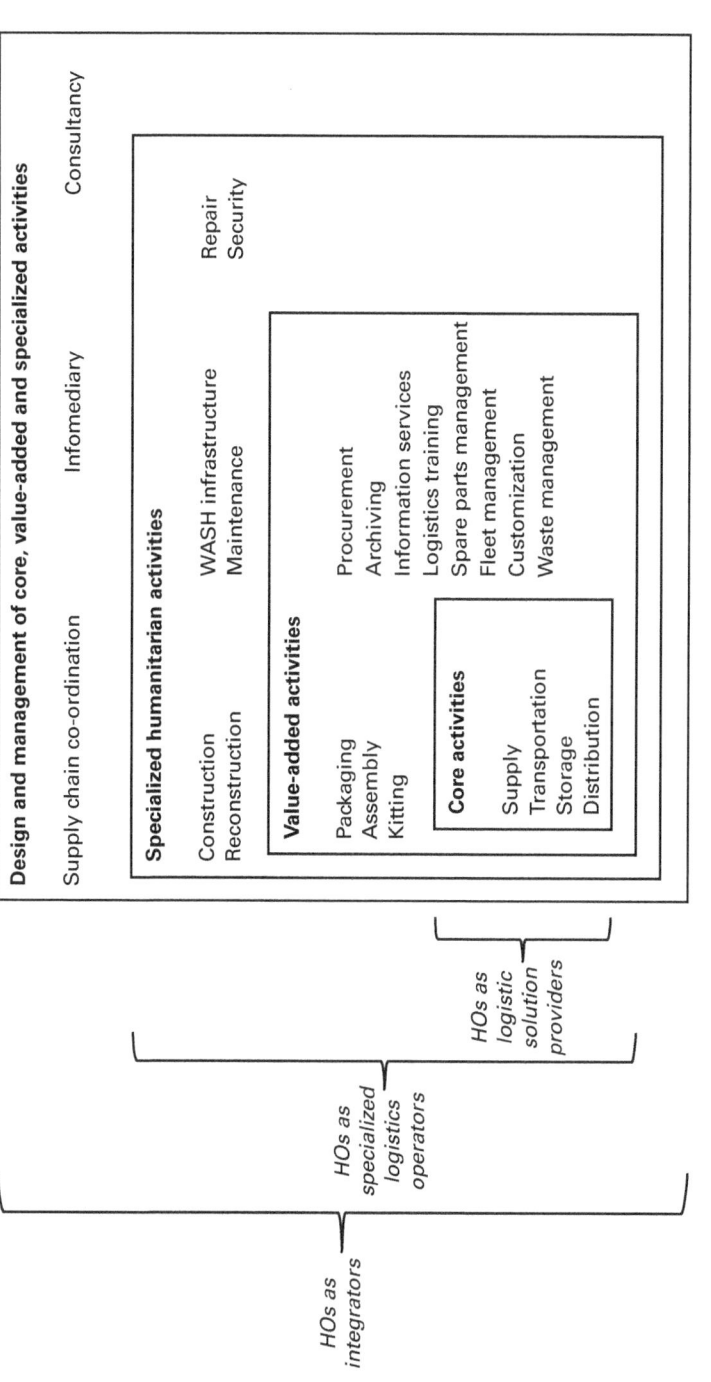

Design and management of core, value-added and specialized activities

Supply chain co-ordination Infomediary Consultancy

Specialized humanitarian activities

Construction WASH infrastructure Repair
Reconstruction Maintenance Security

Value-added activities

Packaging Procurement
Assembly Archiving
Kitting Information services
 Logistics training
 Spare parts management
 Fleet management
 Customization
 Waste management

Core activities

Supply
Transportation
Storage
Distribution

{ HOs as logistic solution providers
{ HOs as specialized logistics operators
{ HOs as integrators

SOURCE Reproduced with kind permission of Vega and Roussat (2019) the boundaries of logistics to include activities that are found in the humanitarian sector but not likely to be performed by commercial LSPs (see Figure 3.2). Hence, HSPs are well suited to carry out activities on behalf of their peers:

- As logistics intermediator, they can design, manage or execute logistics operations (transporting and storing diverse physical flows).
- As logistics solution providers, they can provide value-added logistics services such as 'information management, inventory management, training and capacity building, co-packing and co-manufacturing' (Vega and Roussat, 2019, p. 5)
- And/or as specialized logistics operators, including the aforementioned services, they are able to provide specific logistics services associated with building and reconstruction and sanitation.
- This work also highlights that HSPs can act as integrators or fourth party logistics (4PL) for other organizations, managing external providers that can be carriers for transportation (Gossler et al., 2019) or local providers for supply (Sigala and Wakolbinger, 2019).

Some examples of service provision by and for humanitarian organizations include simple logistical support to local humanitarian actors by Action Against Hunger, national and international non-governmental organizations by ICRC, or the wider humanitarian community by WFP. More value-added services include warehousing activities provided by, for example, Humanity & Inclusion or ICRC to other organizations, UNICEF providing procurement services to over a hundred partners during the Ebola crisis, or GAVI facilitating UPS transportation services for UNHCR and UNICEF during that same emergency. HSPs also provide customized solutions like warehousing and inventory management training for other organizations and governments, and have developed expertise in areas that are unique to the humanitarian context (e.g. water treatment or reconstruction) that no commercial provider would be able to offer. This positions HSPs as multi-flow, multi-activity and multi-client service providers that places them as worthy competitors to their commercial counterparts.

In their article, Heaslip et al. (2018) describe the international humanitarian aid delivery to the last mile, as a chain in which the many organizations (logistics service providers) are included. Accordingly, logistics services can be provided upstream between suppliers and humanitarian organizations at the headquarters and downstream between the headquarters and the local chapter, teams in the field or other local organizations before customs, as well as further downstream after customs, which is still separate from deliveries in the last mile between the local chapter, teams on the field or local organizations and the beneficiaries. The existing literature has mainly studied the characteristics of the two most common providers of logistics services and their offerings. However, very little has been published about different organizational forms that contribute to the array of logistics services offered to humanitarian organizations.

A continuum of actors

Commercial providers and humanitarian organizations acting as providers have been presented as being distinct entities or approaches – namely 'business' for the former and 'humanitarian' for the latter (Nurmala et al., 2017). Their respective specificities have been pointed out as levers or impediments either to perform in logistics or to adapt to humanitarian contexts. Bealt et al. (2016) note the existence of reciprocal misperceptions and lack of trust (Christopher and Tatham, 2011) between humanitarian and commercial actors, the former preferring to regulate their own operations while LSPs are likely to provide knowledge and expertise (Cozzolino et al., 2017). Beyond that private–public opposition, studying other organizational forms reveals the existence of perfectly hybrid actors, either commercial companies providing logistics services free of charge or humanitarian organizations charging for their services but in a not-for-profit model. The following are a few examples of such organizations.

Bioport

Bioport is a dual structure performing two different activities: one, a non-profit organization founded in 1994 by Dr Charles Mérieux (founder of Bioforce Institute) that provides logistics services to HOs; and another non-profit organization created in 1998 that works on social integration, helping unemployed people to get a future job while training them in forklift driving or order picking. While the latter structure was transformed into a limited liability company in 1999, the logistics activities are still performed under the non-profit structure. Bioport sells logistics services – mainly international freight forwarding – to local or international actors, namely NGOs (e.g. Humanity & Inclusion), foundations, local charity associations, local and regional authorities, schools and universities.

Distribute Aid

Distribute Aid is a non-profit NGO that provides logistics services to connect donors with frontline aid organizations, aiming to increase the efficiency and impact of on-the-ground organizations by helping them collaborate on aid shipments and handle complex aid planning, shipping and distribution. It was established in 2018 as a charitable organization in Sweden, and has a staging hub in Cambridge, UK from which donations and relief items are palletized, stored and loaded for shipment to grassroots organizations. Distribute Aid assists every point in the supply chain: assessing needs, gathering supplies at home or from in-kind donors, coordinating shipments and advising on distributions. As of 2021, Distribute Aid proposes two regular shipments to support aid groups in Lebanon and France.

Crown Agents

Crown Agents (CA) is a not-for-profit business that focuses on accelerating self-sufficiency and prosperity of communities, businesses, institutions and countries. The UK-based international organization works in 50 countries providing different services that include training, institutional efficiency, fund management, programme design, delivery and evaluation, supply chain and inspections. CA was established in 1833 by the British government. During the late 1960s and the early 1990s, CA managed the UK Government's humanitarian deployment logistics and procurement capacity, responding to crises around the world. In 1997 it became a private company owned by a non-profit-making foundation. Today CA have 16 offices around the globe that provide logistics-related and other services to UN agencies, national governments and private sector organizations.

Direct Relief

Direct Relief (DR) is an international humanitarian aid organization seeking to improve the health and lives of people affected by poverty or emergencies. The American-based organization works in 80 countries providing medical resources to community-based institutions and organizations. DR was established in 1948 in the aftermath of the second world war as a non-for-profit organization providing post-war assistance to enable people to help themselves. In the late 1950s and early 1960s, DR decided to focus on the provision of medical assistance to local facilities, serving disadvantaged populations living in medically under-served communities throughout the world.

Riders for Health

Riders for Health was created in the 1990s by Andrea and Barry Coleman and their lifelong friend Randy Mamola, a racing rider. The association was born out of the world of motorcycle racing and takes its roots from two observations the trio made in Africa: the awful condition of the roads and the number of abandoned vehicles. Considering that around 70 per cent of people in sub-Saharan Africa live in rural zones, reaching them properly for health services appeared crucial. Riders for Health improves access to healthcare for those populations by providing transportation services, including vehicle management, training and support services, healthcare products distribution or the door-to-door collection of samples, to national governments and health care delivery organizations in Africa.

These actors suggest that, rather than opposing commercial firms and humanitarian organizations one should consider the actors playing an active part in humanitarian supply chains by positioning them on a continuum: from an LSP (commercial firm) exploiting a business, an LSP (maybe the same one) acting free of charge as part of its CSR strategy, then hybrid actors (firms in the social and solidarity economy, organizations charging fair prices to keep their activities alive) and finally HSPs, humanitarian organizations acting as logistics providers with free-of-charge logistics services (see Figure 3.3). Arguably, this vision brings previous approaches back together and may help to demonstrate that commercial and humanitarian mindsets can be combined and, indeed, are actually combined by the hybrid actors we studied. The existence of such actors and the resulting complexity of the chains reinforce the need to coordinate or (to align with the commercial sector) to orchestrate a humanitarian supply chain network in a new choreography (Grange et al., 2019).

Summary and conclusion

This chapter is aimed at presenting and characterizing the actors that provide logistics services for humanitarian relief. As such, it fits in a global research agenda aimed at understanding the roles that different actors can play in humanitarian logistics. First, extant literature makes a stark distinction between the not-for-profit humanitarian environment and the for-profit, commercial logistics service providers. According to this literature, these follow different institutional logics, although there can be more than these two extremes at play.

FIGURE 3.3 A continuum of logistics service providers for humanitarian relief

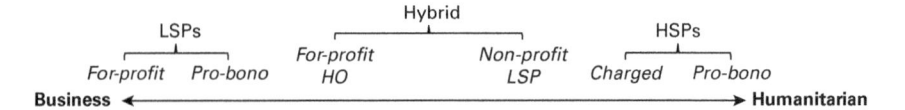

```
              LSPs                    Hybrid                 HSPs
                              For-profit    Non-profit
         For-profit  Pro-bono    HO            LSP      Charged  Pro-bono
Business ◄─────────────────────────────────────────────────────► Humanitarian
```

Rather, the combination of the business logic of commercial entities with the humanitarian or development logic of humanitarian supply chains gives rise to interesting constellations of (institutional) logics. Such new constellations enable the rise of new types of actors, some of which were introduced in this chapter.

Identifying and exploring those rather new (or previously unstudied) actors in humanitarian supply chains confirms the intertwining of logistics and non-logistics services in the humanitarian chain. This is in line with Annala et al.'s (2019) insight on the constellation of logics giving rise to various supply chain configurations, and in this case to new, hybrid types of actors. Previous literature referred to in this chapter has identified specific aid services mixed with logistic ones. Here again, we observe that interconnection with, for example, the freight forwarder for NGOs, Bioport, managing personal possessions for expatriate aid workers or Riders for Health developing ambulance activities along with their focus on fleet management. As HOs and LSPs, hybrid actors have 'expanded their definition of logistics' (Vega and Roussat, 2019, p. 947) to a growing body of activities (Liu and Lyons, 2011), customised to their clients and contexts (Leahy et al., 1995). Future research should explore such expansion and contribute to this ever-growing body of knowledge.

Note

1   See ungm.org/ASR (archived at https://perma.cc/BB5D-P487)

References

Abidi, H, de Leeuw, S and Klumpp, M (2015) The value of fourth-party logistics services in the humanitarian supply chain, *Journal of Humanitarian Logistics and Supply Chain Management*, 5(1), pp 35–60

Akbari, M (2018) Logistics outsourcing: A structured literature review, *Benchmarking: An International Journal*, 25(5), pp 1548–1580

Annala, L, Polsa, P, Kovac, G (2019) Changing institutional logics and implications for supply chains: Ethiopian rural water supply, *Supply Chain Management*, 24(3), pp 355–376

Anderson, E J, Coltman, T, Devinney, T M and Keating, B (2011) What drives the choice of a thirdparty logistics provider?, *Journal of Supply Chain Management*, **47** (2), pp 97–115

Baharmand, H, Comes, T (2019) Leveraging partnerships with logistics service providers in humanitarian supply chains by blockchain-based smarts contracts, *IFAC PapersOnLine*, **52**(13), pp 12–17

Baharmand, H, Comes, T and Lauras, M (2017) Managing in-country transportation risks in humanitarian supply chains by logistics service providers: Insights from the 2015 Nepal earthquake, *International Journal of Disaster Risk Reduction*, **24**, pp 549–559

Bealt, J, Barrera, J C F. and Mansouri, A (2016) Collaborative partnerships between logistics service providers and humanitarian organizations during disaster relief operations, *Journal of Humanitarian Logistics and Supply Chain Management*, **6** (2), pp 118–144

Christopher, M, and Tatham, P (2011) Introduction in M Christopher and P Tatham (eds.), *Humanitarian logistics: Meeting the challenge of preparing for and responding to disasters,* pp 1–14, Kogan Page, London

Cohen, L (2016) The outsourcing decision process in humanitarian supply chain management: Evaluation through the TCE and RBV principles, RIRL2016, 11th Research Meeting in Logistics and Supply Chain Management, Lausanne, Switzerland.

Cozzolino, A, Wankowicz, E and Massaroni, E (2017) Logistics service provider's engagement in disaster relief initiatives: An exploratory analysis, *International Journal of Quality and Service Sciences*, **9**(3/4), pp 269–291

Dufour, E, Laporte, G, Paquette, J and Rancourt, M-E (2018) Logistics service network design for humanitarian response in East Africa, *Omega*, **74**, pp 1–14

Gossler, T, Sigala, I F, Wakolbinger, T (2019) Applying the Delphi method to determine best practices for outsourcing logistics in disaster relief, *Journal of Humanitarian Logistics and Supply Chain Management*, **9**(3), pp 438–474

Grange, R, Hesalip, G, McMullan, C (2019) Coordination to choreography: the evolution of humanitarian supply chains, *Journal of Humanitarian Logistics and Supply Chain Management* **10**(1), pp 21–44

Heaslip, G (2013) Services operations management and humanitarian logistics, *Journal of Humanitarian Logistics and Supply Chain Management,* **3** (1), pp 37–51

Heaslip, G (2014) The increasing importance of services in humanitarian logistics, in *Humanitarian Logistics: Meeting the Challenge of Preparing for and Responding to Disasters*, 2nd ed., pp 115–127, Kogan Page, London

Heaslip, G (2016) Service triad case study, in *Supply Chain Management for Humanitarians – Tools for practice*, Haavisto, I, Kovács, G and Spens, K (eds.), pp 656–65, Kogan Page, London.

Heaslip, G, Kovács, G and Grant, D B (2018) Servitization as a competitive difference in humanitarian logistics, *Journal of Humanitarian Logistics and Supply Chain Management*, 8(4), pp 497–517

Hertz, S and Alfredsson, M (2003) Strategic development of third-party logistics providers, *Industrial Marketing Management*, 32(2), pp 139–149

Jahre, M, Jensen L-M, Listou, T (2009) Theory development in humanitarian logistics: A framework and three cases, *Management Research News*, 32(11), pp 1008–1023

Kim, S, Ramkumar, M and Subramanian, N (2019) Logistics service provider selection for disaster preparation: A socio-technical systems perspective, *Annals of Operations Research* 283, pp 1259–1282

Kovács, G (2014) Where next? The future of humanitarian logistics, in Christopher, M and Tatham, P (eds), *Humanitarian Logistics: Meeting the Challenge of Preparing for and Responding to Disasters*, 2nd ed., pp 275–285, Kogan Page, London.

Kovács, G and Spens, K (2007) Humanitarian logistics in disaster relief operations, *International Journal of Physical Distribution and Logistics Management*, 37(2), pp 99–114

Langley, C J Jr and Infosys (2020) 2020 24th annual third-party logistics study: the state of logistics outsourcing

Leahy, S E, Murphy, P R and Poist, R F (1995) Determinants of successful logistical relationships: A third-party provider perspective, *Transportation Journal*, 35(2), pp 5–13

Liu, C-L and Lyons, A C (2011) An analysis of third-party logistics performance and service provision, *Transportation Research Part E*, 47, pp 547–550

McLachlin, R and Larson, P D (2011) Building humanitarian supply chain relationships: Lessons from leading practitioners, *Journal of Humanitarian Logistics and Supply Chain Management*, 1(1), p 3

Nurmala, N, de Leeuw, S, Dullaert, W (2017), Humanitarian-business partnership in managing humanitarian logistics, *Supply Chain Management: An International Journal*, 22(1), pp 82–94

Quinn, J.B. (1999), Strategic outsourcing: Leveraging knowledge capabilities, *Sloan Management Review*, 40(4), pp 9–21

Selviaridis, K and Spring, M (2007) Third party logistics: a literature review and research agenda, *The International Journal of Logistics Management*, 18(1), pp 125–150

Sigala, I F, Wakolbinger, T (2019) Outsourcing of humanitarian logistics to commercial logistics service providers: An empirical investigation, *Journal of Humanitarian Logistics and Supply Chain Management*, 9(1), pp 47–69

Schmenner, RW (1986) How can service businesses survive and prosper?, *Sloan Management Review*, 28(3), pp 21–32

Schmenner, RW (2004) Service Businesses and Productivity, *Decisions Sciences*, **35** (3), pp 333–347

United Nations Office for Project Services (2021) The 2021 Annual statistical report on UN procurement, www.ungm.org/Shared/KnowledgeCenter/Pages/ASR (archived at https://perma.cc/T644-D2ZT)

Vega, D and Roussat, C (2019) Towards a conceptualization of humanitarian service providers, *International Journal of Logistics Management*, 30(4), pp 929–957

Vega, D and Roussat, C (2015) Humanitarian logistics: the role of logistics service providers, *International Journal of Physical Distribution & Logistics Management*, 45(4), pp 352–375

Zacharia, Z G, Sanders, N R and Nix, N W (2011) The emerging role of the third-party logistics provider (3PL) as an orchestrator, *Journal of Business Logistics*, **32** (1), pp 40–54

04

Risky business revisited: Disasters within disasters

PAUL D LARSON

Everything is fine, until the moment when it is not. And when that moment comes it can be very quick and very bad.
AIMERY MBOUNKAP, AS TOLD TO JONATHAN HARR, REPORTING FROM EASTERN CHAD, NOVEMBER 2007

ABSTRACT

Reported attacks on humanitarian aid workers have grown in the last 25 years. Such attacks were always less than 100 per year before 2003. Starting in 2018, there has not been fewer than 399 attacks per year. These disasters within disasters seem to have surged with the global 'war on terror' and a rising number of intrastate conflicts around the world. This chapter reviews a broad selection of literature on humanitarian security, with special reference to aid workers. It also classifies all attacks in 2021 reported in the Aid Worker Security Database into the following five categories: personal, criminal, political, collateral and unknown. This motivates a call for mission-specific security strategies, ranging from obliviousness to several proactive strategies to withdrawal of the mission. Based on the 'security triangle', the proactive approaches – acceptance, protection and deterrence – involve increasing loss of neutrality and autonomy, compromising the humanitarian principles. The chapter ends with a call for more research into attacks on aid workers to support mission-specific security strategies.

The death of two of our colleagues in an attack by the Myanmar army has devastated us at Save the Children. They, as well as at least 35 other people – including four children – were killed in the attack. Our colleagues, aged 28 and 32, were first-time parents, with babies a few months old, returning to the office on December 24 after attending to the needs of children ... The military forced people out of their cars, arrested some, killed many, and burned their bodies. (Waaijman 2022)

This quotation describes one of the most recent episodes of carnage in the humanitarian world – of disasters within disasters. According to Stoddard (2020), despite having *protected status* under international humanitarian law, 'aid workers have fatality rates exceeding those of uniformed military and law enforcement personnel'. Figure 4.1 depicts the increase in attacks on humanitarian aid workers, in number of incidents and workers affected (AWSD 2021). Both metrics have surged since 1997, rising to peaks of nearly 500 workers murdered, kidnapped or injured in 2013 and 2019.

Alexander and Parker (2021) argue that aid agencies operated rather freely during the 1990s, with few security limitations, even in the most dangerous places. Back then, local people perceived humanitarian aid workers and international non-governmental organizations

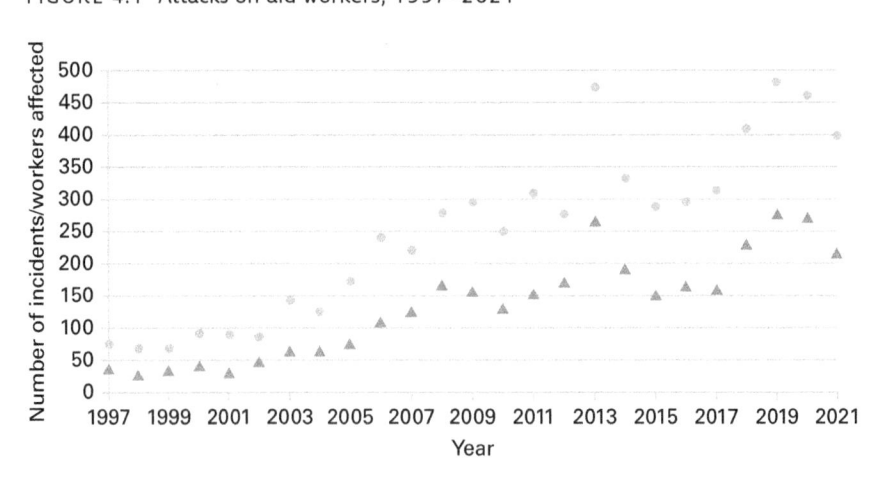

FIGURE 4.1 Attacks on aid workers, 1997–2021

(NGOs) to be neutral, autonomous and independent of political agendas. Unfortunately, it all changed after 2000. Especially after 9/11 and the American response, perceived neutrality and autonomy began to fade away. Attacks on aid workers began to rise.

Inspired by these dreadful developments, this chapter looks at attacks on aid workers as disasters within disasters. The chapter is organized into three more sections. First, there is a review and classification of a selection of relevant literature. This is followed by a framework for matching security strategies and tactics to situations in the field. Finally, the chapter closes with a brief agenda for further research.

Review and classification of the literature

In this section, a selection of literature is reviewed and classified by: (1) causes and consequences of attacks; and (2) security management strategies.

To identify academic literature on the topic, an electronic database covering a vast array of disciplines was used. A search for scholarly/peer reviewed articles with the terms 'humanitarian' and 'security' in their abstracts yielded over 5,000 hits, after elimination of exact duplicates. When 'aid workers' was added as an additional search term, the number of hits fell to 95. After screening for exact duplicates, 46 relevant articles remained. These are the primary sources in the literature review to follow. Several additional articles and reports were found among the references of the 46 pieces from the database.

As shown in Figure 4.2, the earliest article found was published in 2000. While there appears to be a slight increase in number of articles over the years, the maximum number of articles in any given year is four. It seems interest in this topic among academics has not quite matched the growing number of incidents and workers affected in the field.

The articles were also classified according to field of study. As can be seen from a scan of the reference list, this topic is extraordinarily multidisciplinary. The most prevalent field represented by this literature is

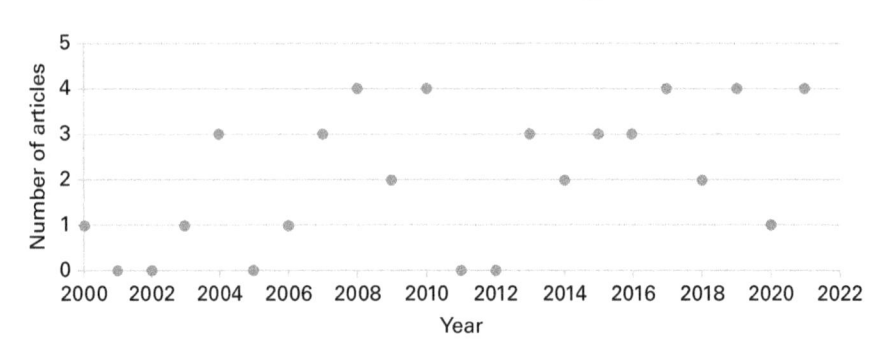

FIGURE 4.2 Articles on humanitarian aid worker security, 2000–2021

medicine/public health, closely followed by risk/security/disaster management. Next, the third most prevalent field is humanitarian/ development aid. Additional, notable areas of study include conflict and international relations, peace-keeping, gender, law and sustainability. Interestingly, as recently noted by Larson (2021), the humanitarian logistics and supply chain management literature has seldom studied matters of aid worker security.

Causes and consequences of attacks on aid workers

Literature on aid worker security addresses causes and consequences of the accelerating rate of attacks. Based on their study of attacks on aid workers from 1997 to 2014, Hoelscher, Miklian and Nygård (2017) found the 'presence and severity of armed conflicts' in a country tended to increase attacks. Stoddard et al. (2017) confirm the link between active conflict and threat of violence against humanitarians. Like attacks on aid workers, the numbers of armed conflicts in the world have been rising nearly every year, especially since 2010 (Pettersson, 2021).

Aid workers are also prone to numerous health issues ranging from Ebola virus (Haggman, Kenkre and Wallace, 2016) to altitude sickness (Bayer, 2017). Another cause of targeted attacks is their increasing lack of neutrality (Katz and Wright, 2004), as perceived by local populations. Neutrality is one of the four humanitarian principles, along with humanity, impartiality and independence (OCHA,

2012). Guided by neutrality, humanitarians do not 'take sides in hostilities or engage in controversies of a political, racial, religious or ideological nature'.

Stoddard and Harmer (2006) connect rising violent attacks on aid workers to challenges of preserving neutrality in the context of the 'global war on terror', and its counter-insurgency military operations. Integrated missions (Harmer, 2008), i.e. combining military and humanitarian operations, make it hard for NGOs to maintain perceived neutrality. For instance, the so-called provincial reconstruction teams (PRTs) in Afghanistan obscured the distinction between humanitarians and anti-insurgency combatants. In 2010–2011, Mitchell (2015) found that NGOs operating in provinces with PRTs endured more security incidents, but only in provinces with teams led by nations other than the United States. The cloak of neutrality comes off as aid agencies work aside foreign military forces (Lischer, 2007).

Martin (1999) includes 'NGO competition and culture' among the causes of attacks on aid workers. At times, aid agencies compete to be first on the scene, where the action is. Hasty entry can yield greater exposure. In addition, there may be a tendency for aid worker 'culture' to reject the discipline needed to implement effective security procedures.

Turning to consequences, Houldey (2019) reports that stress-related problems, including post-traumatic stress disorder (PTSD), are increasing along with the growth in attacks on aid workers. With a focus on operations in Kenya, she suggests that stress is induced by NGO policies and practices, as well as threats from external actors. Corey et al. (2021) identify the following inter-related consequences of exposure to trauma: anxiety, depression and PTSD. Earlier, Curling and Simmons (2010) compared the sources of stress experienced by aid workers based on their gender, location (at headquarters versus in the field) and origin (national versus international staff). Attacks on aid workers overwhelmingly occur in the field.

Attacks also make recruitment and retention of aid workers more challenging. Only 40 per cent of a large sample of Médecins sans Frontières (MSF) workers re-enlisted for a second mission. Interestingly, personal reasons and origin (from developed nations) seems to have had more of an effect on retention than the security situation (Korff et al., 2015). Of course, local (as opposed to international) aid workers may face the same security situation on and off the job.

Stoddard et al. (2017) observe that poor security conditions and threat of attack greatly compromise the ability of humanitarians to reach those in need of aid. Prior reports describe the depressing impact of security problems on the delivery of humanitarian aid in Iraq (Kapp, 2003) and Somalia (Wakabi, 2009). In Afghanistan, MSF withdrew after the murder of five of its workers, temporarily ceasing operations (Ahmad, 2004).

Security management strategies

In response to increasing attacks on aid workers, humanitarian agencies have used a variety of security management strategies, ranging from effective indifference to the risk to total withdrawal of operations, as noted above.

Martin (1999) outlines a security protocol structured around the 'security triangle' consisting of these three elements: (1) acceptance, (2) protection and (3) deterrence. The first item refers to acceptance of aid workers by the local community including law enforcement. Protection, which might compromise acceptance, utilizes tactics (e.g. communications equipment, perimeter security, helmets and vests); policies (e.g. curfews) and procedures (e.g. training); and NGO collaboration to provide security for aid workers. Finally, deterrence includes diplomacy, in consort with the United Nations (UN) and other diplomatic entities; deployment of guards; and coordination with military forces. In a recent blog post, Skelly (2021) advocates acceptance as the best way to remain true to the humanitarian principles.

Humanitarian agencies can operate under the principles of neutrality, impartiality and independence and adopt a strategic *indifference* to attacks on aid workers, as a risk of doing business in certain places (Childs, 2013). Due to the obvious limitations of obliviousness to violence, a variety of *in-house tactics* have emerged, such as building protective walls, hiring (un)armed guards, installing barbed wire and providing staff with security training (Eckroth, 2010). While Rondeau

(2009) stresses the role of training to assure safety and security for those working to support humanitarian missions of the United States, Rowley, Burns and Burnham (2013) advocate including security awareness in aid worker job descriptions. Providing security using in-house tactics implies a need to consider security during recruitment, hiring and on-boarding of aid workers.

Another option is to *outsource* security management to private military and security providers (Maisel, 2016). G3 Security Limited (https://www.g3security.co.uk/), headquartered in London, is an example of a commercial security provider. This option has budget implications, along with possible donor relations implications. It might also move an NGO further away from neutrality and independence, in the minds of beneficiaries and other local folks (Larson, 2021). Renouf (2007) discusses the possible role of private security companies for protecting aid workers, including strengths and weaknesses of these providers, along with their impact on humanitarian missions.

Further down the path to non-neutrality is to *collaborate* on an operational basis with military units, e.g. United Nations peacekeeping forces or national armies or foreign military units. Security of humanitarian missions has been a discussion topic for more than 20 years. In response to various security issues arising during complex emergencies involving refugees, Pilch (2000) proposes the possibility of 'standby forces' having a mandate to support and protect humanitarian workers and refugees. Lischer (2007) identifies several forms of 'militarized charity', such as soldiers as humanitarians and/or aid workers as government agents. She argues that as humanitarians work closer alongside intervention forces, their neutrality evaporates in the minds of local people.

Attacks on aid workers due to their perceived connection to Western governments have been especially rampant in a small group of nations, including Afghanistan, Somalia and Sudan (Rubenstein, 2013). In an analysis of 67 nations that endured intrastate conflict between 1997 and 2018, Kisangani and Mitchell (2021) observe that aid workers are more likely to be attacked in nations with integrated UN missions. In Afghanistan, it was suggested that national security forces be trained to protect aid workers (Ahmad, 2004).

Finally, humanitarian agencies can *withdraw*, i.e. cease operations and evacuate people, in response to attacks against aid workers (Runge, 2004). Operational withdrawal is *reactive* if it is based on an actual attack and *proactive* if it is based on threat of attack or anticipation of attack. Following the murders of five of its aid workers on a single day, MSF withdrew from Afghanistan (Ahmad, 2004; Katz and Wright, 2004). Similar withdrawal strategies have been in play during the deteriorating security situation in the Darfur region of Sudan (Reeves, 2007). Sometimes withdrawal is partial, e.g. pulling out all international aid workers and replacing them with local or national staff (Haver, 2007).

The approaches beyond indifference and withdrawal put increasing distance between aid workers and the people they serve. These approaches also compromise neutrality and independence. In a study of NGO security management in high-risk environments, Guidero (2022) identified several relevant factors, including the specific location, logistics, the organizational mandate, perceptions of aid recipients and the actions of other NGOs and local government agencies. The following section builds on this work, with a framework for security evaluation and management.

Figure 4.3 extends the security triangle to a security pentagon by adding more reactive security strategies to the mix. The triangle elements are all proactive and all three items could involve in-house

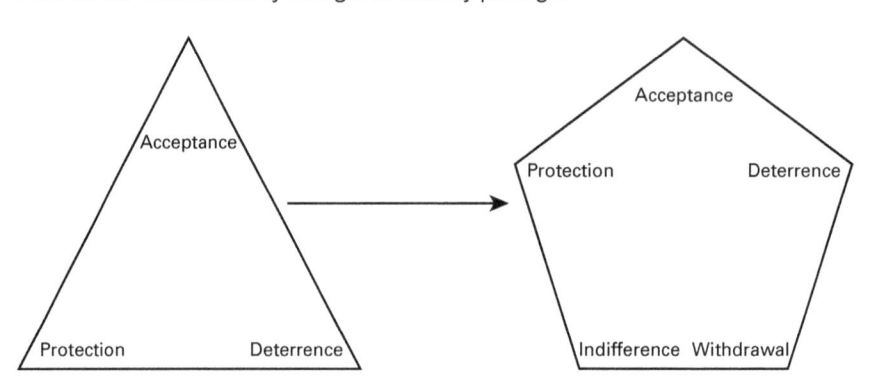

FIGURE 4.3 From security triangle to security pentagon

security tactics, outsourcing security and/or broader forms of collaboration. However, the two new elements – indifference (which assumes local community acceptance) and withdrawal – are largely reactive.

Security evaluation and management

Security evaluation and management (SEAM) starts with a thorough assessment of the local situation. *What types of attacks are likely to occur?* The author studied all attacks on aid workers in 2021, as listed in the Aid Worker Security Database (AWSD, 2021), with special focus on the 'details' column. Attacks can be classified into the following five categories, based on the perpetrators' motivations: personal, criminal, political, collateral and unknown. Following are several examples of each type of attack and country of occurrence, drawn from AWSD (2021) details.

Attacks motivated by *personal* reasons appear to be the smallest category of attacks. These attacks occur in relatively few countries. Effective strategies and tactics to reduce this type of attack might involve staff training and awareness.

> One NGO staff member was killed by a family member in an interpersonal dispute … (South Sudan)
> One NGO staff member was shot and wounded over a personal dispute … (South Sudan)
> One INGO guard was shot and killed while walking near an INGO warehouse. Reports indicate the incident was a revenge killing and not an attack on the INGO. (South Sudan)
> One UN national humanitarian staff member was assaulted and injured during a dispute with a housing tenant. (Sierra Leone)

Criminal attacks (e.g. assault, rape and theft) seem considerably more common than personal attacks. Dealing with these attacks could also involve staff training and awareness, as well as cooperation with local law enforcement, if available. Criminal attacks against aid workers happen across a wide range of countries.

> One female UN humanitarian staff member was attacked and robbed by unknown men while withdrawing money ... (Namibia)
>
> A female NGO staff member was abducted by a taxi driver and sexually assaulted ... (Afghanistan)
>
> One national male UN humanitarian agency staff member was injured after an unknown criminal broke into his house ... (Liberia)
>
> Two INGO national staff members, a doctor and a nurse working in an ambulance, were ambushed by an unknown criminal gang ... (El Salvador)
>
> Twelve international staff members of a Christian INGO were kidnapped by the criminal group 400 Mawozo, while on route from an orphanage visit ... (Haiti)
>
> One female NGO staff, on her way home from duty, was stopped by two 'security' personnel (who) directed the staff member to remove her trousers due to an alleged dress code for women in the camp. The staff member was assaulted when she refused ... (South Sudan)

The previous example of a criminal attack suggests there is a grey-area between criminally-inspired and politically-motivated attacks. Attacks motivated by political matters are the most common type of attack, and they seem to be largely limited to countries in the midst of armed conflict. These are the sorts of attacks for which some sort of formal, collaborative security arrangement might be considered. Since international humanitarian law fails to require combatants to guarantee the safety of aid workers, Seatzu (2017) proposes the creation of international agreements between NGOs and parties at war.

> One male humanitarian NGO staff member was reportedly kidnapped by Boko Haram while traveling ... (Nigeria)
>
> One INGO staff member was killed during a complex attack in Mogadishu. Official statements report Al Shabaab militants exploded a car bomb and then stormed the building with guns... (Somalia)
>
> Four female NGO workers were shot and killed and a male NGO driver was wounded ... reports indicate the aid workers were ambushed on the road while delivering services for women in the area. No group has claimed responsibility, but reports point to the Pakistani Taliban ... (Pakistan)

> Three NGO aid workers were reportedly detained and beaten by the Myanmar military because they were providing medical assistance to injured protestors … (Myanmar)
>
> One NGO staff member was shot and killed at the conclusion of a food distribution event … Reports indicate the shooting motivation was the alleged food distribution 'to the Junta' … (Ethiopia)
>
> Two UN agency staff members were killed after their aid convoy was ambushed by UPC militants … (Central African Republic)
>
> One Red Crescent paramedic was injured by the glass from a broken windshield after Israeli Defense Forces shot at the clearly marked ambulance … (Occupied Palestinian Territories)
>
> The director of an NGO all-girls school was abducted by Taliban militants from his home. He was held for two days and tortured in regards to his work with the organisation … (Afghanistan)

Another type of attack in areas of ongoing armed conflict leads to *collateral* damage. Aid workers are not the intended targets, though sometimes the target may not be clear. Negotiations with competing combatants in a conflict could help reduce the frequency and severity of these attacks.

> One humanitarian staff member was reportedly killed by a stray bullet during a clash between ISWA militants and the Nigerian Armed Forces. (Nigeria)
>
> One male aid worker, head of an INGO office, died after being shot in crossfire. (Ethiopia)
>
> One NGO staff member was killed along with his family, when his home was hit by a Russian military shell … (Syria)
>
> One NGO staff member was reportedly killed in crossfire during community clashes … (South Sudan)

Finally, the motivation behind some attacks is *uncertain*, often because it is unclear whether they are driven by criminal or political ambitions. Security evaluation and management efforts should include investigation into the cause of past security incidents.

One male INGO national staff member was shot and killed for unknown reasons. (Democratic Republic of the Congo)

A marked INGO ambulance transporting patients was attacked by a group of unknown armed men. Three staff on board and three patients were assaulted and detained in the hot sun for several hours before being allowed to continue. (Mali)

Two other important questions for NGOs in risky situations are: who are our stakeholders? How do security risks affect each stakeholder group? These stakeholders include beneficiaries or recipients of aid, staff at headquarters, international and national aid workers in the field, donors, other NGOs, host government agencies and current or potential security partners. As demonstrated above, aid workers face security issues and uncertainties in the field, along with shrinking access to the beneficiaries they seek to serve (Asgary and Lawrence, 2014).

In addition, NGOs should evaluate their exposure to various security events. Exposure follows from a variety of factors, ranging from largely uncontrollable (e.g. local culture and extent of armed conflict) to controllable (e.g. organizational security strategy and tactics). Other factors, like perceived neutrality of the NGO in the minds of local people, may be partially controllable.

SEAM recognizes the importance of developing and implementing strategies to deal with aid worker security (Fast, 2010). The ultimate purpose is to match security strategy and tactics to situations. The traditional reactive strategy of assuming all is well, then withdrawing if terrible things happen serves neither beneficiaries nor aid workers well. Proactive approaches, as outlined in the security triangle (Martin, 1999), follow from a thorough analysis of any given situation. Under what conditions will we enter a dangerous area? Under what conditions will we exit? Such questions should be asked before deployment, rather than after.

Creating policies and standards (Bollettino, 2008; Rowley et al., 2013) enables a proactive approach to security management. Perhaps there are times and places for relative indifference to security threats. Gathering intelligence, developing an early warning system and having a plan are activities that require resources. Being proactive can distract aid workers from the primary job of assisting people in need. Still, as they go about their risky business, NGOs and other humanitarian agencies should do whatever they can to protect their aid workers.

Research agenda

For more than 20 years, researchers have been analysing attacks on aid workers. Sheik et al. (2000) analysed 382 aid worker fatalities between 1985 and 1998. They found that most of these deaths were due to intentional violence, which peaked during the 1994 crisis in Rwanda. Also focusing on fatalities, King (2002) studied reports from the ReliefWeb database from 1997 to 2001. Nearly half of all the non-accidental deaths occurred during ambushes of transport vehicles. Three local aid workers were killed for every expatriate aid worker killed.

More recently, Hoelscher et al. (2017) analysed attacks on aid workers at the country level from 1997 to 2014 and Morokuma and Chiu (2019) used AWSD data to study security incidents involving aid workers from January 1997 to December 2016. The latter piece categorized incidents as occurring in healthcare versus non-healthcare settings and identified a need for 'high-quality' data from the field to support improvements in security. Extending the time period further, Kisangani and Mitchell (2021) analysed aid worker attacks in 67 countries suffering under intrastate conflict between 1997 and 2018.

What appears to be missing from these previous investigations is inclusion of NGO security policies and practices, linked to each attack on aid workers. Future work is needed to expand the datasets beyond location, type of incident, outcome and brief details, to include coding of security strategies and tactics being used by the affected NGOs and other agencies. Linking incident outcomes to security approaches, as well as location and other details, will enable benchmarking and facilitate improvement.

There is also a need for more case-study research on security in the field. Previous examples include a study of security practices in Haiti (Beerli, 2018) and interviews with security managers on aid agencies' use of security technology (Kalkman, 2018). Ultimately, each attack on aid workers is a case worthy of careful study for the purpose of reducing these horrific disasters within disasters.

References

Ahmad, K (2004) Aiding the aiders, *The Lancet*, **364**(9442), pp 1303–1304, https://doi.org/10.1016/S0140-6736(04)17209-7 (archived at https://perma.cc/83D7-WVJF)

Alexander, J and Parker, B (2021) Then and now: 25 years of aid worker (in)security, *The New Humanitarian*, (25 February), https://www.thenewhumanitarian.org/feature/2021/2/25/then-and-now-25-years-of-aid-worker-insecurity (archived at https://perma.cc/B8TN-JGMC)

Asgary, R and Lawrence, K (2014) Characteristics, determinants and perspectives of experienced medical humanitarians: a qualitative approach, *BMJ Open*, **4**(12), pp 1–14, https://doi.org/10.1136/bmjopen-2014-006460 (archived at https://perma.cc/43US-ZGAB)

AWSD (2021) Humanitarian outcomes, aid worker security database https://aidworkersecurity.org/ (archived at https://perma.cc/82KY-RBRQ)

Bayer, R G (2017) Altitude illness, *Disaster Prevention & Management*, **26**(1), pp 55–64, https://doi.org/10.1108/DPM-07-2016-0146 (archived at https://perma.cc/X8K2-UC95)

Beerli, M J (2018) Saving the saviors: Security practices and professional struggles in the humanitarian space, *International Political Sociology*, **12**(1), pp 70–87, https://doi.org/10.1093/ips/olx023 (archived at https://perma.cc/S85N-QUEG)

Bollettino, V (2008) Understanding the security management practices of humanitarian organizations, *Disasters*, **32** (2), pp 263–279, https://doi.org/10.1111/j.1467-7717.2008.01038.x (archived at https://perma.cc/UZ6D-ZMLX)

Childs, A K (2013) Cultural theory and acceptance-based security strategies for humanitarian aid workers, *Journal of Strategic Security*, **6**(1), pp 64–72, https://doi.org/10.5038/1944-0472.6.1.6 (archived at https://perma.cc/288W-HLF3)

Corey, J, Vallières, F, Frawley, T, De Brún, A, Davidson, S and Gilmore, B (2021) A rapid realist review of group psychological first aid for humanitarian workers and volunteers, *International Journal of Environmental Research and Public Health*, **18**(4), p 1452, https://doi.org/10.3390/ijerph18041452 (archived at https://perma.cc/9CG2-WJ6Y)

Curling, P and Simmons, K B (2010) Stress and staff support strategies for international aid work, *Intervention: Journal of Mental Health and Psychosocial Support in Conflict Affected Areas*, **8**(2), pp 93–105, https://www.interventionjournal.com/sites/default/files/Stress_and_staff_support_strategies_for.2.pdf (archived at https://perma.cc/K7NU-MPPE)

Eckroth, K R (2010) Humanitarian principles and protection dilemmas: Addressing the security situation of aid workers in Darfur, *Journal of International Peacekeeping*, **14**(1/2), pp 86–116, https://doi.org/10.1163/187541110X12592205205694 (archived at https://perma.cc/25G9-2E8C)

Fast, L A (2010) Mind the gap: Documenting and explaining violence against aid workers, *European Journal of International Relations*, **16**(3), pp 365–389, https://doi.org/10.1177/1354066109350048 (archived at https://perma.cc/C3FC-ND9R)

Guidero, A (2022) Humanitarian (in)security: Risk management in complex settings, *Disasters*, **46**(1), pp 162–184, https://doi.org/10.1111/disa.12457 (archived at https://perma.cc/ND65-MZ7B)

Haggman, H, Kenkre, J and Wallace, C (2016) Occupational health for humanitarian aid workers in an Ebola outbreak, *Journal of Research in Nursing*, **21**(1), pp 22–36, https://doi.org/10.1177/1744987116630578 (archived at https://perma.cc/T96F-2BXD)

Harmer, A (2008) Integrated missions: A threat to humanitarian security? *International Peacekeeping*, **15**(4), pp 528–539, https://doi.org/10.1080/13533310802239824 (archived at https://perma.cc/N5J7-254N)

Harr, J (2009) Lives of the Saints, *The New Yorker*, (January 5), pp 47–59

Haver, K (2007) Duty of care? Local staff and aid worker security, *Forced Migration Review*, **1** (28), pp 10–11

Hoelscher, K, Miklian, J and Nygård, H M (2017) Conflict, peacekeeping, and humanitarian security: Understanding violent attacks against aid workers, *International Peacekeeping*, **24** (4), pp 538–565, https://doi.org/10.1080/13533312.2017.1321958 (archived at https://perma.cc/67G5-DM78)

Houldey, G (2019) Humanitarian response and stress in Kenya: Gendered problems and their implications, *Gender & Development*, **27**(2), pp 337–353, https://doi.org/10.1080/13552074.2019.1615281 (archived at https://perma.cc/35EX-K87F)

Kalkman, J P (2018) Practices and consequences of using humanitarian technologies in volatile aid settings, *Journal of International Humanitarian Action*, **3**(1), https://doi.org/10.1186/s41018-018-0029-4 (archived at https://perma.cc/YC4F-AW9T)

Kapp, C (2003) Plight of Iraqi civilians worsens, according to UN, *The Lancet*, **361**(9364), p 1190, https://doi.org/10.1016/S0140-6736(03)12970-4 (archived at https://perma.cc/W7ND-QH94)

Katz, I T and Wright, A A (2004) Collateral damage – Médecins sans Frontières leaves Afghanistan and Iraq, *New England Journal of Medicine*, **351**(25), pp 2571–2573

King, D (2002) Paying the ultimate price: An analysis of aid worker fatalities, *Humanitarian Exchange*, **21**(July), Humanitarian Practice Network, Overseas Development Institute, London, pp 15–16, https://odihpn.org/publication/paying-the-ultimate-price-an-analysis-of-aid-worker-fatalities/ (archived at https://perma.cc/WPQ8-86LK)

Kisangani, E F and Mitchell, D F (2021) The impact of integrated UN missions on humanitarian NGO security: A quantitative analysis, *Global Governance*, **27**(2), pp 202–225, https://doi.org/10.1163/19426720-02702005 (archived at https://perma.cc/473A-2SNJ)

Korff, V P, Balbo, N, Mills, M, Heyse, L and Wittek, R (2015) The impact of humanitarian context conditions and individual characteristics on aid worker retention, *Disasters*, **39**(3), pp 522–545, https://doi.org/10.1111/disa.12119 (archived at https://perma.cc/9TKA-5TRP)

Larson, P D (2021) Security, sustainability and supply chain collaboration in the humanitarian space, *Journal of Humanitarian Logistics and Supply Chain Management*, **11**(4), pp 609–622, https://doi.org/10.1108/JHLSCM-06-2021-0059 (archived at https://perma.cc/RF8X-DYEF)

Lischer, S K (2007) Military Intervention and the Humanitarian 'Force Multiplier,' Global Governance: A Review of Multilateralism and International Organizations, **13** (1), pp 99–118, https://doi.org/10.1163/19426720-01301007 (archived at https://perma.cc/GXV9-4ZNR)

Maisel, N (2016) Strange bedfellows: Private military companies and humanitarian organizations, *Wisconsin International Law Journal*, **33**(4), pp 639–666

Martin, R (1999) NGO field security, *Forced Migration Review*, **4**(April), pp 4–7, https://www.fmreview.org/security-at-work/martin (archived at https://perma.cc/22YW-CK5G)

Mitchell, D F (2015) Blurred lines? Provincial reconstruction teams and NGO insecurity in Afghanistan, 2010–2011, *Stability: International Journal of Security and Development*, **4**(1), https://doi.org/10.5334/sta.ev (archived at https://perma.cc/4A3H-UY6J)

Morokuma, N and Chiu, C H (2019) Trends and characteristics of security incidents involving aid workers in health care settings: A 20-year review, *Prehospital and disaster medicine*, **34**(3), pp 265–273, https://doi.org/10.1017/S1049023X19004333 (archived at https://perma.cc/69XK-78KK)

OCHA (2012), What are humanitarian principles? OCHA on Message: Humanitarian Principles, (June), https://www.unocha.org/sites/unocha/files/OOM_Humanitarian%20Principles_Eng.pdf (archived at https://perma.cc/V99N-69V9)

Pettersson, T (2021) *UCDP/PRIO armed conflict dataset codebook*, Version 21.1, Uppsala Conflict Data Program, Department of Peace and Conflict Research, Uppsala University, Sweden and Centre for the Study of Civil Wars, International Peace Research Institute, Oslo, https://ucdp.uu.se/downloads/ucdpprio/ucdp-prio-acd-211.pdf (archived at https://perma.cc/HN9E-K9TL)

Pilch, F T (2000) Security issues and refugees: Dilemmas, crises, and debates, *Refuge*, **19**(1), pp 25–34, https://doi.org/10.25071/1920-7336.22069 (archived at https://perma.cc/E7GR-DY9K)

Reeves, E (2007) Genocide without end? The destruction of Darfur, *Dissent*, **54**(3), pp 8–13, https://doi.org/10.1353/dss.2007.0055

Renouf, J S (2007) Do private security companies have a role in ensuring the security of local populations and aid workers? Universites d'Automne de l'Humanitaire, (September), https://www.researchgate.net/publication/265572536_Do_Private_Security_Companies_Have_a_Role_in_Ensuring_the_Security_of_Local_Populations_and_Aid_Workers (archived at https://perma.cc/47VD-66ZG)

Rondeau, D L (2009) Providing a safe and secure environment: in support of America's humanitarian missions, *Journal of Counterterrorism and Homeland Security International*, **15**(2), pp 18–22

Rowley, E, Burns, L and Burnham, G (2013) Research review of nongovernmental organizations' security policies for humanitarian programs in war, conflict, and postconflict environments, *Disaster medicine and public health preparedness*, **7**(3), pp 241–50, https://doi.org/10.1001/dmp.2010.0723 (archived at https://perma.cc/LQ6N-3UXC)

Rubenstein, L S (2013) A way forward in protecting health services in conflict: moving beyond the humanitarian paradigm, *International Review of the Red Cross*, **95**(890), pp 331–340, https://doi.org/10.1017/S1816383113000684 (archived at https://perma.cc/PB99-RWPP)

Runge, P (2004) New Security threats for humanitarian aid workers, Social Work and Society, **2** (2), https://doaj.org/article/123d8003b9bf467482e97d1f44a86f1b (archived at https://perma.cc/M2MZ-FPLJ)

Seatzu, F (2017) Revitalizing the international legal protection of humanitarian aid workers in armed conflict, La Revue des droits de l'hommeOpenAIRE, **11**, https://doi.org/10.4000/revdh.2759 (archived at https://perma.cc/JZC9-NF2R)

Sheik, M, Gutierrez, M I, Bolton, P, Spiegel, P, Thieren, M and Burnham, G (2000) Deaths among humanitarian workers, *British Medical Journal*, **321**(7254), pp 166–168, https://doi.org/10.1136/bmj.321.7254.166 (archived at https://perma.cc/ZS2M-M3V9)

Skelly, J (2021) Community Acceptance: A cornerstone of humanitarian security risk management, (16 February), https://gisf.ngo/blogs/community-acceptance-a-cornerstone-of-humanitarian-security-risk-management/ (archived at https://perma.cc/SQR3-CJVG)

Stoddard, A and Harmer, A (2006) Little room to maneuver: The challenges to humanitarian action in the new global security environment, Journal of Human Development, 7(1), pp 23–41, https://doi.org/10.1080/14649880500501146 (archived at https://perma.cc/DEN5-U7NL)

Stoddard, A, Jillani, S, Caccavale, J, Cooke, P, Guillemois, D and Klimentov, V (2017) Out of reach: How insecurity prevents humanitarian aid from accessing the neediest, *Stability: International Journal of Security & Development*, **6**(1), pp 1–25, https://doi.org/10.5334/sta.506 (archived at https://perma.cc/NL9D-MQG8)

Stoddard A (2020) 'Tracking the Toll: Measuring Violence against Aid Workers,' in *Necessary Risks: Professional humanitarianism and violence against aid workers*, Palgrave Macmillan, Cham, https://doi.org/10.1007/978-3-030-26411-6_1 (archived at https://perma.cc/X9PN-FHRC)

Waaijman, G (2022) For a 2022 in which the attacks on aid workers end, *El Pais*, (10 January), https://elpais.com/planeta-futuro/red-de-expertos/2022-01-10/por-un-2022-en-el-que-acaben-los-ataques-hacia-los-cooperantes.html (archived at https://perma.cc/SKN6-Z9QF)

Wakabi, W (2009) Fighting and drought worsen Somalia's humanitarian crisis, *The Lancet*, **374** (9695), pp 1051–1052, https://doi.org/10.1016/S0140-6736(09)61687-1 (archived at https://perma.cc/UAY2-XUE5)

05

The journey from a patchy to a comprehensive supply chain in UNHCR (2005–2015)

SVEIN J HÅPNES

ABSTRACT

This chapter outlines the process of transforming and improving the supply chain management (SCM) function in UNHCR, from 2005 to 2015. Being an operational agency of the UN, UNHCR in 2005 had a minimal SCM function, unable to support the objectives of the organization. The chapter describes the process, hurdles and tools used on the way towards a more rightsized and capable SCM function. The chapter concludes with the main lesson identified and key success criteria for the development of the UNHCR SCM function that took place in this period.

Introduction to the United Nations High Commissioner for Refugees (UNHCR)

The United Nations High Commissioner for Refugees (UNHCR) is one of the three larger *operational* United Nations (UN) agencies, together with United Nations Children's Fund (UNICEF) and the World Food Program (WFP). UNHCR provides assistance to refugees globally, except for Palestine refugees who are handled by the UN Relief and Works Agency for Palestinian refugees in the Near East. The organization was created in 1950 to deal with the refugees following World War II in Europe, with a mandate provided in the 1951 Refugee Convention (UN, 1951), further expanding its legal framework and reach with the subsequent 1967 Refugee Protocol (UN, 1967). Notable when comparing UNHCR with other large UN agencies or international non-governmental organizations (INGOs) is the fact that UNHCR is a *mandated* organization. UNHCR is mandated to aid and protect refugees and stateless people, and to assist in their repatriation, integration or third country resettlement. This mandate influences the power balance between administration, budget/finance, legal protection, operations and supply chain management (SCM), which is important to understand when analysing UNHCR, and its approach to SCM.

During the period when Sadako Ogata was the UN High Commissioner for Refugees (November 1990–December 2000), at a time of large crises caused by the wars in former Yugoslavia and the Great Lakes region in Africa, UNHCR became known as an organization with a forceful logistics capability able to deliver the basics to beneficiaries in need. De facto, prior to the forming of the Inter-Agency Standing Committee thematic clusters as a part of the Humanitarian Reform Agenda in 2005, the UNHCR operation in Croatia and Bosnia and Herzegovina was a massive internally displaced people operation. During the war, UNHCR took on the responsibility of serving the basic food needs of approximately 2.7 million internally displaced people inside Bosnia and Herzegovina and Croatia (more than 1.1 million metric tons). UNHCR, with support of the UN Department of Peacekeeping Operation (DPKO), the UN Protection Force (UNPROFOR) and troop contributing countries, also led the Sarajevo airlift, Operation Provide Promise, the longest lasting airlift operation in history. The operation lasted 1,285 days from 3 July 1992 until 9 January 1996, delivering 12,886 airlifts moving 160,677 metric tons of supplies to the besieged city of Sarajevo.

With the retirement of Ogata in December 2000, Rudd Lubbers (in the role from January 2001 to June 2005) was appointed as the next High Commissioner. Following his arrival, there was a gradual change of perspective of what UNHCR's core activities were, moving the focus away from providing assistance and hence SCM. Following a funding crisis in the early 2000s, this resulted in larger personnel cuts and organizational changes, and the SCM capacities and capability built during the Yugoslav and Great Lakes-crises era was demolished. Lubbers' period as Commissioner was cut short, as he was forced to leave the post in February 2005 following a sexual harassment complaint.

Building the understanding of supply chain management in UNHCR (2005–2006)

In May 2005, the former Portuguese Prime Minister António Guterres was appointed as the new High Commissioner. (He occupied the position from June 2005 to December 2015.) Guterres, recently re-elected to his second term as the UN Secretary-General, is known as a founding member of the Portuguese Refugee Council and for his involvement with the independence of Timor-Leste and the transfer of sovereignty of Macau. Guterres took an early interest in strengthening the ability of UNHCR to provide assistance (and hence its capability to deliver) as a part of an attempt to increase the relevance of UNHCR. At the time of the author's arrival in UNHCR (January 2006), the Supply Management Service (SMS) was a unit in the Division of Finance and Administration, perceived and placed to act as a more administrative procurement function. The structure to lead, oversee and advise in respect of logistics and delivery was only three individuals, two officers and one local staff member, hence there was only a very limited capacity to manage and lead SCM operations in support of emergency responses.

In March 2006, a massive flooding hit the Sahrawi refugee camps in Tindouf, Southern Algeria. Being a core mandate situation, the onus was on UNHCR to rapidly deliver much needed relief supplies.

The patchy response exposed many of the shortfalls and weaknesses of the SCM capabilities in UNHCR. Relief supplies were eventually made available from the surplus at the 'Iraqi stockpile' in Amman, Jordan, and transport secured through a UN Office for the Coordination of Humanitarian Affairs (OCHA) call for donated flights using the Oslo Guidelines (UN OCHA, 2007). As a result, UNHCR was only able to produce a rather slow and uncoordinated response to this sudden onset emergency. A review undertaken after the response uncovered many shortfalls, the three most prominent SCM findings being the lack of: rapid access to stocks of relief items; financial mechanisms for funding of transport; and availability of trained SCM staff to manage the operations on the ground. In May the same year, civil unrest broke out in Timor-Leste, and although not a mandated refugee situation, the High Commissioner decided that UNHCR should respond. Based on the learnings from Tindouf, the High Commissioner early took the decision to release (dead) stock from the Iraq and Dubai stockpiles. Equally as important, the High Commissioner decided to advance funds from the UNHCR reserves based on projections of donations, enabling the SMS to respond rapidly. It also allowed for the building of a cost-efficient follow-on air/sea supply chain through a transit hub in Darwin, Australia, that was harmonized with the growing local receiving and logistics capacities in Dili, Timor-Leste.

In July/August 2006, the UNHCR SCM came under its ultimate test following the Israeli invasion of Lebanon. Coinciding with the roll-out of the Humanitarian Reform Agenda launched in 2005, the response to the war in Lebanon became the first real test of the cluster approach in a man-made emergency. As was the case for other UN agencies, UNHCR struggled to deliver under its mandate in Lebanon, mainly because of the security situation and the UN-imposed security regulations (phase IV and Minimum Operating Security Standards requirements). The strict security regulations forced larger parts of the humanitarian community to be stranded at the Mövenpick Hotel in Beirut. Regardless, compared with the UNHCR emergency responses early in 2006, the Lebanon response in SCM terms was an improvement, as trained and experienced logistics personnel were deployed and able to operate from a rented warehouse in Beirut port. Equally, UNHCR was able to operate regular relief flights into Beirut throughout the conflict, using Jordanian, Portuguese, Belgian and German Air Force C-130/160s operating under bilateral agreements out of Amman, Jordan. Through diplomatic channels, UNHCR was able to provide sufficient security and safety guarantees allowing these aircrafts to operate. Space was also offered to other humanitarian agencies through the logistics cluster to enable the movement of essential medicines for other UN agencies and NGOs.

Establishing the need to improve supply chain management in UNHCR (late 2006)

In the UNHCR internal review (UNHCR, 2006a) following the response in Lebanon, a set of SCM recommendations was highlighted. The recommendations urged UNHCR to prepare: in-house procedure for emergency declaration; rosters with staff in protection and shelter; trained staff and material resources to undertake local procurement; establish/operationalize delivery of relief items from emergency stockpiles; undertake a joint review of the functioning of the logistics cluster with WFP; develop its own internal logistics capacities; and strengthen the contingency planning capacity. It had become evident to UNHCR's senior management that there was an urgent need to revamp the SCM capability in order to improve the organization's capacity to deliver emergency assistance. Based on the lessons identified in 2006, a decision to move the SCM functions from Division of Finance and Administration to the Division of Operational Support was taken. As a part of the preparations for this move, an internal paper (UNHCR, 2006b) analysing how to best re-organize was developed. The paper stressed the need for: rapid and robust emergency intervention processes; integrated planning procedures; emergency response implementation based on regular processes and human resources (do the same, just faster); a centrally-controlled emergency stockpile management; internal pre-funding arrangements for sourcing of delivery means; and focused selection and training of SCM personnel. This led to the establishment of the '72-hours principle', ensuring first delivery of core relief items anywhere in the world within this time frame. This was achieved by establishing an internal process for rapid: access to funds, release from global stocks and chartering of air assets.

While the need to improve UNHCR SCM capacities started to mature with senior management, the organization also came under pressure from donors to reduce its staffing overhead in Geneva. As a result, the High Commissioner launched a review of core functions at headquarters. An initial step was an external review of the SCM functions undertaken by INSEAD (2006). Their main findings were that UNHCR, as an operational agency, needed to drastically reinforce its capacities within logistics. Furthermore, they observed an absence of cross-functional (integrated) planning, resulting in the SCM functions not being properly aligned with the assistance and protection functions in UNHCR. This happened at the same time as UNHCR started to operationalize its role as lead in the protection cluster, and as co-lead in the shelter and the camp coordination and camp management (CCCM) clusters. With these clusters (which also includes core relief items such as tents, blankets, water containers, etc.) being logistics-heavy, this reinforced the need for improvements and increased capacity of the SCM function. Thus, SMS became actively involved in the development of the mentioned clusters, as well as the logistics cluster (LC), but insufficient human resources were available for the tasks. However, SMS' involvement and participation in the logistic cluster's Logistics Response Team training and the humanitarian logistics certification became a very important tool for UNHCR SCM staff due to the lack of their own inhouse SCM training opportunities.

One of the more important learnings was the growing understanding that the one-size-fits-all approach, with decentralized control within all functions, did not support the ability of the SCM to deliver emergency responses. It became clear to the High Commissioner and senior management that the compartmentalization of ownership and control of emergency stocks and the absence of financial instruments to deal with the funding time lag and lead times seriously hampered UNHCR's ability to respond to emergencies forcefully and in a timely manner. In the case of the UNHCR SCM, there was a need to move the point and place of decision-making of the '6W' question, who wants what, where, when and why?, to the strategic SCM level to mitigate against uncertainty and long lead times. As stated in the UN Office of Internal Oversight Service (OIOS) audit report (UN OIOS, 2007), 'while UNHCR is not a logistical agency, much of what it does depends on the critical support role of logistics to UNHCR's protection and assistance activities'.

The internal 'battle' for change (2007)

While the level of internal recognition of the need to improve inhouse SCM capacities grew, the donor pressure to reduce staffing levels at the Geneva headquarters (HQ) continued to increase. To review the possibility of relocating certain HQ functions to lower cost locations, PricewaterhouseCoopers (PwC) was engaged to undertake an outposting (relocation) feasibility study (PwC, 2007). The PwC study focused on direct costs savings 'due to the moral imperative (of UNHCR) to minimize head office and costs, to maximize resources available for its beneficiaries in the field'. As a result, the so-called *back-office functions* as carried out in the Division of Human Resources Management and Division of Finance and Administration became the main target. As the mandate of the PwC study had been prepared prior to SMS' move to the Division of Operational Support, it also became a part of the targeted functions due to its former placement in Division of Finance and Administration. Although PwC did not directly question the rationale behind UNHCR including the SCM functions in the review, they advised that a decision of relocation of SMS needed to take account of not only the direct staff costs, but also of the ability of SMS to influence (operational) procurement and logistics cost. PwC estimated that 36 per cent of UNHCR's total budget in 2005 was spent on SCM and SCM activities, and it recommended that UNHCR implement additional steps to reduce SCM risks that would be created if the SCM HQ was to be away from the other operational HQ functions based in Geneva.

As the review undertaken by PWC was very much focused on reducing staff cost in the Geneva HQ, it was difficult to argue for the potentially even greater improvement and efficiency opportunities created by the ability of the SCM function to influence operational decisions. As earlier reviews were omitted in the elaborations, the SCM focus of operational effects and cost-efficiency was not given significant weight in comparison to the higher staff cost in Geneva. Equally, as the organization had relative low maturity in understanding SCM, it was difficult to argue for the benefits of integration and close collaboration with other operational functions (or, indeed, the understanding of SCM as an operational function). As all prior reports had concluded that there was a need to strengthen the SCM, this would require increased staffing, which would be very difficult in Geneva at the time. In effect, the final decision in 2007 was to relocate larger parts of Division of Human Resources Management, the finance part of Division of Finance and Administration and the SMS part of the Division of Operational Support (later changed to the Division of Emergency, Security and Supply) to Budapest, Hungary as of March 2008. The key factors of choosing Budapest were a beneficial package of low office and local staff costs. In isolation, this would allow for rightsizing of the strategic SCM structure to meet the *actual* requirements of the organization. As a result, an enlarged staffing table including functions of central SCM emergency response and central emergency stock management was approved for Budapest. Although the potential benefit of keeping an SCM liaison officer function in Geneva was discussed, the final decision left the SCM function completely unrepresented at the Geneva HQ. An overview of the main events and implications for UNHCR SCM is presented in Table 5.1 below.

TABLE 5.1 Summary of important developments and documents 2005–2007

Time	Description	Impact
June 2005	António Guterres appointed High Commissioner for Refugees	This led to a 10-year period with a new view of the importance of SCM for UNHCRs mandate delivery
March 2006	Flooding in Tindouf	UNHCR emergency response operations exposing the lack of and shortfalls in SCM capacities and capabilities
May 2006	Unrest on Timor-Leste	
July/August 2006	The war in Lebanon	
October 2006	Real-time evaluation of UNHCR's response to the emergencies in Lebanon & Syria	Making the need for improvements of the SCM function clear to senior management of UNHCR
November 2006	INSEAD report; considerations for UNHCR's Supply Chain	An independent external review highlighting the need for improvements to the UNHCR SCM
December 2006	How to best organize the Supply Chain Management function (in UNHCR)	An internal study on how to best reorganize UNHCR's SCM to ensure improved capacity to deliver
April 2007	PwC; UNHCR Outposting Feasibility Study	Creating the possibility to grow, but leaving the SCM with no representation at the Geneva HQ

Establishing in Budapest, building the new supply chain organization (2008–2015)

During the local recruitment in Budapest, it became evident that, due to the competitive salaries of the UN, the interest was high. For most positions, the selected candidates would hold a relevant MSc degree. However, no analysis exposing the limited availability of personnel with prior experience of humanitarian work or SCM operations in less-developed parts of the world had been undertaken. As a result, the training programmes for the new Division of Human Resources Management, Division of Finance and Administration and SMS staff had to include sections trying to mitigate this situation. As of March 2008, SMS started operations in Budapest with an increased staff structure. As no local staff transferred from Geneva to Budapest, it took most of 2008 and 2009 to become fully operational and to begin to gain benefits from the enlarged staff structure.

To help to maintain the momentum of change, SMS commenced a cooperation with the Fritz Institute to get an impartial evaluation of the SCM function in UNHCR. The core findings of the Institute's 2008 baseline study (Fritz Institute, 2008) were that the UNHCR SMS had: strong capabilities of *emergency response*, which negatively impacted the ability of timely delivery to beneficiaries in *ongoing operations*; a strong audit mentality negatively impacting the ability to deliver, and with no ERP (enterprise resource planning) system support; lack of visibility of the delivery chain from a strategic level. Further, the study benchmarked the UNHCR SCM as *immature* in comparison with Oxfam, WFP, MSF and UNICEF. It also noted the lack of understanding of the operational support role of SMS, urging the need for 'supply to be given a place at the High Commissioner's table' and to increase the vision from pure supply to SCM. Based on the scale and complexity of UNHCR operations, the study highlighted the need for an SCM Division in UNHCR. In their recommendations, the Institute listed six critical areas for improvement: organizational will; supply chain strategy; processes; information integration and tools; people (staff development); and relationships (with other organizations/entities). Another effect of the 2008 report was the commencement of a process of mapping and systemizing the competencies required for all positions within the UNHCR SCM structure. This was done to ensure that an objective review of the competencies of candidates for SCM positions would take place, replacing the previous more subjective approach. This work was concluded in 2012 with the approval of the UNHCR Supply Chain Competency Framework document (UNHCR, 2012a), together with a new process whereby all proposed candidates would be vetted and approved by the Head of SMS prior to recruitment.

For the SMS management at the time, it became evident that a longer-term vision and change strategy, anchored with senior management, would be the only viable approach forward. As such, the development of a supply chain strategy aligned with other operational functions based in Geneva became a priority. The first SMS Strategic Plan (UNHCR, 2010) was established in late 2010, clearly indicating the direction for the UNHCR SCM for the next three years. Equally as important was the establishment of a Business Analysis Section given, amongst others, the task of overseeing the implementation of the plan and measuring progress. Due to the increased capacity and data analysis of the need for change, the first strategy was replaced in late 2012 with the Supply Chain 2015 Plan (UNHCR, 2012b). Based on the current state of SCM in UNHCR, and the recommendations in the Institute's 2008 report, the plan had three main pillars: reliable delivery (for both emergency and ongoing operations); improved emergency response; and human resources capacities. The strategy focused on increasing the ability of SMS to establish the future needs for core relief items for emergency response purposes (pre-funding) rather than waiting for orders based on the budget processes. Furthermore, the relevance of the SCM function in UNHCR was to be enhanced through: improved planning; building SCM competencies of SMS staff; better trained SCM emergency response staff; and global stockpiles prepared and capable of rapid emergency response.

To improve the functioning and support role of the UNHCR global stock management network, SMS reached out to the Lund University in Sweden. This resulted in a joint warehouse network project (UNHCR, 2012c) with the objectives of: validating existing and potentially new warehouse locations; optimizing the stock level per core relief items; evaluating inclusion of new items as core relief items; establishing a replenishment strategy (per warehouse per commodity); defining the emergency (safety) stock levels per location and commodity; and proposing commodities and locations for potential for 'white stock' with suppliers. The project had limited resources but produced very useful insights that enabled SMS to establish new emergency stock locations and secure commitment by the High Commissioner to fund the global stock levels sufficient to respond comprehensively to 600,000 persons of concern at any time. The global stock management inventory levels has, as a result of heightened crises levels, at times been temporarily increased up to 750,000 persons of concern, funded from the UNHCR reserves. While the analysis of the project was a one-off, SMS continued the cooperation with Lund University, enlarging it with the Norwegian Business School in Oslo and Northeastern University in Boston, United States. This cooperation culminated in the development of an UNHCR purpose-built optimization model, based on 'must-meet-demand' criteria (Jahre et al., 2016). The aim of the model was to maximize the advantages/benefits of the regular flow of core relief items for ongoing operations in support of both ongoing and emergency response operations.

Halfway into the implementation period of the Supply Chain 2015 strategy in 2013, it was decided that UNHCR would need to upgrade its ERP-system entitled the Managing System for Resources and People (MSRP). For the SCM function this was a golden opportunity, as the existing version of MSRP implemented and rolled out in 2005–2008, only consisted of very basic finance/administrative SCM functionality and provided little SCM value-added. Although the main rationale behind the MSRP project was to upgrade the core and financial modules that would soon become unsupported by Oracle/PeopleSoft, this was also a unique opportunity to improve the SCM function. By increasing the scope of processes supported as well as the range and quality of the data collected, this presented significant opportunities for better strategic oversight and decision-making based on facts rather than assumptions. However, complicating the picture, just prior to the MSRP upgrade project, SMS had been divided into two services: Supply Management Logistics Service (SMLS) and Procurement Management and Contracts Service (PMCS), without the appointment of *one* overall SCM Director. The Business Analysis Section, supporting both SMLS and PCMS, took lead in establishing a joint MSRP Upgrade Strategy (UNHCR, 2014a). The strategy was developed by defining the actual system needs for a seamless UNHCR end-to-end supply chain, based on four principles: *one* supply chain; stay 'vanilla' (reduce customization of the ERP-system as much as possible); support improvement;

and enable and simplify. As a result, the SCM organization was able to secure support for the use of approximately two-thirds of the $17 million upgrade project budget for enhancement of supply chain functionality.

The final module selection included many functions and SCM functionality that previously had not been covered at all, covered only partially, or managed off-line. The new functional areas included were Demand Planning, Strategic Network Optimization, (e) Procurement, (e) Supplier Relationship, Real Estate (and Project), Asset Lifecycle, and Transportation Management, in addition to new modules within Governance, Risk and Compliance, and Enterprise Performance Management. In sum, the new ERP system landscape would drastically increase the ability to rapidly establish and maintain a UNHCR internal Humanitarian Recognized Logistics Picture (HRLP). While approaching the implementation period of the MSRP Upgrade Project, an initiative to improve fleet management in UNHCR also took off. An attempt had been made already in 2006, based on strong recommendations from the UNHCR internal Policy Development and Evaluation Unit (UNHCR, 2006c) and the UN OIOS (2007). These reports had highlighted the impact of the absence of Fleet Management System (FMS) capabilities in UNHCR, resulting in: fleet sizes not driven by operational needs; life-cycle waste in the range of $20,000 per vehicle; and procurement of vehicles costing in the range of $3–4,000 more than needed. Regardless of the reports and experiences available from the International Federation of Red Cross and Red Crescent Societies (IFRC) fleet management system implementation in 2002, little resources for fleet management in UNHCR were made available at the time. Lacking funding, senior management support and staff resources, the initiative stranded after the development of a fleet administration tool in the MSRP-system.

With a strengthened SCM function filled with competent staff, greater attention from senior management and acknowledgement of the dependency of proper fleet management as an enhancer to operations, the time was ready for a revamped fleet management system proposal. Learning from experiences from both the IFRC and WFP fleet management projects, as well as using expertise from Fleet Forum, an enhanced solution suiting the UNHCR operational needs was developed, the Global Fleet Management project. A UNHCR Fleet Strategy 2014–2018 (UNHCR, 2014b), detailing the actions needed to reach optimal fleet management in 2018, was developed and signed off by senior management. The implementation of a centrally-controlled internal lease (and insurance) programme (ILP) became a great success. The Year One report of INSEAD (INSEAD, 2014) showed remarkable results: 11 per cent reduction of fleet size; 21 per cent drop in average age; reduced procurement costs of 21 per cent; and an income of $10 million from centrally-controlled disposal (up from $1.2 million in 2012), providing a potential annual saving of $5 million in 2014. That said, these rather impressive figures must also be interpreted as being a result of a long overdue clean-up of historical poor fleet management in UNHCR. Further adding to the workload related to change management, a cooperation project with company GS1, aimed at establishing a humanitarian logistics barcode standard, was launched at the same time. As the Director of SMLS had previously been the Chief Executive Officer of GS1, it was understandable that UNHCR took a lead role in this interagency project, taking capacity and focus away from the already initiated inhouse SCM improvement initiatives/projects. The main events and implications for the UNHCR in this period are summarized in Table 5.2.

The development in forcible displaced in the world and UNHCR budgets (2005–2015)

Before analysing any potential effects and/or benefits of the change to UNHCR's SCM functions, it is interesting to relate it to the development in the surrounding environment in the same time period. In the 10-year period that is the focus of this chapter there were two main changes to the external environment: a dramatic increase in the number of people in need and an increased pressure from the donors for accountability and cost-efficient delivery. As shown in Figure 5.1, the number of individuals forcibly displaced rose from approximately

TABLE 5.2 Summary of important developments and documents 2008–2015

Time	Description	Impact
March 2008	SCM moved to Budapest, along with Finance and Human Resources	Giving an opportunity for a rightsized SCM structure, diminishing the ability to influence operational decisions
November 2008	Fritz Institute; baseline report	Establishing the baseline of the UNHCR SCM, allowing for factual measuring the improvements of the function
November 2010	SMS Strategic Plan 2011–2013	Establishing the framework for strategic and longer-term change, used to secure support from senior management
May 2012	UNHCR Supply Chain Competency Framework	Establishing competency requirements for all SCM posts, supporting the professionalization of the function
May 2012	Strategic Warehouse Network Analysis, a cooperation with Lund University	Creating the base for a long-term GSM strategy, building an enhanced foundation for UNHCR emergency response
April 2012	Supply Chain 2015	Building on the first strategy, supported by analysis of SCM data, emphasizing the focus areas for the next three years
Mid-2013	The UNHCR's MSRP upgrade decided	Creating an opportunity to get a revamped ERP-system, better covering and supporting UNHCR SCM operations
Mid-2013	The former SMS is split into SMLS (logistics) and PMCS (procurement)	The creation of two equal services with no SCM Director appointed is working against the one SC principle
February 2014	Fleet strategy 2014–2018	Building on IFRC and WFP FMS solutions, resulting in the launch of UNHCR's enhanced fleet management/ILP solution in 2014
May 2014	MSRP (ERP-system) Upgrade Strategy, for supply chain management	To ensure a one SC approach, taking maximum benefit and effect of the MSRP upgrade project
2013–2015	Cooperation with BI Norwegian Business School, Lund University and Northeastern University	To provide a purpose-built optimization model for UNHCR's SC, with the potential to be included in MSRP

FIGURE 5.1 UNHCR displacement and funding statistics 2005–2015

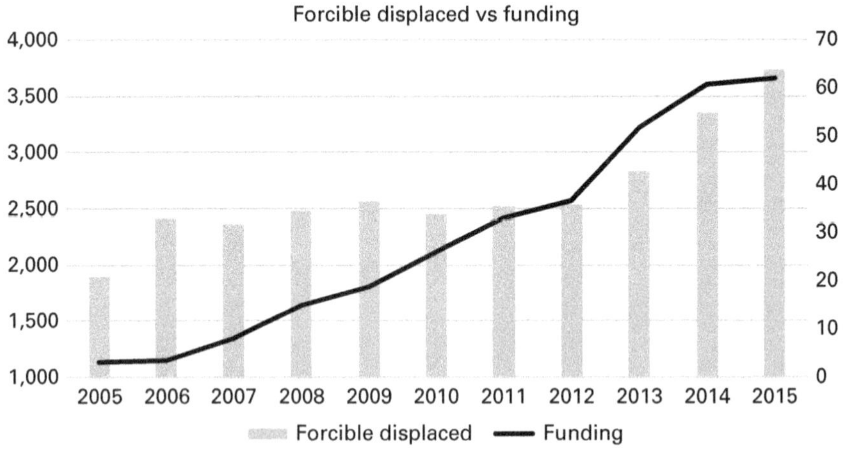

Forcible displaced vs funding

SOURCE www.unhcr.org

21 million in 2005, to a staggering 63.9 million in 2015. Sadly, the figure is continuing to grow, and based on the latest UNHCR statistics, it reached 82.4 million in 2020, further extending UNHCR's operations to a new continent – South America.

Interestingly, in the same period and despite donor pressure and potential fatigue, there has been a continuous and increased interest from donors to continue to fund UNHCR operations. Although the displacement figures went down in 2007, and only slightly increased until 2012, there was a steady and growing interest in the funding of the organization. Equally, following the sharp increase in displaced at the time of the Syria crises, the response from donors tended to be aligned with needs. Indeed, it is interesting to observe how the increased willingness to provide funding follows the trend of the improvements made in UNHCR's ability to deliver assistance and its SCM function.

Among other framework elements influencing the UNHCR SCM, it is important to mention the Supply Chain Manual, UNHCR Chapter 8. Many of the changes in the SCM function such as, for example, the growth of cash assistance meant that new processes and procedures needed to be defined. Often the quality, flexibility and speed of implementation in bureaucratic organizations such as UNHCR will either be limited or enhanced by the ability to update internal processes and regulations given their inherent audit focus. One of the clear weaknesses of Chapter 8 of the manual is that it encompasses both policy and procedures. What are the implications? The downside is that even minor changes to procedures will require a policy change and, as such, trigger a lengthy internal approval process involving all policy owners (divisions at HQ). On the positive side, the process of approval across the organization will improve the knowledge of SCM, its operations and procedures. But which factor outweighs which? In the case of UNHCR, this leads to the procedural framework often lagging years behind the de facto changes in the environment, support systems and need for updated or new procedures in the field. The worst example was in 2013, after a much needed and long overdue *complete* overhaul of Chapter 8, the approval process in UNHCR HQ took well over a year.

Improvements made to the UNHCR supply chain management function (2005–2015)

The 2015 Fritz Institute report analyses (across each area) the improvements to UNHCR SCM against the Institute's 2008 baseline report. The general observation is that UNHCR has made significant progress in all six highlighted areas of importance: organizational will; supply chain strategy; processes; information integration and tools; people (staff development); and relationships (with other organizations/entities). Although there is a strong will among senior management to improve, there is a shortfall in the organizational will in terms of creating the basis for *real* integrated planning. Based on the author's own experiences from both HQ and the field, in a similar way to the 'physical distance' between SMLS/PMCS in Budapest and the HQ in Geneva, there is a 'distance' between SCM and operations/programme in the field. The SCM function is still not perceived as core to operations, and therefore not consistently involved in preparedness, response and implementation planning across the organization.

In terms of processes, tools and information, a unique opportunity was available through the full-scale revision of the UNHCR SCM Manual (Chapter 8) in parallel with the MSRP upgrade. However, it is clear that the speed of updating and revising the SCM manual must be drastically increased, in order to keep pace with the dynamic changes to humanitarian operations, requirements from donors, etc. UNHCR SCM needs to drastically reduce the cycle time from identification/initiation of the need to change to its implementation across the organization. In an audit-focused environment such as the UN, the speed of implementation is often linked to the pace of policy and procedure approval at a central level. Another effect of the Fritz Institute 2008 baseline report was the establishment of a Business Analysis Section responsible for, amongst other activities, providing SCM data analysis to all actors. To have *one* central point of supply chain data analysis has proven to be very valuable, in particular following the split of the SCM function into the two services, SMLS and PMCS. Upgrading MSRP with SCM functionality supporting a more consistent capturing of the functioning of the end-to-end supply chain provides the basis for better and more timely available information to decision-makers at all levels. The crucial role of Business Analysis as an 'impartial' internal body measuring the SCM performance, mapping the causes/effect across functional borders and levels, is invaluable.

As for the MSRP upgrade, opportunities were lost due to the lack of one central owner/director of the supply chain. Nor was the opportunity of implementing a UNHCR custom built optimization model in the Demand Planning, Strategic Network Optimization module capitalized upon. This reduced the ability to achieve the full benefit from the revamped MSRP when managing emergency response and ongoing operations supply pipelines as *one*, mutually supporting each other. In addition, many of the selected modules were not implemented, or not properly implemented using business process re-engineering methodologies, to maximize the effects of the new functionality.

TABLE 5.3 Review of UNHCR's supply chain organization, 2013 figures

Item/organization	ICRC	IFRC	MSF-OCB	Oxfam GB	UNHCR	UNICEF	WFP
Annual budget USD million	1,200	365	360	605	3,230	3,900	4,400
Procurement USD million (% of budget)	340 (28%)	220 (60%)	90 (25%)	303 (50%)	953 (29.5%)	2,839 (73%)	1,600 (36%)
HQ vs. local procurement	45 / 55	45 / 55	56 / 44	40 / 60	40 / 60	76 / 24	41 / 59
Total personnel strength	12,000	1,800	8,000	5,000	8,600	12,000	14,000
SCM personnel (% of total)	800 (7%)	125 (7%)	400 (5%)	350 (7%)	391 (4.5%)	876 (7%)	3,000 (21%)

SOURCE Fritz Institute, 2015

As shown in Table 5.3, UNHCR SCM staffing was still at a low level when compared with similar organizations. This in particular was due to UNHCR's high degree of local procurement, which is a staff-heavy activity. Often, the deciding factor in terms of SCM staffing levels is the capability of the field Senior Supply (Chain) Officer and the understanding of SCM by the representative, and not the actual requirements of the field operation. As objective factors are still not determining the SCM staffing structures, this hampers the timely build-up of local procurement, logistics and delivery capacity. Attempts have been made in including the SMLS/PMCS in the process of budgeting and approving country level staffing proposals, but with varying degrees of success. That said, based on the author's own field experiences in Jordan/the Middle East in 2012–13, it is possible to get acceptance for a rightsized SCM structure. Similarly, adequate staffing levels for the demanding SCM operations in Jordan, Iraq, Lebanon, Egypt and Turkey, were approved for the 2013 operating budgets. In the case of Jordan, an SCM operation able to sustain an arrival rate of up to 8,000 people a day to the Al Za'atari camp needed to be established. In physical terms, this resulted in a massive national and international pipeline of goods representing up to 37 long haul truck deliveries to the camp a day. The delivery, large engineering efforts, contracts to sustain camp operations, etc., resulted in a spend in the range of up to $450,000 per day. To manage this wide array of SCM tasks, a mixture of long- and short-term local staff, regular and emergency assigned international staff, and secondees from standby partners, such as the Norwegian Refugee Council (NRC) and Danish Refugee Council (DRC) and RedR Australia, allowed for the SCM operation to be managed in an efficient and accountable manner.

A mechanism for the pre-funding of delivery means and advance funding for the global stock management (to mitigate long lead times) has been established and is providing valuable benefits to the first phase emergency response. The mechanisms are overseen and controlled by the Assistant High Commissioner, Operations, ensuring a short chain of command. Since its initiation in 2006, the SCM organization has consistently delivered on its 'first delivery within 72-hours' promise. However, benchmarking by the Fritz Institute against a similar mechanism in WFP indicates that there is room for improvement but, as demonstrated under the Syria crises, the existing system can support larger emergency interventions over time. UNHCR is continuing to expand the global stock management network and to establish regional support hubs. The fleet management system (UNHCR, 2018) based on the mandatory internal lease programme has been a great success, providing cost-efficiencies and operational benefits to field operations. One of the key success factors was the openness to learn from IFRC and WFP experiences, and to shape a program that was tailormade to UNHCR's operational needs (Kunz et al., 2015). Through the establishment of regional support hubs, and by collaborating with other UN-agencies, UNHCR has become the lead agency in the process of the retirement and sales of used vehicles. As noted in the OIOS Audit (UN OIOS, 2016), Global Fleet Management facilitated an annual saving of $11 million in 2017, and delivered the promised gradual reduced leasing costs to operations (33.84% in 2014 versus 19% in 2018, less than the purchase price on a 5-year perspective). The Global Fleet Management is well integrated into UNHCR operations, has a clear mandate, strategy and business model.

As the UNHCR SCM is characterized by a high degree of local procurement, the strategic procurement (PMCS) needs to focus its efforts on guidance, capacity building and establishment of global/regional frame agreements in support of local operations. Some progress has been made, as in improved procurement competency of field SCM personnel through Chartered Institute of Procurement and Supply (CIPS) training and certification, but it is still hampered by the compliance and order placement focus. Transactions that potentially can be better handled by field SCM personnel are still undertaken at the strategic level. Focusing the capacities of the strategic level PMCS on strategic procurement, procurement strategy, supplier management and development and performance monitoring would be beneficial for the performance of the end-to-end supply chain in UNHCR. As an example, in 2017 UNHCR still did not have a global/

regional frame agreement for the implementation of cash programmes. As contracts for cash distribution are complex, require specialist knowledge and normally involve larger global or regional financial institutions, this is an area where the strategic level PMCS could have made great impact.

Equally as important is the procurement oversight mechanisms, in UNHCR named the Committee of Contracts (CoC), to oversee the procurement process for transactions above a certain threshold. In UNHCR the threshold for local review and approval is low and equal for all operations. As a result, larger country operations must submit most of their procurement cases for review at the HQ Committee of Contracts creating a time lag. By comparison, UNICEF has taken a risk management approach, setting individual thresholds based on the competency and seniority of the SCM personnel and personnel appointed to the local review committee. This has resulted in a control framework better matching the dynamics of each individual operation, whilst managing the risks and securing the need for compliance and accountability. In terms of compliance and audit focus, progress has been made, but there is still a need to shift further towards a more risk management-based approach. With the tools implemented as a part of the MSRP Upgrade project, the system support is there, but strategic direction needs to be given. Based on the author's own field experiences in Jordan 2012–13 and Greece 2016–18, it is believed that it is possible to balance between operational delivery, accountability, compliance and risk management. That said, this is dependent on the acceptance of UN and external audit bodies taking an intent-based audit approach, rather than undertaking paragraph coherence audits. Overall, as shown in the UNHCR Statistics in Table 5.2, UNHCR has been able to maintain the confidence of the donors. In a more difficult and demanding donor market, it seems that the improved ability of UNHCR to deliver assistance, consistently, transparently, effectively and efficiently with its improved SCM functions has had a substantial share in the increased donor confidence.

Key lessons identified

The right people, with the right skills, in the right place!

The improved ability of the SCM function to deliver has generated an acceptance of the need for a more rightsized and professionalized UNHCR SCM personnel structure in the field and at headquarters. The move to Budapest gave an opportunity to increase the SCM staff structure, and to create the functions needed to be able to manage the end-to-end supply chain. Similarly, the competency framework set the professional requirements for each SCM post, paired with a centralized functional vetting process, although it must be stated that the SCM function is still struggling to be accepted as a core operational support function to operations and protection. The lack of objective tools and/or influence of SCM managers in local HR planning processes hampers the ability to further align the field-based structure to match the actual requirements on the ground. This is a particular problem in smaller and more underfunded operations with less senior, or non-, SCM officers in place.

Out of sight, out of mind!

Based on the starting point described in the initial reviews and reports, considerable progress has been made in improving the capacity and capability of the UNHCR SCM function. Most improvement objectives have been reached either in full or in part, but the main objective of streamlining the SCM function as an integrated strategic support function has not been reached. The move to Budapest gave an opportunity to increase the staff structure but resulted in less direct collaboration with other operational functions at the HQ level. Hence, UNHCR has still not implemented a *true* integrated planning approach. The establishment of two parallel services in Budapest, PMCS and SMLS, with *no* SCM Director, further diminished the SCM function's ability to coordinate influence in the decision-making processes. UNHCR has still not established an SCM division, and the function does not have a seat at the table of the High Commissioner!

Lasting change takes consistent efforts over time!

The framework of revolving 3-year strategic SCM plans has proven its value. The same methodology has been applied for larger change initiatives such as the Global Fleet Management and MSRP Upgrade projects. The process of developing the strategic plans and establishing commonly agreed goals is the most feasible route to ensure commitment towards the described end-state by SCM personnel and from senior decision makers. This will also enable the organization to better maintain focus on a steady course over time, avoiding being derailed by short-term individual motives.

Separate between policy and procedures!

As shown in the case of UNHCR Chapter 8 overhaul (the SCM manual), there is a need to differentiate and separate change management procedures for policy and procedures. The procedures should be owned by the strategic management of the function and not the senior management of the organization, as procedures need to be dynamically updated based on the ever more rapid changes in the operational environment. Meanwhile, the senior management should focus on policy, ensuring that all functions are coordinated when supporting the implementation and achievement of the objectives of the organization.

Act on facts, not feelings!

The MSRP Upgrade, providing SCM modules, functionality, and process support, had the potential of being a game changer through timely capturing of consistent supply chain data at all levels. To sustain the change initiatives, the establishment of a central business analysis function in SCM has proven invaluable. The ability to extract and present facts relating to the functioning of the supply chain and the effects of the different initiatives and changes has been successfully used by SCM managers in discussions with senior managers. This has provided opportunities, support and funding of the change process(es). As shown with the failure of Global Fleet Management in 2006, and its success in 2014, change takes time and must be based on undisputable facts that demonstrate benefits for the organization and the *people of concern.*

The decentralized nature of humanitarian organizations

As noted by Kunz et al. (2015), the decentralized nature of most humanitarian organizations with high decision power delegated to field offices, works against the SCM principles that have made global stock management and Global Fleet Management a great success for UNHCR. To convince the organization and its senior managers of the benefits of a centrally-controlled SCM coupled with capable decentralized management of delivery and implementation has taken significant time and effort. Equally, to build the required SCM capability, capacity, strength and systems enabling a truly global end-to-end supply chain requires persistency and focus. In a mandate-oriented organization with a strong field mindset, this has been even more challenging.

Cooperation with academic institutions

As shown in some of the UNHCR's change initiatives, utilization of the knowledge and capacity of academic institutions can be very useful. That said, to ensure beneficial outcomes and products, there is a need to invest in the cooperation (resources) and carefully establish one's own requirements and expected output. As demonstrated with global stock management, the SCM optimization model and the Global Fleet Management project, UNHCR has taken great advantage of working and collaborating with academia, although unfortunately has not always been able to take the full advantage out of it.

Do the right thing because it is the right thing to do!

In bureaucratic organizations as UNHCR, it is easy to focus on rigid process adherence, to ensure accountability and reporting in line with donor expectations. But following this perspective, there is an inherent risk of losing sight of the accountability towards the people of concern. Seen from a SCM perspective, one of the more effective tools would be to continue the path towards intention-based auditing, supported by a balanced risk management approach, as this would allow for more flexible methods when delivering SCM support in rapidly changing operations.

Focus on building ability and agility!

In a more volatile world, the relevance of operational organizations such as UNHCR is becoming more dependent on the ability and agility of the SCM function to support operations. As such, the demand for professional qualifications and skills is increasing. Furthermore, with growing pressure on funding, the ability and mechanisms for collaboration with other humanitarian actors must be in place, as no organization can afford to have all the required functions or capacities themselves. Hence, a network approach that requires the capability of information gathering systems to manage and analyse the information and cooperation across organizations is becoming more important. The success of *all* organizations is dependent on the ability of organizations to jointly and quickly establish a Humanitarian Recognized Logistics Picture, and in coordination deliver the required support to all sectors irrespective of their particular home organization.

Summary and conclusion, UNHCR SCM 2015 onwards

To maintain the momentum of change and improvements to the SCM function in UNHCR will be a challenging exercise. The two main factors are the earlier mentioned power balance between administration, budget/finance, legal protection, operations and SCM, coupled with a HR policy leading to a high degree of personnel rotation/turnover – both factors influence the focus within the SCM domain, but equally, the management of other functions fight over the same scarce resources. Guterres left his post as UNHCR High Commissioner at the end of 2015, making it uncertain if the UNHCR top management would maintain its focus on the SCM function as an essential operational support function. Guterres' personal interest was instrumental in SCM change from 2005 to 2015.

With the appointment of Filippo Grandi as the new High Commissioner on 1 January 2016 with a five-year tenure (extended in 2021 by two-and-a half years), it is not certain whether he will continue the previous strategic course in relation to the SCM function. Equally, due to the previously mentioned rotation policy, it is unclear whether the individuals appointed to key leadership positions in other functions would understand the central elements of the successful transformation of the UNHCR SCM. Indeed, there are some signs that parts of the positive changes are eroding with, as an example, a regionalization and decentralization program in 2018–2020 resulting in a reduction of the SCM workforce in Budapest by approximately one third, without any significant additions to local/regional SCM staffing. On the other hand, there are some positive signs as evidenced by the appointment of a new Head of SMLS, coming from a similar position in an NGO, and a new Director of Division of Emergency, Security and Supply focusing on SCM capabilities.

As with many ERP-projects, the UNHCR MSRP upgrade project also hit delays and problems during its implementation. This has resulted in some of the SCM modules not having been implemented to the full extent of their potential, or not implemented at all. Most prominently are the cloud-based modules that were only partially implemented, such as (e) Procurement, (e) Supplier Relationship, and Transportation Management, this having great impact due to the disruptions on global supply chains caused by the Covid-19 pandemic. That said, a new cloud-based ERP solution is planned to be implemented in 2023, including Procurement, Logistics and Inventory Management modules, whilst a Shipment Tracking/Transportation Management module is going live in 2022. UNHCR is currently working on including sustainability and greening of the supply chain, including for the fleet management system/ILP programme, as a part of the current SCM strategy. A review and update of the UNHCR SCM manual (Chapter 8) is also well underway and is planned to be published soon. As pointed out earlier, the establishment of a Business Analysis Section was instrumental to the development of such changes due to its role in measuring the functioning of UNHCR's global supply chain and in overseeing the implementation of key SCM improvement initiatives. Sadly, this function was dismantled during 2016 and 2017 and has only just recently been reactivated, albeit to a lesser degree.

The most prominent issue is that SCM is still not considered a *strategic core function* and thus given the importance that it requires. In other words, the function still does not have a seat at the Commissioner's table, nor has *one* overall SCM Director been appointed, and the implementation of the concept of *true* integrated planning at all levels is still pending. Located away from the main HQ, the SCM function in Budapest is dependent on the focus, prioritizations and supply chain understanding of the Director of the Division of Emergency, Security and Supply (who typically does *not* have an SCM background), to fight the SCM cause in Geneva.

References

Fritz Institute (2008) Mizushima, M, Coyne, J, de Leeuw, S, Kopczak, L R and McCoy, J, UNHCR SCM: Assuring Effective Supply Chain Management to Support Beneficiaries, 4 November

Fritz Institute (2015) M. Mizushima, L.R. Kopczak, S. de Leeuw, K. Echavarri-Queen, and J. Mcdonald: UNHCR's Supply Chain Review 2015; 26 March

INSEAD (2006) Humaitæd consulting: Z. Zacca, R.M. Tomasini, and L. van Wassenhove: Considerations for UNHCR's Supply Chain, 28 November

INSEAD (2015) Humanitarian research Group; N. Kunz and L van Wassenhove: UNHCR Global Fleet Management, Year One (2014) Baseline report

Jahre, M, Kembro, J, Rezvanian, T, Ergun, O, Håpnes, S J and Berling, P (2016) Integrating supply chains for emergencies and ongoing operations in UNHCR, *Journal of Operations Management* **45**, pp 57–72

Kunz, N, van Wassenhove, L, McConnell, R and Hov, K (2015), Centralized Vehicle Leasing in Humanitarian Fleet Management: The UNHCR case, *Journal of Humanitarian Logistics and Supply Chain Management,* 5(3), pp 387–404

PriceWaterhouseCoopers (2007) UNHCR outposting feasibility study, final report (UNHCR Internal), 16 April

UN (1951) Convention relating to the status of refugees, 28 July

UN (1967) Protocol relating to the status of refugees, 31 January

UNHCR (2006a) Policy development and evaluation service: Real-time evaluation of UNHCR's response to the emergency in Lebanon and Syria, July–September 2006, October

UNHCR (2006b) Supply management and emergency Service sections: M. Bisau and S.J. Håpnes; The integration of SMS into DOS; How to best organize the Supply Chain Management functions, to ensure improved emergency logistics delivery capacity, 12 December

UNHCR (2006a) Policy development and evaluation service: Real-time evaluation of UNHCR's response to the emergency in Lebanon and Syria, July–September 2006, October

UNHCR (2006b) Supply management and emergency Service sections: M. Bisau and S.J. Håpnes; The integration of SMS into DOS; How to best organize the Supply Chain Management functions, to ensure improved emergency logistics delivery capacity, 12 December

UNHCR (2012a) Supply management service; Supply Chain Competency Framework, 19 May

UNHCR (2010) Supply Management Service; SMS Strategic Plan 2011–2013, 16 November

UNHCR (2012b) Supply Management Service; Supply Chain 2015, 6 April

UNHCR (2012c) Supply Management Service in cooperation with Lund University: Project proposal; Strategic Warehouse Network Analysis, 15 May

UNHCR (2014a) Supply Management Logistics and Procurement Management and Contracts Services: MSRP Upgrade Strategy: Supply Chain Management, 30 May

UNHCR (2006c) Policy development and evaluation unit: Evaluation of the utilization and management of UNHCR's light vehicle fleet, February

UNHCR (2014b) Supply management logistics and procurement management and contracts services: Fleet strategy 2014–2018, February

UNHCR (2005–2015) People of concern statistics and financial reporting to the executive committee of the high commissioner's programme

UNHCR (2018) Evaluation Service, Report ES/2018/13: Evaluation of UNHCR's Global Fleet Management December

UN OCHA (2007) Oslo guidelines: Guidelines on the use of foreign military and civil defense assets in disaster relief, revision 1.1, www.unocha.org (archived at https://perma.cc/43UY-9DEA)

UN OIOS (2007) Audit report, Assignment No. AR2006/161/01: UNHCR Fleet Management, 6 July

UN OIOS (2016) Audit Report, Assignment No. AR2015/167/02: Audit of the arrangements for fleet management at the Office of the UNHCR, 8 September

06

Humanitarian supply chain service performance

RUTH BANOMYONG, PUTHIPONG JULAGASIGORN, PAITOON VARADEJSATITWONG AND THOMAS E FERNANDEZ

ABSTRACT

Performance measurement is a necessity for relief organizations and critical for their accountability. Despite service being an important aspect of relief operations, there is a lack of performance measurement relative to service quality perceived by beneficiaries. The purpose of this chapter is to propose an instrument entitled Humanitarian Supply Chain Service Performance (HUMSERVPERF), and to illustrate how HUMSERVPERF could be used in a humanitarian operation. An instrument used to assess service quality in commercial supply chains was adapted based on suggestions of researchers in the humanitarian literature and further modified by practitioners in relief operations. HUMSERVPERF is proposed as a starting point for further development of performance measurement of relief service quality. A case of a flood relief operation was further employed to demonstrate how to use HUMSERVPERF. Discussions on future directions of HUMSERVPERF are made.

Introduction

The Centre for Research on the Epidemiology of Disasters indicated an increasing trend in economic losses caused by disasters (www.cred.be). Such an increase in the severity, frequency and scale of disasters not only causes loss of lives and properties but also increases competition among relief organizations (Oloruntoba and Gray, 2009; Burkart et al., 2016). Many relief organizations are competing for limited resources, which come from individual donors, private firms, and governments (Kovács and Spens, 2007; Oloruntoba and Kovács, 2015).

Performance measurement is a necessity for relief organizations and critical for their accountability. Performance reports provide baseline information, indicating the success of relief efforts, and need to be submitted to decision makers in relief organizations and donors (Schiffling and Piecyk, 2014; D'Haene et al., 2015). It is evident that donors that are both organizations and individuals want to know how their money is utilized (Shabbir et al., 2007; Banomyong et al., 2019). Some organizational donors use such reports to manage, control and improve relief efforts at the strategic, tactical and operational level (Gunasekaran and Kobu, 2007).

Relief organizations provide goods along with services for their beneficiaries. However, despite service being an important aspect of relief operations, there is a lack of performance measurement relative to service quality, as perceived by beneficiaries (Banomyong et al., 2019; Cardoso et al., 2021). Although some studies measured service level, these were stated in terms of stock efficacy, delivery performance, delivery time and availability and speed (Haavisto and Goentzel, 2015). Performance measurement of relief operations should not only consist of effectiveness and efficiency, but also consider beneficiaries' perceptions, such as satisfaction and kindness of staff (Medina-Borja and Triantis, 2007; Oloruntoba and Gray, 2009; Tatham and Hughes, 2011; Banomyong et al., 2019). If a relief organization cannot satisfy the needs of beneficiaries, this not only affects its performance but also decreases its reputation and potential future funding (Oloruntoba and Gray, 2009). In contrast, good performance can help ensure long-term funding (Schiffling and Piecyk, 2014).

The purpose of this chapter is to develop an instrument used to assess service quality of relief operations, as perceived by beneficiaries. Supported by evidence from the humanitarian literature, an instrument used to assess service quality in commercial supply chains was adapted to fit with the context of relief operations. Practitioners in disaster management and relief operations in Thailand further helped modify the instrument. The adapted instrument was then proposed to assess service quality in the humanitarian context. The instrument is called the Humanitarian Supply Chain Service Performance (HUMSERVPERF) framework. A case of flood in Thailand was employed to demonstrate how HUMSERVPERF could be used.

The next section presents a review of service quality performance measurement from the perspective of commercial supply chains and a review of service quality performance measurement used in the humanitarian literature. HUMSERVPERF, as developed in this chapter, is proposed next and is supported through insights reported by the practitioners, who helped modify HUMSERVPERF. A case study is then presented and this is followed by discussions regarding the challenges and further development of HUMSERVPERF. The final part of this chapter highlights theoretical and practical implementations and limitations.

Service quality performance measurement used in commercial supply chains

Service quality is one of the main streams of performance measurement in the commercial supply chains and Parasuraman et al.'s (1985, 1988, 1991) works have been considered as seminal papers in the field (Varadejsatitwong et al., 2021). Such seminal works are aimed at assisting the field to move forward from a focus on technical aspects (effectiveness and efficiency) and include customers' perceptions on what the service should be and how it is conveyed.

Parasuraman *et al.* (1985) insisted that the quality of goods is not sufficient to understand *service* quality. They defined service quality as the gap between customer's expectation of a service and their

TABLE 6.1 Ten SERVQUAL's dimensions

Dimension	Definition
Reliability	Consistency of performance and dependability
Responsiveness	Willingness of employees to provide service
Competence	Essential skills and knowledge for performing the service
Access	Approachability and ease of contact
Courtesy	Politeness, respectfulness and friendliness of contact personnel
Communication	Keeping customers informed and helping them understand essential information
Credibility	Firm's trustworthiness, believability and honesty
Security	Customers feel safe in conducting transactions with the firm
Understanding Customers	Understanding the customers' needs
Tangibility	Physical evidence of the service

SOURCE Parasuraman et al. (1985)

perception of actual service. Based on this concept, they proposed 10 SERVQUAL's dimensions as shown in Table 6.1.

Parasuraman et al. (1988) further refined dimensions included in Table 6.1. Competence, courtesy, communication, credibility and security were combined into 'assurance', while access and understanding/knowing the customers were combined into 'empathy'. Finally, Parasuraman et al. (1991) proposed five SERVQUAL's dimensions consisting of 22 items (Table 6.2). Because each item measures two perspectives (expectations and perceptions), there are 44 items used when asking customers to rate SERVQUAL. The resultant assessment of service quality is calculated by comparing the difference between the expectations and the perceptions.

However, SERVQUAL has been criticized in relation to its theoretical assumption, validity and reliability (Ladhari, 2009). Cronin and Taylor (1992) strongly argued that the conceptualization and operationalization of SERVQUAL is inadequate, as customers provide their own ratings by automatically comparing their expectations and perceptions. Therefore, the measures regarding expectations can be discarded, and the measures regarding perceptions should be used alone. Cronin

TABLE 6.2 Five of SERVQUAL's dimensions

Dimension (items)	Definition
Tangibility (4)	Physical facilities, equipment and appearance of personal
Reliability (5)	Ability to perform the promised service dependably and accurately
Responsiveness (4)	Willingness to help customers and provide prompt service
Assurance (4)	Knowledge and courtesy of employees and their ability to inspire trust and confidence
Empathy (5)	Caring, individualized attention the firm provides its customers

SOURCE Parasuraman et al. (1988)

TABLE 6.3 SERVQUAL vs. SERVPERF

Instrument	SERVQUAL	SERVPERF
Assumption	Service quality equals to expectations minus perceptions	Service quality equals to perceptions only
Dimension	Reliability, responsiveness, tangible, assurance, empathy	Reliability, responsiveness, tangible, assurance, empathy
Measures	22 (expectation) + 22 (perception)	22 (perception)
Contribution	Dominance in the service literature	Rival concept
Prediction power	Lack behind	Outperformed
Applications	Identifying service quality shortfalls	Assessing overall service quality
Adaptation	Require more adaptation	Require less adaptation

SOURCE The authors complied from Cronin and Taylor (1992), Brady et al. (2002), Jain and Gupta (2004) and Carrillat et al. (2007)

and Taylor (1992), thus, proposed service quality measurement as a performance-based framework (SERVPERF). SERVPERF consists of 22 items, which are similar to the questions appear in SERVQUAL's dimensions. Table 6.3 presents the differences and similarities between SERVQUAL and SERVPERF.

Zeithaml et al. (1996) suggested that whether to use the SERVQUAL or the SERVPERF model depends on a study's purpose: the perceptions-only operationalization is appropriate for a study that aims to explain the variance in dependent constructs; the perceptions-minus-expectations is appropriate for a study that aims to diagnose service shortfalls. With this in mind, the purpose of this chapter is to develop an instrument that can be used to assess service quality of relief operations. We aim to adapt instruments that originated from commercial supply chains and investigate whether or not such a modified instrument could be used in the context of relief operations. Therefore, in our adaptation of the SERVQUAL or the SERVPERF used in this chapter we observed results to be similar, as both instruments share the same descriptions of items.

The SERVQUAL and the SERVPERF model is often employed with three dependent constructs: overall service quality, satisfaction and behavioural intention (Cronin and Taylor, 1992; Cronin et al., 2000; Ladhari, 2009). Cronin et al. (2000) also demonstrated that service quality leads to satisfaction and to behavioural intention directly and indirectly through satisfaction.

The next section presents a review of service quality performance measurement used in the humanitarian literature.

Service quality performance measurement in relief operations

Current measures within the humanitarian context are various such as flexibility, resource efficiency, cost, accuracy, process adherence, delivery time, defect rate, coverage and equity, utilization, innovation and learning and quality of life and well-being (Oloruntoba and Gray, 2009; Haavisto and Goentzel, 2015; Behl and Dutta, 2018). Oloruntoba and Gray (2009) further noted that service measurements in the humanitarian literature focus on metrics such as the number of tons delivered, number of containers landed, number of aid-users fed, and number of relief staff arriving promptly. Haavisto and Goentzel (2015) observed that service level was another important measure but this was stated in terms of stock efficacy (Van der Laan et al., 2009), delivery performance, delivery time (Schulz and Heigh, 2009), availability and speed (De Leeuw, 2010). Other researchers indicated that measurements often neglect the satisfaction and perceptions of beneficiaries (Tatham and Hughes, 2011; Banomyong et al., 2019; Cardoso et al., 2021). Tatham and Hughes (2011) highlighted that relief organizations should consider 'other "soft" or intangible aspects that impact the perception by the recipient of the quality of the aid they receive' (p. 11).

Medina-Borja and Triantis (2007) suggested that performance measurement of a relief organization is composed of three dimensions: effectiveness (outcomes or impacts of relief efforts such as the number of death and homelessness as well as improved skills and knowledge of recipients), efficiency (input compared to output), and beneficiaries' perceived service quality (satisfactions and perceptions towards issues such as timeliness and kindness of staff). Furthermore, relief operations involve not only the delivery of goods, but also other aspects such as the behaviour of field staff and the reputation of the relief organization (Oloruntoba and Gray, 2009; Schiffling and Piecyk, 2014; Burkart et al., 2016; Banomyong et al., 2019; Heaslip and Kovács, 2019).

Although previous studies indicated that there is a lack of service quality performance measurement as perceived by beneficiaries, we observed that a few studies have investigated this topic, albeit from different perspectives. Sheu (2014) adapted variables from the literature including Parasuraman et al. (1988) to propose a dimension called 'perceived humanitarian logistics service quality', which includes empathy, trustworthiness, accessibility and responsiveness, to capture service quality; while dependent variables include satisfaction and other types of beneficiaries' perceptions such as benefits to physical recovery, their hope for future, their willingness to survive and increased positive thinking. Nolte et al. (2020) employed SERVQUAL's dimensions as a theoretical underpinning to explore insights into the quality of public services provided during the refugee crisis in Germany and have demonstrated the application of SERVQUAL's dimensions. Recently, Cardoso et al. (2021) conducted a systematic literature review to derive a set of service quality categories in the context of house rebuilding. One of their categories is called 'assistant', which relates to beneficiaries' satisfactions with regards to response time, reliability of the information and quality of products/services provided by a government-contracted construction company. The beneficiaries' satisfaction was used as a predicted variable in relation to each category.

There are three conclusions derived from the review. First, it is observed that although previous studies adopted different approaches to investigate service quality, some of them adapted the whole or elements of SERVQUAL's dimensions to define service quality perceived by beneficiaries. Furthermore, it seems that all studies used beneficiaries' satisfaction as a predicted variable of service quality. Finally, the humanitarian literature points out that no study employs SERVQUAL or SERVPERF to develop an instrument used to assess service quality in a relief operation. Before adapting such instrument in the relief context, two issues need to be addressed: customers in relief operations and a predicted variable of service quality. These two issues are critical when adapting an instrument from the commercial supply chains because the characteristics of humanitarian operations are very complex and differ from the commercial context. The next section, therefore, discusses these issues.

Customers in relief operations

Applying the concept of customers in the humanitarian context is problematic (Oloruntoba and Gray, 2009; Heaslip and Kovács, 2019). In the commercial context, service quality is perceived by customers who need goods or services, place orders to obtain such goods or services, and pay for the cost of such goods or services (Besiou and Van Wassenhove, 2021). In the humanitarian context, customers are various and their roles differ from one to another. Generally, donors pay money without receiving goods or services, while beneficiaries are passive receivers of goods or services without placing orders or paying money (Heaslip and Kovács, 2019; Besiou and Van Wassenhove, 2021).

However, most studies we have reviewed used the term 'customers' when referring to both donors and beneficiaries (Beamon and Balcik, 2008; Oloruntoba and Gray, 2009; Burkart *et al.*, 2016). Some researchers further identify other stakeholders such as local communities, military and implementing partners as the customers of a relief organization (Oloruntoba and Gray, 2009; Heaslip and Kovács, 2019). This is because defining who customers are depends on the context of relief operations under investigation and on who the service provider is (Oloruntoba and Gray, 2009; Schiffling and Piecyk, 2014). In this chapter, beneficiaries are the focus due to the fact that they are the reasons for the existence of relief organizations and are those who often neglected by researchers (Oloruntoba and Gray, 2009; Tatham and Hughes, 2011).

The second issue is related to the predicted variable of service quality. Satisfaction and dissatisfaction have been tested in previous studies and would appear to be the most appropriate measures to be used as the predicted variable of service quality (Shabbir et al., 2007; Oloruntoba and Gray, 2009; Sheu, 2014; Carsodo et al., 2021). Satisfaction is defined as the ability to satisfy or to fulfil specific beneficiaries' needs (Oloruntoba and Gray, 2009). Behavioural intentions are, however, more complex than satisfaction. In the commercial context, if customers have favourable behavioural intentions towards a service provider's ability, they will express this in terms of (1) remaining loyal to the provider through repurchasing and (2) spending more with the provider (Zeithaml et al., 1996).

In the humanitarian context, beneficiaries hope that encountering disasters and engaging with relief organizations are a one-off experience. Thus, ideally, beneficiaries do not want to repeat a disaster (Kovács and Spens, 2007; Kovács, 2014). Consistently, therefore, the aim of humanitarian logistics is *not* to gain repeat custom from beneficiaries (Oloruntoba and Gray, 2009). The loyal customer concept can thus only be applied in respect of donors as customers (Shabbir et al., 2007), but not beneficiaries (Heaslip et al., 2018). For this reason, the concept of loyalty and repeat purchases, which are often-used measures in commercial supply chains (Banomyong and Supatn, 2011), are not appropriate for the relief context. Therefore, in this chapter, we will adapt SERVPERF's dimensions and satisfaction to fit the relief context. The method used to achieve this adaption is presented in the following section.

Developing HUMSERVPERF questionnaire

SERVPERF, as an instrument, was adapted to fit with the relief context. The method used is presented in Figure 6.1, which was adapted from Banomyong and Supatn (2011). The steps for developing HUMSERVPERF consist of (1) adapting SERVPERF to serve relief operations, (2) improving HUMSERVPERF to serve in disaster situations, and (3) validating HUMSERVPERF.

In the first step, several items that appeared in SERVPERF were modified: 'XYZ' was changed to 'the relief organization', 'employees'

FIGURE 6.1 Developing the HUMSERVPERF

	Adapting the SERVPERF to fit relief operation context
Step 1	• Modify questions in SERVPERF • Add new items based on humanitarian literature • Propose the first iteration of HUMSERVPERF

	Improving the HUMSERVPERF
Step 2	• Translate the first iteration of HUMSERVPERF into Thai language using a bilingual translator • Conduct email interviews (attached with the Thai version of first iteration of HUMSERVPERF) with the regional DDPM director; ask the director for comments and recommendations • Revise the first iteration of HUMSERVPERF (English and Thai version) using the bilingual translator • Propose the second iteration of HUMSERVPERF • Conduct semi-structured interview with two local volunteers (Thai and English proficient) • Ask the two local volunteers to comment and recommend on the second iteration of HUMSERVPERF • Ask the two local volunteers to validate translation of HUMSERVPERF between Thai and English • Revise the second iteration of HUMSERVPERF (English and Thai) using bilingual translator • Propose the third iteration of HUMSERVPERF

	Validating the HUMSERVPERF
Step 3	• Submit the third iteration of HUMSERVPERF to the regional DDPM director, the two local volunteers and two DDPM frontline staff (a deputy of DDPM-SAO, director and a chief officer) • Conduct interviews with all respondents and ask them to comment and recommend • No further comment was added from all respondents • The HUMSERVPERF is ready to be tested

DDPM denotes Department of Disaster Prevention and Mitigation
DDPM-SAO denotes Department of Disaster Prevention and Mitigation Sub-district Administrative Organization

SOURCE The authors

was changed to 'employees/volunteers', 'service' was changed to 'relief service', and 'you' and 'customer' were changed to 'beneficiaries'. Negative words were also reversed back to positives for ease of understanding.

Based on the review of the humanitarian literature, eight new items were added. Two items relating to staff's knowledge and skills were included, as these are important relief staff characteristics (Kovács and Spens, 2009; Oloruntoba and Gray, 2009). Two items relating to the accountability and reputation of the organization were included, as these are important to relief organizations (Beamon and Balcik, 2008; Oloruntoba and Gray, 2009). Another three items concerned information about emergency and safe locations, how instructions appeared on relief supplies and equipment, as well as effective communication during disaster events, as all these are critical for relief operations (Kovács and Spens, 2009; Banomyong et al., 2019). These seven new items were added to 'assurance' since it is the combined dimension of communication, competence and credibility, as suggested in the original SERVQUAL (see Table 6.1). A supply quantity measure was added, as current measures in the literature did not cover the needs of beneficiaries, regarding the adequacy of relief supplies (Oloruntoba and Gray, 2009). Regarding the dependent variable, questions regarding overall service quality and satisfaction were adapted from Parasuraman et al. (1985) and Cronin and Taylor (1992). The first iteration of HUMSERVPERF was thus developed.

In the next step, we improved the first iteration of HUMSERVPERF by asking practitioners in disaster management and relief operations in Thailand to assist further in modifying the first iteration of HUMSERVPERF. In Thailand, the Department of Disaster Prevention and Mitigation (DDPM) is the main relief organization (relief service provider) working under the Ministry of Interior. Figure 6.2 illustrates the organizational structure of DDPM. The DDPM has direct responsibility for handling disasters, preventing disaster damages and losses and mitigating calamities caused by manmade or natural disasters. The DDPM has regional offices which are responsible for monitoring the government budgets and the performance of local

FIGURE 6.2 Organizational structure

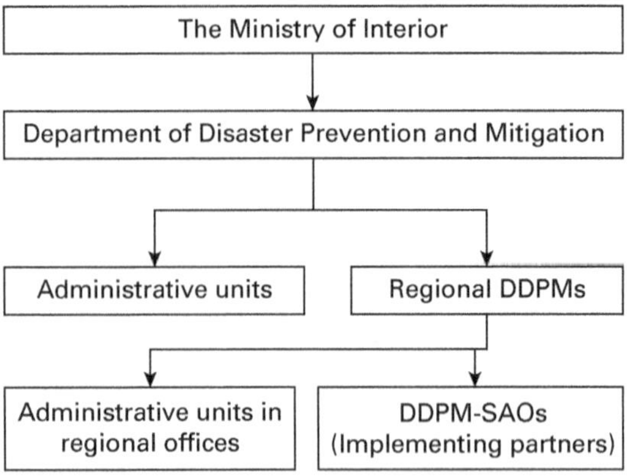

```
The Ministry of Interior
        ↓
Department of Disaster Prevention and Mitigation
        ↓                              ↓
Administrative units            Regional DDPMs
        ↓                              ↓
Administrative units in         DDPM-SAOs
regional offices                (Implementing partners)
```

SOURCE Adapted from DDPM's website, https://www.disaster.go.th/

DDPMs (the DDPM sub-district administrative organizations or DDPM-SAOs). DDPM-SAOs are, therefore, the frontline units or implementing partners who are helping beneficiaries in need.

The first iteration of HUMSERVPERF was mainly improved by three practitioners: a regional director of the DDPM in Thailand, a local volunteer and a foreign volunteer. The two volunteer respondents were both proficient in English and Thai language and had many years' experience in working with relief organizations including with a local DDPM during disaster events such as floods, storms, poverty and refugee crises. The regional director participated to help modify the questionnaire, while the two volunteer respondents further contributed to the modification and translation of the questionnaire. All respondents were asked to provide comments regarding the appropriateness of the items in the proposed HUMSERVPERF and provide recommendations for further revision. Questions involved were: *Are these items making sense? Which one should be revised, how should it be and why?*

After multiple iterations, the final version of HUMSERVPERF (in English and Thai) was sent to all three respondents. Furthermore, the last iteration of HUMSERVPERF (in Thai) was sent to two DDPM frontline staff (a deputy of a DDPM-SAO and the chief rescue officer of a DDPM-SAO). Interviews were then undertaken to confirm if anything else needed to be added, but no further modifications were made. The HUMSERVPERF was, therefore, ready to be tested.

HUMSERVPERF questionnaire

HUMSERVPERF is presented in Appendix 6.A. HUMSERVPERF consists of two parts: dependent variables and independent variables. The dependent section consists of four questions relating to overall service quality and satisfaction. The independent section comprises 29 questions which are intended to assess six dimensions of service quality: tangibility (3 items), reliability (5 items), responsiveness (4 items), assurance (11 items), empathy (5 items), and supplies quantity (1 item). Five-point Likert scales anchored at 1 (lowest) to 5 (highest) are used in specifying the scale of HUMSERVPERF's enquiries. Practitioners were also invited to provide reasons to support their modifications (Table 6.4).

The following section summarizes the discussions around the respondents' recommended revisions. First, SERVPERF considers the importance of promises of on-time and reliable service (Parasuraman et al., 1985) but, in the case of relief operations, these are almost impossible to commit to. It is acknowledged that dealing with disasters is complex and unpredictable (Van Wassenhove, 2006). In a refugee crisis, relief staff will avoid any promises in respect of service provision to beneficiaries and reliability that is considered relevant to service quality as both are difficult to guarantee (Nolte et al., 2019). In our case, for example, the DDPM was unable to provide promises and time-definite service. What they could provide was a tentative timetable and tentative information about when and where aid would be delivered. The regional DDPM argued that this was because of the difficulty of disaster prediction and other conditional factors such as traffic, road and weather conditions.

Secondly, SERVPERF considers customer personalized attention, as businesses are customer-oriented (Parasuraman et al., 1985; Homburg et al., 2011). If there is a service backlog, a firm can adjust its service staff or can increase its service capacity by hiring more

TABLE 6.4 The changes made by the respondents with supporting reasons

Dimension	SERVPERF	HUMSERVPERF	The respondents' reasons for their revisions
Tangibility	XYZ___ has up-to-date equipment.	The relief organization has up-to-date equipment that is at the ready for use.	Equipment can be up-to-date but, sometimes, are not well-functioning or cannot be used.
	XYZ___'s physical facilities are visually appealing.	(deleted)	Facilities used in a flood relief operation are always dirty. It is impossible to make such facilities visually appeal.
	XYZ___'s employees are well dressed and appear neat.	The relief organization's employees/ volunteers dress appropriate to their relief work.	Dressing well and neat seems to be suit when you are a business staff. Relief staff can only dress appropriate to their relief work.
	The appearance of the physical facilities of XYZ___ is in keeping with the type of service provided.	The physical facilities and equipments are used correctly according to the type of relief service provided.	This item seems to be redundant with the first item of this dimension. What should be added is an emphasis on the correct use of physical facilities and equipments because relief staff can be volunteers who may not well know how to use such facilities.
Reliability	When XYZ___ promises to do something by a certain time, it does so.	When the relief organization strives to do something by a certain time, it does its utmost to respond and fulfil the needs of beneficiaries.	DDPM's staff cannot give promises to beneficiaries due to many external and internal conditions such as weather condition, transport-related condition and our limited staff and supplies.
	XYZ___ provides its services at the time it promises to do so.	The relief organization provides its relief services within an appropriate time frame.	A relief service cannot be promised due to many factors above-mentioned. We can only provide an appropriate time frame such as we will arrive at the disaster area within one-two hours according to the road condition.

(continued)

TABLE 6.4 (Continued)

Dimension	SERVPERF	HUMSERVPERF	The respondents' reasons for their revisions
Responsiveness	Employees of XYZ___ are too busy to respond to customer requests promptly.	Employees/ volunteers of the relief organization are available to respond to beneficiaries' requests when on duty.	It needs to add 'when on duty'. Although a local DDPM is available for 24 hours, the DDPM's staff and volunteers are limited and available only when they are on duty.
Assurance	Employees of XYZ___ are polite.	Employees/ volunteers of the relief organization are polite, humble and strive to maintain the dignity of the beneficiaries.	Beneficiaries had lost so much. Being humble with dignity and empathy are critical characteristics for relief staff.
Empathy	XYZ___ does not give you individual attention.	The relief organization gives beneficiaries personal-centered care.	DDPM's staff and volunteers are limited. It is impossible to give beneficiaries with personalized attention. What we can do is give adequate care to those in the most need.
	Employees of XYZ___ do not give you personal attention.	Employees/ volunteers of the relief organization give beneficiaries personal attention when appropriate.	
	XYZ___ does not have operating hours convenient to all their customers.	The relief organization has operating hours that serve beneficiaries well.	'Convenient time' does not seem to apply to the relief context. Beneficiaries need our help regardless of their convenient time. Relief services should be performed at our operating hours.

SOURCE The authors

staff (Oliva, 2001). In a relief operation, the number of relief staff is, however, far less than the number of beneficiaries. Relief staff cannot always provide beneficiaries with personalized attention, as this can burn up their physical and psychological energy (Nolte et al., 2019). In our case, the limited number of DDPM staff presents a challenge to relief operations that deal with multiple sudden-onset disasters that occur simultaneously in different locations. What the DDPM could offer was that relief care had to be prioritized to those in most need. 'It is important that staff focus more on the affected areas that require the most help,' was stated by the chief.

Thirdly, in the commercial context, courtesy is important to companies while rude manner can jeopardize firms (Helms and Mayo, 2008). In the humanitarian context, beneficiaries may be vulnerable to exploitation and misunderstanding and should be treated with respect and dignity (Soliman, 2010). In Thailand, DDPM staff were government officers and were ordered to behave to beneficiaries with dignity to avoid any negative sentiments. The regional DDPM director noted that protection of the right and dignity of beneficiaries is one of their goals. 'Beneficiaries lost so many things and what we have to do is to treat them with dignity,' as stated by the director. All practitioners in our case emphasized the importance of courtesy, respectfulness and the dignity of beneficiaries, and strongly recommended that these issues be included in HUMSERPERF. Although it is challenging for academia to force governments to develop policies that respect the dignity of beneficiaries (Oloruntoba and Banomyong, 2018), protecting beneficiaries rights with regards to these issues may be exercised through service quality performance measurement.

Finally, SERVPERF considers physical appearance as critical to service delivery (Parasuraman et al., 1985). The main goal of a relief organization, however, is to save life (Kovács and Spens, 2007). Relief organizations' key concern is whether or not physical goods are in a good condition, while the aesthetics of such physical goods does not play an important role (Nolte et al., 2019). In our case, the reality was that the facilities used in a flood relief operation and staff uniforms could not always be clean and be visually appealing.

Nevertheless, a key concern from the DDPM was that staff and volunteers should dress appropriately to carry out their relief work.

To illustrate how HUMSERPERF can be used, a case study of flood was employed and is presented in the next section.

CASE STUDY

The case of flood

A case study method is frequently used in the research of humanitarian supply chains (Behl and Dutta, 2019). Flood was selected as an illustrative case in this illustration due to the fact that a quarter of great natural catastrophes over the world are caused by hydrological-related disasters (Wirtz et al., 2014).

Flooding occurs almost every year in Mae Sot district, Tak province, Thailand and the 2020 flood was used as a case study. Residents were affected, losses of buildings and businesses in the areas were clearly observed (Maneechote, 2021). The DDPM at Mae Sot (DDPM regional office No.8) was the main government agency involved in relief efforts. Collaborating with the DDPM regional office, DDPM-SAOs were the primary relief organization helping beneficiaries.

The case selected in this chapter was of a flooded area in Tak province, Thailand, where beneficiaries' houses were under water for days and food shortages lasted for weeks. This area was located far away from the local DDPM-SAO and this was a reason why help arrived late. Furthermore, there was only a one lane dirt road accessing this location and the road itself was rendered impassable by the floodwaters, which cut off the village from the city centre.

The Thai version of HUMSERVPERF questionnaire was modified to be appropriate for the flood context. We included one question asking about behaviour intentions in terms of intention to receive future relief service. Although the humanitarian literature suggests this was not appropriate, we were keen to investigate whether service quality, satisfaction and behavioural intention received similar perception scores. The question related to behaviour intentions in the original SERVPERF was modified to: 'In the next disaster, I would hope to receive further aid from this organization'. It was anticipated that beneficiaries' satisfaction levels were likely to be low while, at the same time, their intention to receive future relief service can be high. This might be because beneficiaries had no choice due to the fact that the DDPM-SAO was their only hope.

A semi-structured interview with HUMSERVPERF questionnaire was carried out with eight beneficiaries living in the flooded areas. All questionnaires were completed. Due to the small sample size, the result could not be tested statistically. Mean scores were calculated to demonstrate the beneficiaries' perceptions of relief service quality.

Beneficiaries' perceptions towards relief service quality

The beneficiaries fully understood the questions in HUMSERVPERF and agreed that HUMSERVPERF was both valid and useable and could reflect their perceptions towards the relief service quality. In Appendix 6.B, the result of HUMSERVPERF in the case of flood is reported. As expected, the scores of five dimensions, on average, were low and overall service quality (2.43, 2.29) and satisfaction (2.43, 2.57) were also low. However, the score on behaviour intention was high (3.86). This indicates that beneficiaries perceived that the service provided by DDPM-SAOs staff could not satisfy their needs, but they would still require help in the case of future disasters. This confirms that the beneficiaries' satisfaction (dissatisfaction) seems not to have an influence on the beneficiaries' behavioural intentions.

The result of interviews with this group of beneficiaries revealed that performing good (bad) services could influence their satisfaction (dissatisfaction). One beneficiary said, '[F]lood occurs every year. What I have to do is waiting[sic] for [DDPM] to help. It does not mean that I want to repeat this [disaster] again but I have no choice and DDPM is our only hope.' Regarding the five dimensions, the beneficiaries had fewer concerns on the appearance of physical facilities, equipment, and staff/volunteers, as one respondent said: 'Who would care of uniforms and cleanliness of equipment? Just come to save us no matter the state of your clothes.' Reliability and Responsiveness in terms of promise and prompt service were important to them, as one said: 'if [DDPM staff] promised us that they will come today, they should do. It was terrible for us to live in the water for one more day'. Regarding Assurance, the beneficiaries believed in the skills and knowledge as well as trust in the relief staff. Regarding Empathy, they understood that the staff had to help others who were in greater need more than them.

Applications of HUMSERVPERF

It is evident that HUMSERVPERF can be used to measure relief service quality. Beneficiaries' responses to the HUMSERVPERF questionnaire were good. Questions put to beneficiaries were clear, and a simple mean could be calculated. Such mean scores may be used as a baseline for a relief organization to improve its relief services in future disasters.

Practitioners in humanitarian organizations often develop their own performance measurements, which are focused on tangible aspects (D'Haene et al., 2015). However, the intangible aspects of relief service have been rarely focused in the humanitarian literature (Tatham and Hughes, 2011). HUMSERVPERF was adapted from the SERVQUAL instrument, which aims to measure the intangible aspect of service quality. The measures in HUMSERVPERF are, therefore, intangible in nature and focus on beneficiaries' satisfactions of relief service. We suggest that HUMSERVPERF should be another common instrument used to assess a relief organization's performance.

In addition, HUMSERVPERF should also be used to supplement already developed hard measures such as efficiency and effectiveness. Measuring beneficiaries' satisfaction levels from both a tangible and intangible perspective should benefit a relief organization. Reporting beneficiaries' satisfactions from both aspects may increase a relief organization's capability to raise funds or ensure a long-term funding. HUMSERVPERF should become a mechanism for beneficiaries to inform whether or not their needs have been met. Rather than focusing on financial efficiency and effectiveness, donors may use HUMSERVPERF to observe whether funds and resources provided can meet beneficiaries' requirements.

Summary and conclusion

This chapter proposed HUMSERVPERF by adapting an instrument used to assess service quality in the commercial supply chains. Practitioners in relief operations helped modify HUMSERVPERF. New items were added as suggested in the humanitarian literature. HUMSERVPERF was proposed as a starting point for further development of performance measurement of relief service quality. There is a need to purify the items and to collect more data to validate all of HUMSERVPERF's dimensions. We need to investigate whether or not all items in HUMSERVPERF could be classified into the five SERVQUAL dimensions. In addition, a correlation test is needed to validate the relationship between relief service dimensions and beneficiaries' satisfaction.

HUMSERVPERF in this chapter was developed with the purpose of assessing beneficiaries' perceptions towards the relief service provided by the DDPM. We do not know whether descriptions in HUMSERVPERF's items will remain the same if it is applied to other relief organizations.

In addition, in most relief supply chains, donors are involved in providing fund to relief organizations. It will be interesting for future research to investigate whether HUMSERVPERF can be used to motivate external donors' funding. Future research should investigate which items in HUMSERVPERF are of concern to such donors and whether the results of HUMSERVPERF can influence donors' decisions on future funding. Lastly, a case of flood relief operation was further employed to demonstrate how to use HUMSERVPERF. We invite humanitarian scholars to test HUMSERVPERF in other disasters as well as other phases of disaster relief.

Acknowledgement

We would like to acknowledge the Kuehne Foundation for their partial funding of the research described in this chapter.

References

Banomyong, R, Julagasigorn, P, Varadejsatitwong, P, and Piboonrungroj, P (2019) The humanitarian supply chain assessment tool (HumSCAT), *Journal of Humanitarian Logistics and Supply Chain Management*, **9**(2), pp 221–249

Banomyong, R and Supatn, N (2011) Selecting logistics providers in Thailand: A shippers' perspective, *European Journal of Marketing*, **45**(3), pp 419–437

Beamon, B M and Balcik, B (2008) Performance measurement in humanitarian relief chains, *International Journal of Public Sector Management*, **21**(1), pp 4–25

Behl, A and Dutta, P (2018) Humanitarian supply chain management: A thematic literature review and future directions of research, *Annals of Operations Research*, **283**, pp 1001–1044

Besiou, M and Van Wassenhove, L N (2021) System dynamics for humanitarian operations revisited, *Journal of Humanitarian Logistics and Supply Chain Management*, **11**(4), pp 599–608

Brady, M K, Cronin Jr, J J, and Brand, R R (2002) Performance-only measurement of service quality: A replication and extension, *Journal of Business Research*, **55**(1), pp 17–31

Burkart, C, Besiou, M, and Wakolbinger, T (2016) The funding – humanitarian supply chain interface, *Surveys in Operations Research and Management Science*, **21**(2), pp 31–45

Cardoso, B, Fontainha, T, Leiras, A, and Cardoso, P A (2021) Performance evaluation in humanitarian operations based on the beneficiary perspective, *International Journal of Productivity and Performance Management*, ahead-of-print

Carrillat, F A, Jaramillo, F, and Mulki, J P (2007) The validity of the SERVQUAL and SERVPERF scales: A meta-analytic view of 17 years of research across five continents, *International Journal of Service Industry Management*, **18**(5), pp 472–490

Cronin Jr, J J and Taylor, S A (1992) Measuring service quality: A reexamination and extension, *Journal of Marketing*, **56**(3), pp 55–68

Cronin Jr, J J, Brady, M K, and Hult, G T M (2000) Assessing the effects of quality, value, and customer satisfaction on consumer behavioral intentions in service environments, Journal of Retailing, **76**(2), pp 193–218

D'Haene, C, Verlinde, S, and Macharis, C (2015) Measuring while moving (humanitarian supply chain performance measurement – status of research and current practice), *Journal of Humanitarian Logistics and Supply Chain Management*, **5**(2), pp 146–161

De Leeuw S (2010) Towards a reference mission map for performance measurement in humanitarian supply chains, in Camarinha-Matos, L M, Boucher, X H Afsarmanesh, H (eds), *Collaborative Networks for a Sustainable World, IFIP Advances in Information and Communication Technology*, 336, pp 181–188, Springer, Berlin

Gunasekaran, A and Kobu, B (2007) Performance measures and metrics in logistics and supply chain management: A review of recent literature (1995–2004) for research and applications, *International Journal Production Research*, **45**(12), pp 2818–2840

Haavisto, I and Goentzel, J (2015) Measuring humanitarian supply chain performance in a multi-goal context, *Journal of Humanitarian Logistics and Supply Chain Management*, **5** (3), pp 300–324

Heaslip, G and Kovács, G (2019) Examination of service triads in humanitarian logistics, *The International Journal of Logistics Management*, **30**(2), pp 595–619

Heaslip, G, Kovács, G, and Grant, D B (2018) Servitization as a competitive difference in humanitarian logistics, *Journal of Humanitarian Logistics and Supply Chain Management*, **8**(4), pp 497–517

Helms, M M and Mayo, D T (2008) Assessing poor quality service: perceptions of customer service representatives, *Managing Service Quality: An International Journal*, **18**(6), pp 610–622

Homburg, C, Müller, M, and Klarmann, M (2011) When should the customer really be king? On the optimum level of salesperson customer orientation in sales encounters, *Journal of Marketing*, **75**(2), pp 55–74

Jain, S K and Gupta, G (2004) Measuring service quality: SERVQUAL vs. SERVPERF scales, *Vikalpa*, **29**(2), pp 25–38

Kovács, G (2014) Where next? The future of humanitarian logistics, in Christopher, M and Tatham, P H (eds), *Humanitarian Logistics: Meeting the Challenge of Preparing for and Responding to Disasters* (2nd ed), pp 275–285, Kogan Page, London

Kovács, G and Spens, K M (2007) Humanitarian logistics in disaster relief operations, *International Journal of Physical Distribution & Logistics Management*, **37**(2), pp 99–114

Kovács, G and Spens, K M (2009) Identifying challenges in humanitarian logistics, *International Journal of Physical Distribution & Logistics Management*, **39**(6), pp 506–528

Ladhari, R (2009) A review of twenty years of SERVQUAL research, *International Journal of Quality and Service Sciences*, **1**(2), pp 172–198

Maneechote, P (2021) Official says Tak flash flood now 'improving', despite continuous reports of flooding, *Thaienquirer*, 29 July, www.thaienquirer.com/30567/official-says-tak-flash-flood-now-improving-despite-continuous-reports-of-flooding/ (archived at https://perma.cc/27N4-6KT5)

Medina-Borja, A and Triantis, K (2007) A conceptual framework to evaluate performance of non-profit social service organizations, *International Journal of Technology Management*, **37**(1–2), pp 147–161

Nolte, I M, Bushnell, A M, and Mews, M (2020) Public administration entering turbulent times: A study of service quality during the refugee crisis, *International Journal of Public Administration*, **43**(16), pp 1345–1356

Oliva, R (2001) Tradeoffs in responses to work pressure in the service industry, *California Management Review*, **43**(4), pp 26–43

Oloruntoba, R and Banomyong, R (2018) Humanitarian logistics research for the care of refugees and internally displaced persons: A new area of research and a research agenda, *Journal of Humanitarian Logistics and Supply Chain Management*, **8**(3), pp 282–294

Oloruntoba, R and Gray, R (2009) Customer service in emergency relief chains, *International Journal of Physical Distribution & Logistics Management*, **39**(6), pp 486–505

Oloruntoba, R and Kovács, G (2015) A commentary on agility in humanitarian aid supply chains, *Supply Chain Management: An International Journal*, **20**(6), pp 708–716

Parasuraman, A, Berry, L L, and Zeithaml, V A (1991) Refinement and reassessment of the SERVQUAL scale, *Journal of Retailing*, **67**(4), pp 420–450

Parasuraman, A, Zeithaml, V A, and Berry, L L (1985) A conceptual model of service quality and its implications for future research, *Journal of Marketing*, **49**(4), pp 41–50

Parasuraman, A, Zeithaml, V A, and Berry, L L (1988) Servqual: A multiple-item scale for measuring consumer perceptions of service quality, *Journal of Retailing*, **64**(1), pp 2–40

Schiffling, S and Piecyk, M (2014) Performance measurement in humanitarian logistics: A customer-oriented approach, *Journal of Humanitarian Logistics and Supply Chain Management*, **4**(2), pp 198–221

Schulz, S and Heigh, I (2009) Logistics performance management in action within a humanitarian organization, *Management Research News*, **32**(11), pp 1038–1049

Shabbir, H, Palihawadana, D, and Thwaites, D (2007) Determining the antecedents and consequences of donor-perceived relationship quality – a dimensional qualitative research approach, *Psychology & Marketing*, **24**(3), pp 271–293

Sheu, J B (2014) Post-disaster relief–service centralized logistics distribution with survivor resilience maximization, *Transportation Research Part B: Methodological*, **68**, pp 288–314

Soliman, H (2010) Ethical considerations in disasters: A social work framework, in Danso, K and Gillespie, D (eds), *Disaster Concepts and Issues: A Guide for Social Work Education and Practice*, pp 223–240, Council on Social Work Education Press, Alexandria, VA

Tatham, P H and Hughes, K (2011) Humanitarian logistics metrics: Where we are, and how we might improve, in Christopher, M and Tatham, P H (eds), *Humanitarian Logistics: Meeting the Challenge of Preparing for and Responding to Disasters*, pp 65–84, Kogan Page, London

Van der Laan, E A, De Brito, M P, and Vergunst, D A (2009) Performance measurement in humanitarian supply chains, *International Journal of Risk Assessment and Management*, **13**(1), pp 22–45

Van Wassenhove, L N (2006) Humanitarian aid logistics: Supply chain management in high gear, *Journal of the Operational Research Society*, **57**(5), pp 475–489

Varadejsatitwong, P, Banomyong, R, and Julagasigorn, P (2021) Developing a performance measurement framework for logistics service providers, in *Proceedings of the International Conference on Industrial Engineering and Operations Management*, Monterrey, Mexico, 3–5 November

Wirtz, A, Kron, W, Löw, P, and Steuer, M (2014) The need for data: Natural disasters and the challenges of database management, *Natural Hazards*, 70(1), pp 135–157

Zeithaml, V A, Berry, L L, and Parasuraman, A (1996) The behavioral consequences of service quality, *Journal of Marketing*, **60**(2), pp 31–46

Appendix 6.A: HUMSERVPERF questionnaire

Independent variables						
Dimension	Questions	Disagree-Agree				
		1	2	3	4	5
Tangibility	The relief organization has up-to-date equipment that is at the ready for use. (T1)					
	The relief organization's employees/volunteers dress appropriate to their relief work. (T2)					
	The physical facilities and equipments are used correctly according to the type of relief service provided. (T3)					
Reliability	When the relief organization strives to do something by a certain time, it does its utmost to respond and fulfil the needs of beneficiaries. (REL1)					
	When beneficiaries have problems, the relief organization exhibits appropriate sympathy and reassuring. (REL2)					
	The relief organization is dependable. (REL3)					
	The relief organization provides its relief services within an appropriate time frame. (REL4)					
	The relief organization keeps records on relief service accurately. (REL5)					
Responsiveness	The relief organization is clear to beneficiaries exactly when relief service is intended to perform. (RES1)					
	Beneficiaries receive prompt relief service from the relief organization's employees/volunteers. (RES2)					
	Employees/volunteers of the relief organization exhibit a willingness to help beneficiaries. (RES3)					
	Employees/volunteers of the relief organization are available to respond to beneficiaries' requests when on duty. (RES4)					

(continued)

(Continued)

Independent variables		
Assurance	Beneficiaries can trust employees/volunteers of the relief organization. (A1)	
	Beneficiaries can feel safe in transacting with the relief organization when receiving relief aid. (A2)	
	Employees/volunteers of the relief organization are polite, humble and strive to maintain the dignity of the beneficiaries. (A3)	
	Employees/volunteers get adequate support from the relief organization to do their relief service well. (A4)	
	Employees/volunteers of the relief organization have passed the relief organization's training. (A5)	
	Employees/volunteers of the relief organization have knowledge and skills related to relief efforts. (A6)	
	The relief organization is subject to a good system of accountability. (A7)	
	The relief organization maintains a good reputation. (A8)	
	The relief organization clearly informs delivery time and locations and emergency meeting points. (A9)	
	The relief organization provides up-to-date and accurate information or any useful information appropriate to the disaster situation. (A10)	
	Beneficiaries are explained or taught clearly or receive instructions about how to use aid and relief equipment. (A11)	
Empathy	The relief organization gives beneficiaries personal-centred care. (E1)	
	Employees/volunteers of the relief organization give beneficiaries personal attention when appropriate. (E2)	
	Employees/volunteers of the relief organization offer empathic understanding of the beneficiaries' needs. (E3)	
	The relief organization has beneficiaries' highest interests. (E4)	
	The relief organization has operating hours that serve beneficiaries well. (E5)	
Supplies Quantity	The relief organization provides relief service that is adequate, appropriate and meets beneficiaries' needs. (Q1)	

SOURCE Adapted from Cronin and Taylor (1992)

Dependent variables					
Overall Service Quality	Poor-Excellent				
	1	2	3	4	5
The quality of relief organization's service is (OSQ1)					
Comparing to my expectation, the quality of the relief organization's service is (OSQ2)					
Satisfaction	Dissatisfied-Satisfied				
	1	2	3	4	5
My feelings towards the relief organization's service can be described as (SAT1)					
My satisfaction towards the relief service is (SAT2)					

SOURCE Adapted from Parasuraman *et al.* (1985); Cronin and Taylor (1992); Banomyong and Supatn (2011)

Appendix 6.B: Result of HUMSERVPERF for the flood context

Dimension	Questions	Score	Dimension	Questions	Score
Reliability	REL1	2.57	Responsiveness	RES1	2.29
	REL2	2.57		RES2	2.29
	REL3	2.43		RES3	2.57
	REL4	2.57		RES4	2.86
	REL5	2.29	Tangibility	T1	2.14
Assurance	A1	2.14		T2	3.00
	A2	2.14		T3	2.14
	A3	2.71	Empathy	E1	2.14
	A4	2.57		E2	2.29
	A5	2.71		E3	2.14
	A6	2.43		E4	2.29
	A7	2.57		E5	2.57
	A8	2.29	Overall Quality	OSQ1	2.43
	A9	2.29		OSQ2	2.29
	A10	2.29	Satisfaction	SAT1	2.43
	A11	2.00		SAT2	2.57
Supplies Quantity	Q1	2.43	Behavioral Intention	BI1	3.86

NOTE Behavioral Intention (BI1) asks respondents that 'In the next disaster, I would hope to receive further aid from this organization'
SOURCE The authors

07

Network design for pre-positioning emergency relief items

GERARD DE VILLIERS

ABSTRACT

The efficiency of emergency relief depends to a large extent on the availability of the right goods at the right time in the right condition and in sufficient quantities at the right place. There is usually very little time to reorder incorrect items or replenish inadequate quantities, but at the same time, the emergency response supply chain should not be burdened with unnecessary stock, especially in the wrong place. Centre-of-gravity analysis has been used very effectively in commercial network design and the purpose of this chapter is to indicate the usefulness of this technique in network design for emergency response. The chapter starts with setting the scene with the need for pre-positioning in war-torn Ukraine, providing Covid-19 context for pre-positioning, defining humanitarian logistics, the planning hierarchy and a discussion of network design. It continues with an overview of disaster event locations and describes the importance of establishing pre-positioning facilities for emergency relief items at the correct locations. The current UNHRD network is discussed and the ESUPS project for pre-positioning introduced. The centre-of-gravity technique is explained and the location of disaster events in Africa used as a case study to suggest how centre-of-gravity analysis can be applied usefully. The chapter ends with brief reference to channel strategies and how each strategy should be aligned with an appropriate logistics network.

Introduction

Commercial supply chains consist of a network of nodes and links between origins and destinations. The nodes consist mainly of intermodal terminals and freight logistics hubs that provide intermediate locations where logistics value is added to the movement of freight. Examples of logistics value added at freight logistics hubs include consolidation or deconsolidation, picking, packing, storing in customised warehousing facilities and related activities to provide for delivering into the market. This is often referred to as the 'last mile' and it is probably the most important link in the supply chain, as it is most expensive due to relatively small loads to be delivered in congested destinations where economies of scale are not possible. Hence it is important to place these facilities at locations that provide cost effective delivery to the market.

Pre-positioning of relief items at cost-effective locations is similarly important in humanitarian supply chains and the same principles apply to having the 'last mile' to the field as cost-effectively as possible for emergency relief. Appropriate network design is needed and although global pre-positioning facilities are currently in place, there is a need for looking carefully at the location of regional facilities to reduce costs of the humanitarian supply chains while increasing the yield of scarce resources committed to emergency relief.

Setting the scene

Following the start of the war in Ukraine on 24 February 2022 the country's security situation deteriorated rapidly, leaving at least 15.7 million people in need of urgent humanitarian assistance and protection. Hostilities escalated across several regions, and the conflict has caused the fastest growing displacement crisis since World War II. The intense military escalation has resulted in loss of life and injuries, as well as massive destruction and damage to civilian infrastructure, interrupting critical services. The humanitarian community prepared for and is rapidly adapting to the unfolding situation, based on the Inter-Agency

> Contingency Plan updated in early 2022 ahead of the onset of the crisis. As anticipated in a worst-case scenario, the violence has prompted a steep escalation in needs and a significant expansion of the areas in which humanitarian assistance is required.
>
> World Food Programme, 2022: 1

The concept of operations continues to state that the instable security situation remains the most significant challenge for the planning and implementation of humanitarian response activities, particularly in eastern and southern areas. Based on ongoing consultations with partner organizations, the logistics cluster has identified the following logistics gaps and bottlenecks in humanitarian relief in Ukraine:

- The current security situation is creating difficulties for planning and executing aid delivery, including the forward movement, and staging of humanitarian cargo. It is anticipated that increasing damage to road infrastructure and limitations of logistics capacity will continue to constrain humanitarian operations.
- While the commercial logistics sector is well developed, many private sector companies relocated to the west due to the conflict. This is leading to limited logistics services currently available in the country's east. There are also limitations on labour force that would otherwise be available, particularly to support the movement of cargo to hard-to-reach areas.
- Fuel shortages are affecting the operational capacity of humanitarian organizations, particularly those in need of light vehicles.
- The conflict has resulted in the closure of all airports, seaports, and border crossings for commercial goods. Shipping lines have also suspended their services from the country and diverted their vessels to other Black Sea ports until further notice.

This narrative from the concept of operations document on the war in Ukraine sets the scene for the discussion on pre-positioning of emergency relief items. It is clear that warehouse facilities that were available for pre-positioning before the war were no longer available or suitable in the country and other regional facilities had to be activated in neighbouring countries. Network design principles as covered in this chapter can be useful in choosing the best location for establishing pre-positioning facilities in the region around such war-torn areas.

Covid-19 context for pre-positioning

Much has been written and will still have to be learnt on how to respond to the Covid-19 pandemic, but it is clear that the impact was highly disruptive and a serious challenge to conventional thinking and practice in supply chains. Sales of certain consumer goods and pharmaceuticals followed a highly erratic pattern and principles of forecasting in inventory management changed overnight. Demand fulfilment became very difficult and emergency items, such as ventilators and distribution of vaccines, dominated supply chains all over the world. Global cooperation will be needed to develop suitable humanitarian supply chains that can deal effectively with demand requirements during future similar pandemics (Niemann and de Villiers, 2022: 4–5).

The UNICEF Supply Division has published a useful report on an assessment of the impact of Covid-19 on global logistics and supplies (UNICEF, 2021). A summary of the general findings of the report at the time include:

- The long-term logistics consequences of the Covid-19 pandemic are continuing to have a negative impact on the shipping industry, with unprecedented major challenges to the delivery of critical supplies, including health technology, medicines, water, sanitation and hygiene supplies to country programmes. Current shipping capacity is extremely tight, with sea carriers preferring to position as many available containers and equipment as possible on the more highly profitable sea lanes between Asia–Europe and Asia–USA. Vessel cancelations continue, with no commitments being made on transit times due to congestion at transhipment hubs. Given tight container fleets, compounded by the shift in trade imbalances, some analysts do not anticipate the normalization of shipping capacity to be resolved over the next 6–12 months.

- Shipping container leasing rates have increased by as much as 300 per cent and higher over the past six month period, coupled with global price increases of many commodities and raw materials, which have direct inflationary pressure on many of the finished products UNICEF and other relief agencies procure through their suppliers.
- As a consequence, following the increased congestion of sea freight carrying capacity at major ports, UNICEF has been observing an increased demand on air freight over the latter part of 2021.
- In order to mitigate the impact of reduced shipping capacity, UNICEF is applying a flexible and agile strategy to overcome some of the difficulties, while it continues to identify alternative solutions.
- UNICEF encourages programmes, countries and partners to plan their procurement as early as possible.

One particular comment in the report regarding pre-positioning is that UNICEF was able to meet emergency response measures using its pre-positioned stocks through its warehouses located in Accra, Copenhagen, Dubai and Panama. This is important and provides useful context for pre-positioning of emergency relief items.

Humanitarian logistics

Humanitarian logistics consists of the same elements or functions of business logistics, such as transport, warehousing, inventory management, procurement, logistics information systems, order management, materials handling, packaging and reverse logistics. 'Customers', though, should rather be replaced by 'beneficiaries', but the definition is in principle the same. Figure 7.1 provides a useful framework that incorporates all components of the humanitarian logistics context.

FIGURE 7.1 Humanitarian logistics and supply chain management

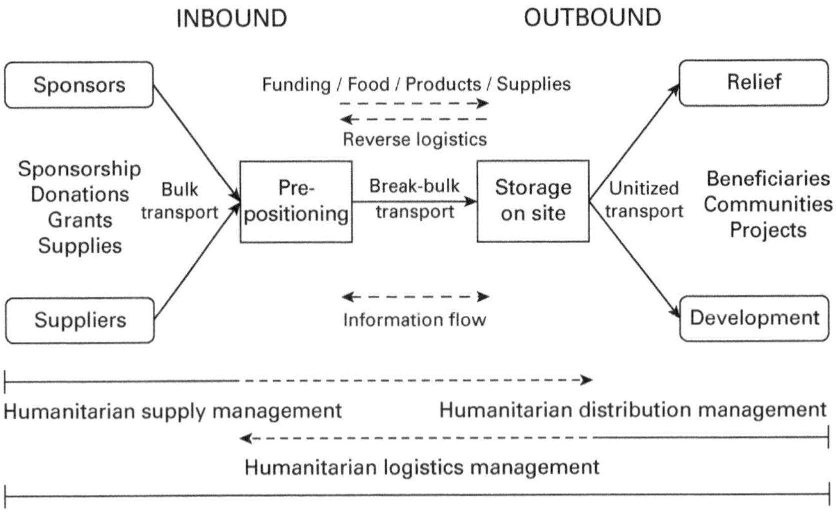

```
                    INBOUND                              OUTBOUND
   ┌──────────┐                                                    ┌──────────┐
   │ Sponsors │      Funding / Food / Products / Supplies          │  Relief  │
   └──────────┘      - - - - - - - →                               └──────────┘
                    ← - - - - - - -
   Sponsorship        Reverse logistics                            Beneficiaries
   Donations   Bulk    ┌──────┐  Break-bulk  ┌────────┐ Unitized   Communities
   Grants    transport │ Pre- │  transport   │Storage │ transport    Projects
   Supplies            │posit.│              │on site │
                       └──────┘              └────────┘
   ┌──────────┐      ← - - - - - →                                 ┌───────────┐
   │Suppliers │        Information flow                            │Development│
   └──────────┘                                                    └───────────┘
   ├────────────────────────────- - - - - - - - - - - - - - →
   Humanitarian supply management      Humanitarian distribution management
                             ← - - - - - - - - - - - - - - - ────────────────┤
                    Humanitarian logistics management
   ├────────────────────────────────────────────────────────────────────────┤
```

SOURCE Niemann and de Villiers, 2022: 417

The upstream (or inbound) side consists of sponsors and suppliers providing sponsorship, donations, grants or supplies, such as food and non-food items. The funding, food, products and supplies move downstream through the supply chain from the inbound to the outbound side while information flows in both directions. Reverse logistics refers to returns, recalls, expired goods and any other packaging material or vehicles that have to be returned from the field or disposed of once the programme or project has been completed.

Planning hierarchy

The planning hierarchy (Stock and Lambert, 2001: 702–704) is very useful in providing a framework for humanitarian logistics planning. Figure 7.2 provides an adapted version of this planning hierarchy to include *inter alia* procurement as well as freight forwarding between the functional and operational levels, plus some minor editing to expand description of some of the key areas.

The importance of this framework for this discussion is the sequence in which planning should happen. It should start at the

FIGURE 7.2 Planning hierarchy

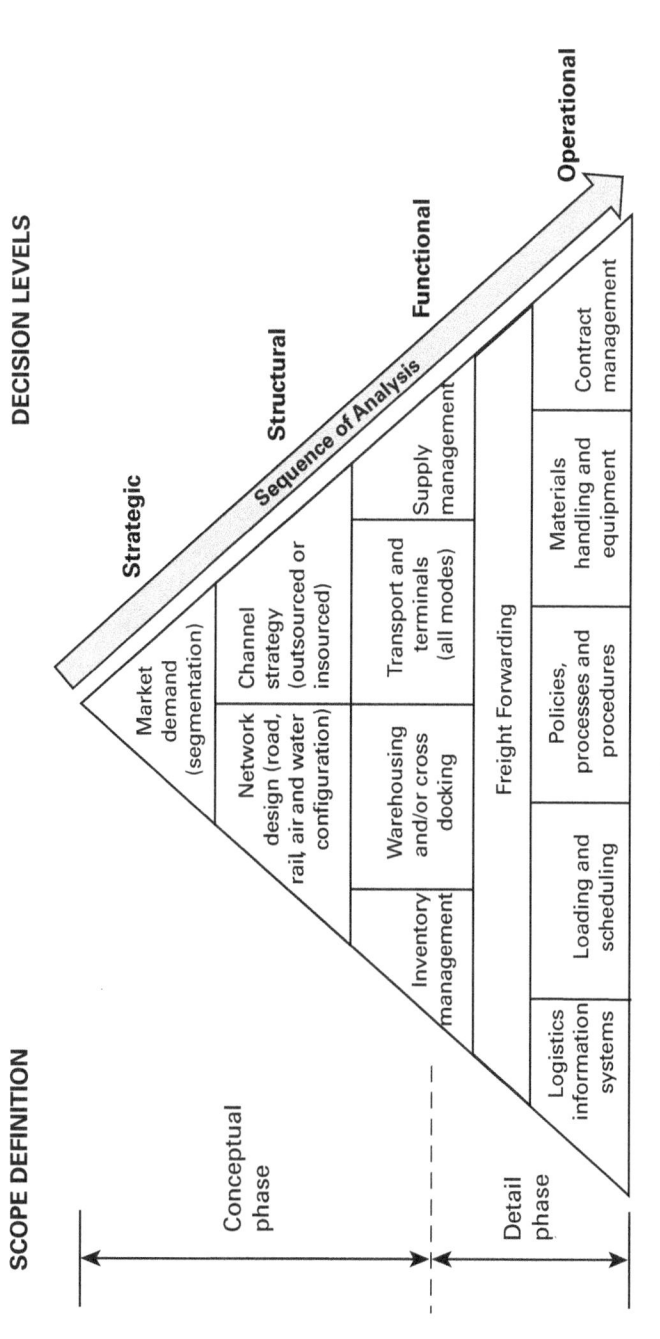

SCOPE DEFINITION   **DECISION LEVELS**

SOURCE Adapted from Stock and Lambert, 2001: 703

strategic level, and follow in a specific order through the structural, functional and operational levels as indicated by the arrow, to ensure structure follows strategy. The pinnacle of the triangle or hierarchy starts with the specific requirements of the customer or the beneficiary in the operational theatre, based on an analysis of the requirements in the field. Once the demand is determined and future growth estimated, the planning process should proceed to network design and channel strategy on structural level. Network design addresses questions such as the number and location of intermediary located facilities, for example, the pre-positioning facilities and centralisation *versus* decentralisation, as discussed in the next section.

Network design

Niemann and de Villiers (2022: 416) suggest that on the outbound side of the humanitarian supply chain, the first stop for food, products or supplies that enters the humanitarian supply chain will highly likely be a pre-positioning facility from where it will be moved to local storage on site, except for direct deliveries immediately after the disaster, when urgency will necessitate a shorter supply chain. The customers on the outbound side are mostly split between relief and development and referred to as beneficiaries, communities or projects. There is usually little or no production-related logistics activity in this supply chain, except maybe 'kitting' of hygiene kits or 'meals ready to eat' (MREs).

The location of these pre-positioning facilities should be determined by appropriate network design techniques such as centre-of-gravity analysis to ensure that the most cost-effective sites are chosen. There are a number of different network design options available but the two most common types are *decentralization* and *centralization*. The first concept of decentralization is presented in Figure 7.3 and indicates a number of global warehouses at the origin and a number of warehouses at the destination.

The second concept of centralization is presented in Figure 7.4 for consolidation of warehousing at the origin and in Figure 7.5 for consolidation of warehousing at the destination.

FIGURE 7.3 Decentralized distribution network

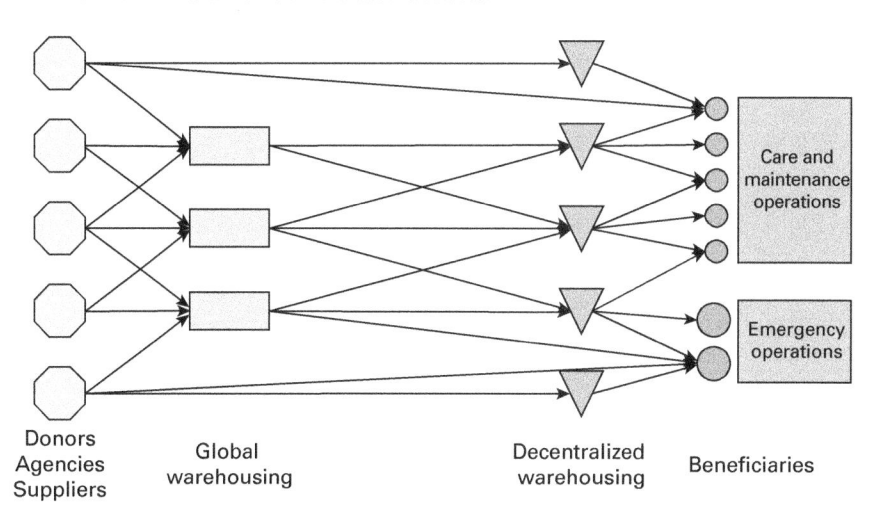

FIGURE 7.4 Centralized distribution network – Consolidated warehousing at the origin

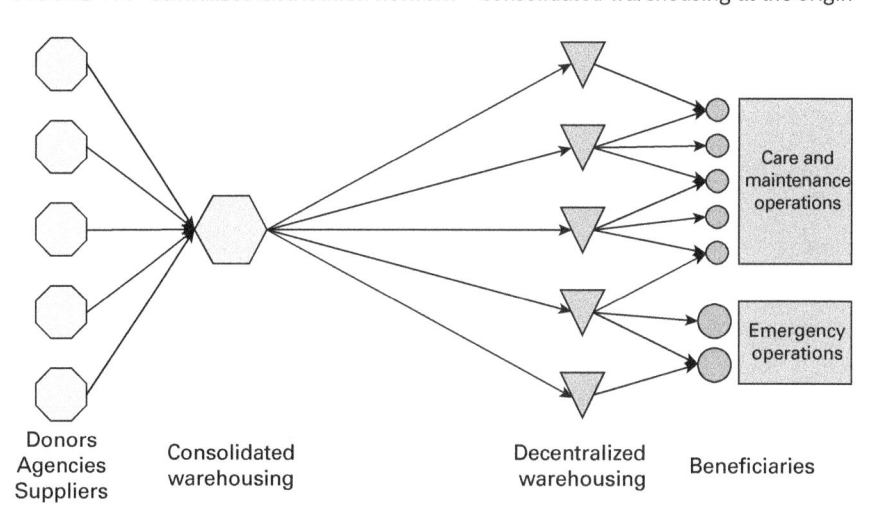

There is no easy answer as to which network is the best and total logistics costs should be used to determine the least cost network design, based on centre-of-gravity analysis. An example of the trade-off between logistics costs is the reality that centralization usually incurs higher delivery transport costs while decentralization incurs lower delivery transport costs due to closer proximity to the market. The opposite is true for warehousing costs as centralization usually

FIGURE 7.5 Centralized distribution network – Consolidated warehousing at the destination

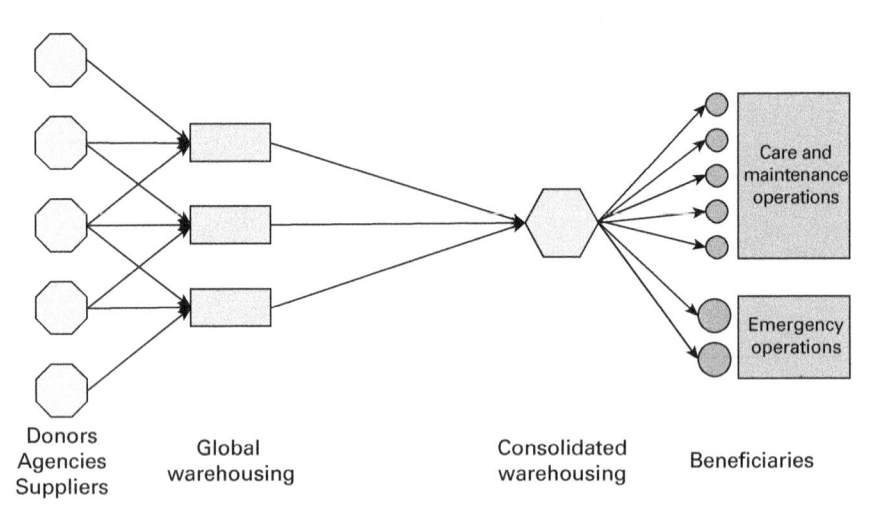

incurs lower warehousing costs while decentralization are due to the higher number of facilities. Inventory carrying costs should also be included in the analyses, and based on the square root law of inventories the costs are likely to reduce significantly with centralization.

The square root law states that total safety stock can be approximated by multiplying the total inventory by the square root of the number of future warehouse locations divided by the current number, as indicated in Equation 1:

$$SS_F = SS_E \sqrt{\frac{n_F}{n_E}} \qquad (1)$$

Where:

SS_F = Future safety stock

SS_E = Existing safety stock

n_F = Number of future facilities

n_E = Number of existing facilities

TABLE 7.1 Advantages and disadvantages of relief procurement

Procurement	Advantages	Disadvantages
Local procurement	• Low transport cost • Prompt deliveries • Local economy support	• Risk strategy to operate solely • Unavailability of enough quantity and quality needed • Create shortage in the local market
Global procurement	• Increase the availability of large quantities of high-quality supplies	• Longer deliver times • Higher transport cost • Supplies not delivered to affected area during the initial critical days due to bidding process
Pre-positioned stock	• Deliver sufficient relief aid within a relatively short timeframe • Less expensive than post-disaster supply procurement • Increase the ability of mobilization • Efficient (low cost, less duplication of efforts, less waste of resource) • Effective (quick response, satisfied demand)	• Financially prohibitive • Complex • Too many uncertainties • Only few can operate • Impossible to depend solely in case of large scale disasters • Capacity limitations

SOURCE Roh and Kim, 2016: 8

Roh and Kim (2016: 7–9) provide an interesting perspective on the inbound side and suggest that once a disaster occurs, humanitarian organizations can acquire relief supplies from three main sources, namely local suppliers, global suppliers and pre-positioned warehousing. The advantages and disadvantages of procuring from these three sources are indicated in Table 7.1.

It is clear from this discussion that the design of the logistics network can have a profound impact on the total logistics costs and hence the yield of scarce resources committed to emergency relief.

Location of disaster events

The design of an appropriate logistics network should start at the location of disaster events and although the exact locations are mostly unknown, it is possible to determine the most likely locations for typical disasters such as earthquakes, as they are mostly linked to geological formations for which the locations are known. Similarly, there are areas or regions on a global scale where hurricanes, typhoons, droughts and other natural disasters seem to happen regularly.

The Emergency Events Database of the Centre for Research on Epidemiology of Disasters (CRED) at the Université Catholique de Louvain (EM-DAT-b, 2017) maintains an extensive and detailed database of disaster events and Figure 7.6 provides a very interesting indication of the location of natural disasters from 1986 to 2015 (EM-DAT-a, 2017).

The same database reports that more recently, the number of 100 million people affected by disasters in 2021 was relatively stable over the 2017–2021 period (EM-DAT, 2021). Peaks in the numbers of total people affected, as reported by EM-DAT, correspond to severe droughts during the 2002 and 2015 monsoon period in India, which each affected more than 300 million people. This is indicated in Figure 7.7.

The impact of such disaster events can be reduced significantly with proper mitigation measures, such as early warning systems, and World Vision has developed a Crisis Country Cluster map, clearly identifying zones in the world where disasters are historically common and what type of disaster to expect (World Vision, 2010: 16). This is indicated in Figure 7.8 and, which shows prone to natural disasters such as hurricanes, typhoons, floods and droughts as well as locations of potential civil unrest and overpopulation.

The monitoring of historical occurrences and information available from research such as the crisis country clusters provide very good indications for where emergency relief items should be pre-positioned for effective deployment to the disaster areas.

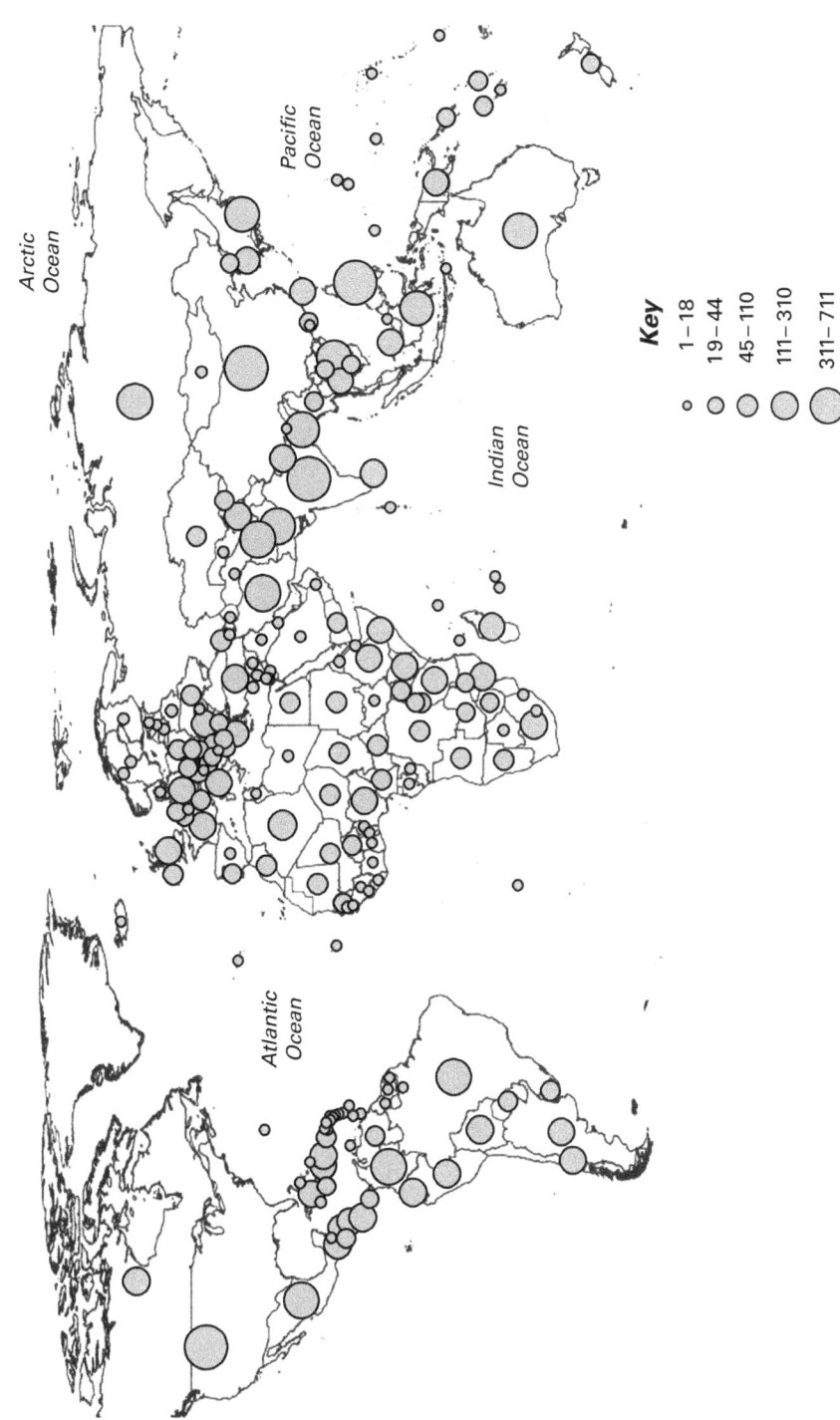

FIGURE 7.6 Location of natural disasters – 1986 to 2015

Key
- 1–18
- 19–44
- 45–110
- 111–310
- 311–711

SOURCE EM-DAT-b, 2017

FIGURE 7.7 Number of disasters by continent and top 12 countries in 2021

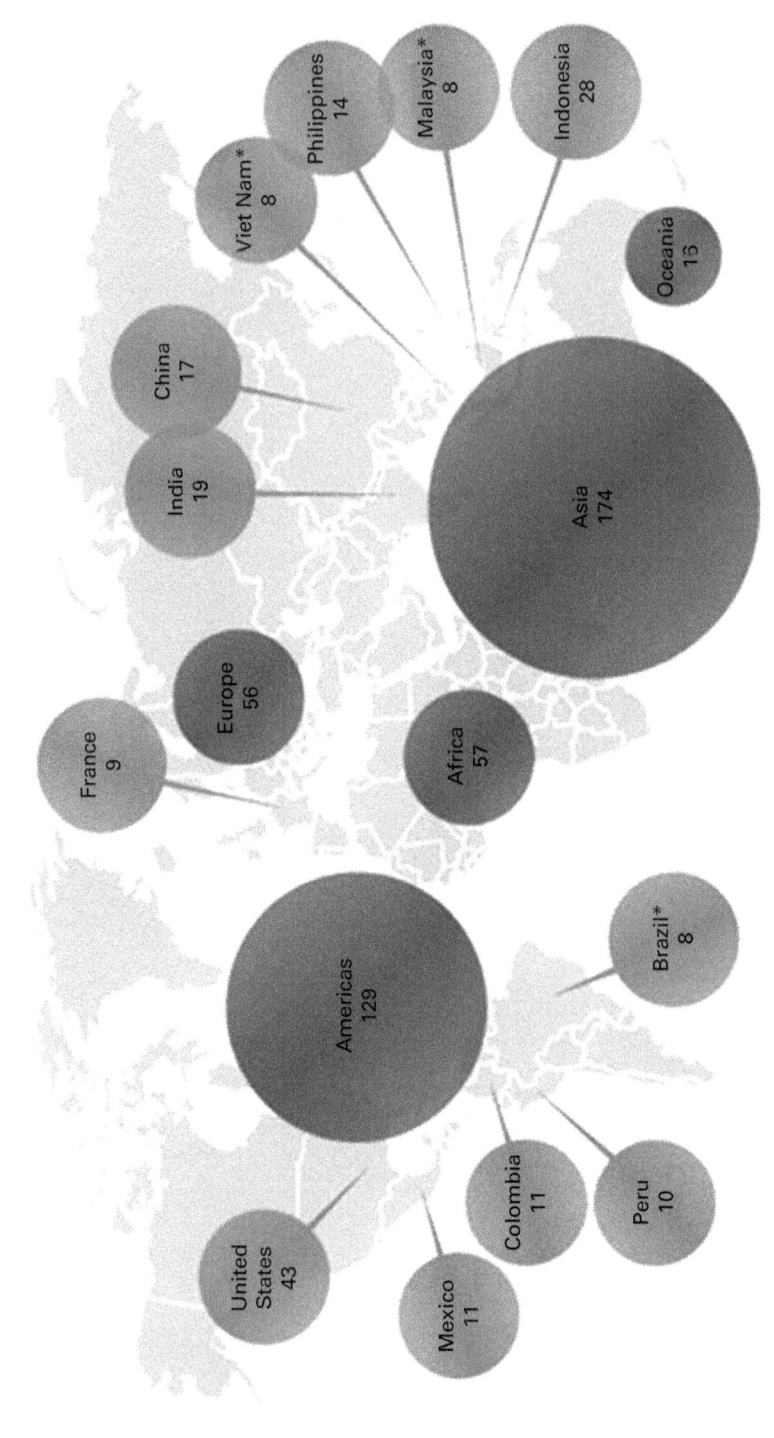

SOURCE EM-DAT, 2021

FIGURE 7.8 Crisis country clusters

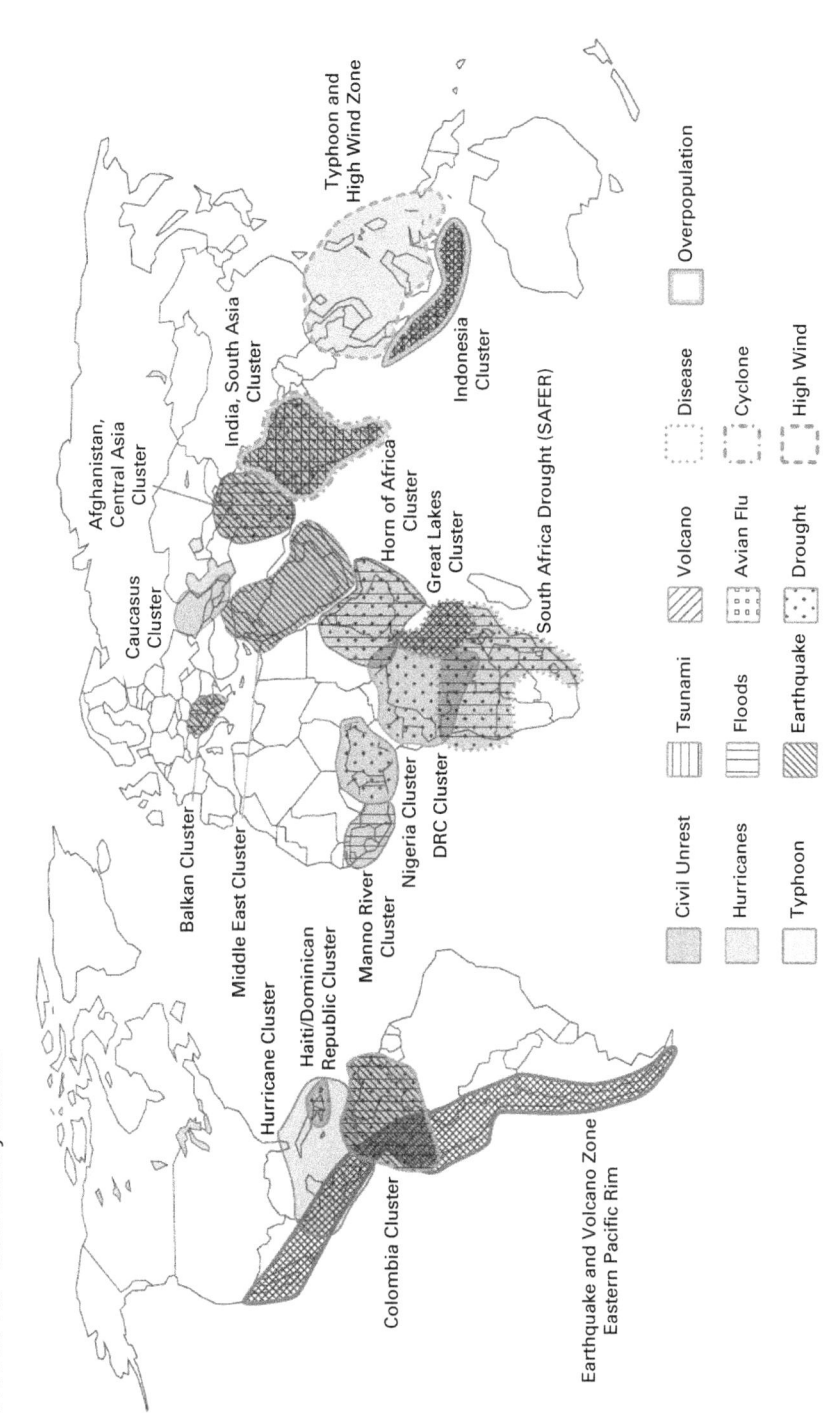

SOURCE World Vision, 2010: Reproduced with permission

Typical emergency relief items to be pre-positioned

Experience in emergency response has over time provided a better indication of the specific items to be stored as well as improved estimates of the required quantities. Duran et al. (2011: 8) found in their work done for CARE that the following relief items should be stored in pre-positioning facilities: food, water and hygiene kits, hot weather tents, cold weather tents and household kits.

De Villiers (2008: 2) mentioned that World Vision carries a similar assortment that includes programme stock and support stock as follows:

- Programme stock: Shelters (tarpaulins, blankets, etc.), water and sanitation (drilling equipment, tanks, etc.) and food (high energy biscuits, etc.).

- Support stock: Electronic equipment (satellite communication devices, computers, etc), security items (bullet proof vests, helmets, etc.), team support (prefabricated offices, staff tents, etc.), transportation (vehicles, etc.), medical equipment (cholera kits, surgical kits, etc.) and storage kits (mobile depot, etc.).

The United Nations Humanitarian Response Depot (UNHRD) network (UNHRD, 2017) was established in 2000 and is managed by the United Nations World Food Programme (WFP). It currently manages around 400 different types of emergency items, ranging from storage tents to ablution units and from refrigerated medicines to ready-to-eat food. The most commonly stored items are prefabricated office and/or accommodation units, tents, storage units, medical supplies and blankets (UNHRD, 2019).

The TLI White Paper (2016: 4) elaborates on the matter and suggests that pre-positioning critical relief supplies in strategic locations can be an effective strategy to improve the capacities in delivering sufficient relief aid within a relatively short lead time, including improvement of logistics infrastructure and processes.

The White Paper mentions that these emergency response facilities will be capable of providing further supporting needful services such as handling and consolidation of humanitarian cargo for distribution of relief goods in the disaster affected areas. The main functional benefits of an established network of distribution centres (DCs) include, according to the research:

- Improvement of capacity of governments and all humanitarian actors to respond to emergencies in a timely and cost-effective manner
- Enablement of timely and coordinated receipt and dispatch of relief assistance via air, sea and surface transport
- Improvement of the immediate availability of relief items, eliminating else needed long lead times for the mobilization of resources, and minimising potential risk of supply disruptions, increasing the overall resilience of the disaster relief supply chain
- Enhancement of capacity building to support operations of repackaging
- Establishment of practical training venues for logistics stakeholders and emergency responders
- Reduction of operational costs.

UNHRD network

The UNHRD network consists of six strategically-located depots that procure, store and transport emergency supplies on behalf of the humanitarian community. The network focuses on emergency preparedness and response and enables the strategic stockpiling of relief items and equipment for its 85 partners including UN agencies, governmental and non-governmental organizations.

The depots provide a full range of comprehensive supply-chain solutions and are located in Italy, United Arab Emirates, Ghana, Panama, Malaysia and Spain, as indicated in Figure 7.9.

By pre-positioning relief items at these locations, UN agencies, governmental and non-governmental organizations can respond

FIGURE 7.9 Location of UNHRD facilities (UNHRD, 2017)

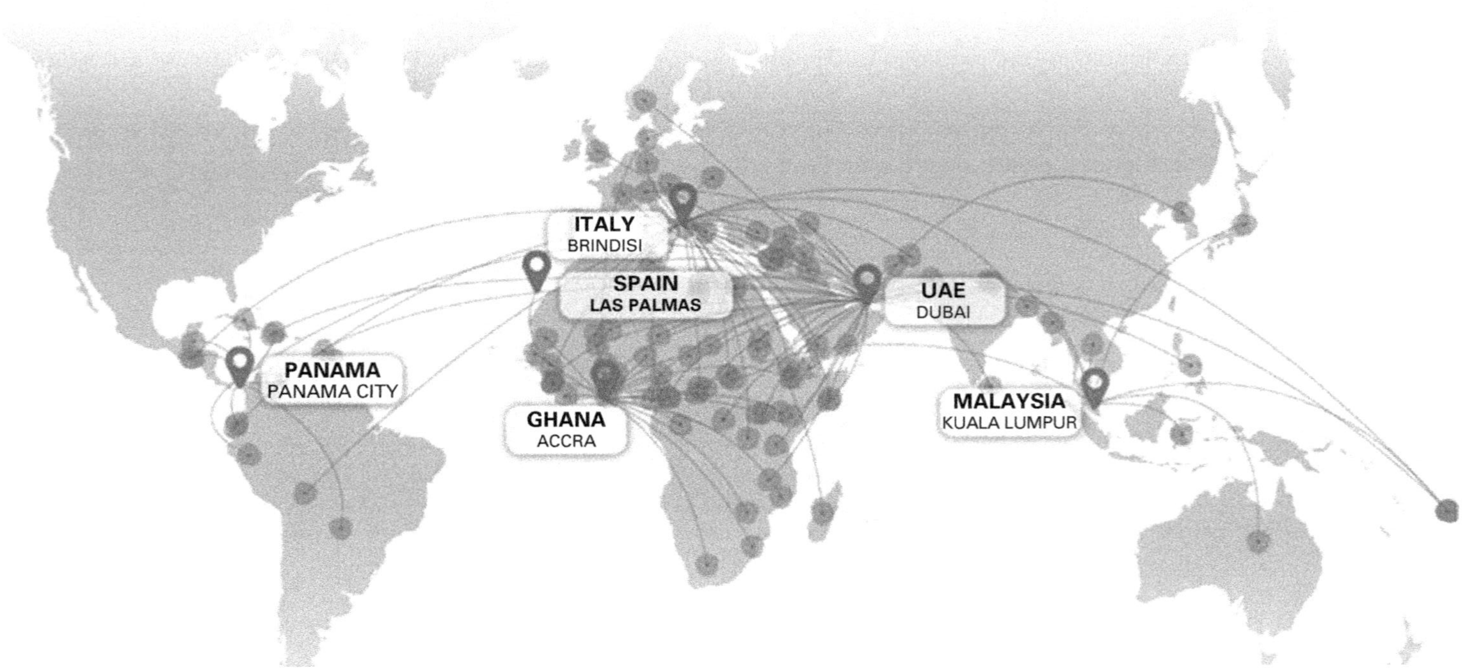

faster and more efficiently to people in need. The mandate of the UNHRD network includes (UNHRD 2009: Slide 3):

- 24/48-hour emergency response
- Support of WFP in meeting its corporate goal of being prepared to respond to four large scale emergencies at any given time
- Support of the emergency response efforts of UN, international, governmental and non-governmental organizations.

ESUPS project

In addition to the global UNHRD network, there are also numerous smaller scale initiatives that contribute to coordination in pre-positioning of relief items. One such example is the Emergency Supply Pre-positioning Strategy (ESUPS, 2022), hosted by Welthungerhilfe in Germany. They work to promote coordinated pre-positioning of emergency supplies and collective logistics strategies, with the goal to reduce gaps and overlaps in emergency preparedness and improve coordination among humanitarian actors. They are currently working in Nepal, Philippines, Indonesia, Vietnam, Madagascar, Colombia and Honduras and ready to start work in Cambodia, Laos and Bangladesh.

They have developed a very useful platform called STOCKHOLM (STOCK of Humanitarian Organisations Logistics Mapping). It offers stock holders access to a visual representation of pre-positioned core relief items. Users are able to filter stock data based on country, province, or any other cross-geographical area; stock data will also be filtered by organization holding the stock, cluster and item. In parallel, ESUPS and its academic partners are carrying out studies to demonstrate the value of collaborative pre-positioning practices.

Centre-of-gravity analysis technique[1]

The design of a suitable logistics network and the development of appropriate channel strategies in multiple markets and multiple suppliers are daunting tasks and some modelling is needed to assist with the complex calculations. Coyle et al. (2008: 542–549) calls centre-of-gravity analysis the 'grid technique' and suggest the technique to be a useful, simplistic heuristic modelling approach for determining the least-cost facility location. This modelling can easily be done by spreadsheets and does not require dedicated optimization software.

Centre-of-gravity analysis is used to determine the least-cost location of a fixed facility (distribution centre, warehouse, depot or terminal) for moving raw materials, finished goods or relief items in the network. The technique assumes sources of raw materials, markets and disaster areas are fixed and that supply and demand are known. The coordinates of each supply and demand point are needed to present the points in geographical format on a map and to calculate the transport costs of all movements.

The technique is shown graphically in Figure 7.10. If all the points are plotted at their respective coordinates on a map and respective supply or demand volumes (represented by weights) are connected through small holes to a central ring on the surface, the ring will come to rest in the centre of gravity (Schoeman, 2017: 40).

The analysis is based on the differential transport costs of raw materials or relief items from the manufacturing facilities or suppliers to one or more fixed facilities, such as distribution centres (primary transport), and finished goods from these facilities directly to destinations in the market, such as retail outlets or used in the field (secondary transport). The result of the analysis provides a theoretical indication of the least cost location of the centralised facility.

Bowersox and Closs (1996: 554–561) provide four analytical techniques for calculating the centre or gravity:

- Ton-centre solution (weight-centre solution)
- Mile-centre solution (distance-centre solution)
- Ton-mile-centre solution (weight-distance-centre solution)
- Time-ton-mile-centre solution (time-weight-distance-centre solution).

FIGURE 7.10 Centre-of-gravity technique

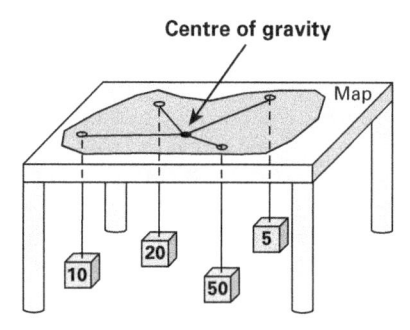

SOURCE Schoeman, 2017: 40

The ton-centre solution (weight-centre solution) is the basic centre-of-gravity calculation and the centre of movement represents the least cost location. The mathematical formula is presented in Equation 2.

$$x = \frac{\sum_{i=1}^{n} x_i F_i}{\sum_{i=1}^{n} F_i}, \quad y = \frac{\sum_{i=1}^{n} y_i F_i}{\sum_{i=1}^{n} F_i} \quad (2)$$

Where:

x, y = Coordinates for the centre of gravity

x_i, y_i = Coordinates for all origins and destinations

F_i = Supply or demand (ton)

The mile-centre solution (distance-centre solution) determines the geographical point that minimizes the combined distance to all points. The ton-mile and time-ton-mile-centre solutions reflect reality most accurately and should be used where possible. They can incorporate a number of variables that need to be considered, such as transport rates ($/ton.km), which are different for primary and secondary transport. Similarly, a differential parameter for congestion in urban areas that reflects the additional time required can be used. This parameter could range from high value for high congestion to a low factor for low congestion. The mathematical formula of a hybrid between the two solutions (adapted from Bowersox and Closs (1996: 554–561)) is shown in Equation 3.

$$x_k = \frac{\sum_{i=1}^{n} x_i F_i R_i C_i / d_i}{\sum_{i=1}^{n} F_i R_i C_i / d_i}, \quad y_k = \frac{\sum_{i=1}^{n} y_i F_i R_i C_i / d_i}{\sum_{i=1}^{n} F_i R_i C_i / d_i} \qquad (3)$$

Where:

x_k, y_k = Coordinates for iteration k (in km)

x_i, y_i = Coordinates for all origins and destinations (in km)

F_i = Supply or demand (ton)

R_i = Transport rate ($/ton.km)

C_i = Congestion factor (relative value)

d_i = Distance between points for each iteration

This calculation requires an iterative process to determine the increasingly improved location, based on the lowest cost. The distance between the locations in each iteration reduces every time a better location is found and the objective is to find the spot where the costs are the lowest. The measurement of distance between two points is done by the formula in Equation 4.

$$d_i = \sqrt{(x_i - x_k)^2 + (y_i - y_k)^2} \qquad (4)$$

Where:

x_k, y_k = Coordinates for iteration k (in km)

x_i, y_i = Coordinates for all origins and destinations (in km)

d_i = Distance between points for each iteration

In summary, centre-of-gravity analysis is based on the differential transport costs from the supply points to one or more distribution centres or terminals (primary transport), and from the distribution centres or terminals directly to the destinations (secondary transport). The result of the analysis provides a theoretical indication of the least cost location of a distribution centre, terminal or pre-positioning facility. Qualitative factors to be considered to accommodate the complexities of the real world include road accessibility, rail accessibility, environmental and geotechnical conditions, land-use and spatial development guidelines. Centre of gravity does not provide the final answer but it provides a scientifically calculated indication of where to start looking for a possible location. Calculations are done for the base year volumes of freight as well as for the anticipated future flows to identify possible trends which might influence the choice of location.

Centre-of-gravity analysis application

It was mentioned earlier that the Emergency Events Database (EM-DAT-b, 2017) contains useful information on various disasters that happened over many years in the world. This section provides a practical application of the centre-of-gravity analysis technique and it was decided to focus on the natural disaster events from this database in Africa from January 2014 to June 2017. These include the following categories of events:

- Geophysical: A hazard originating from solid earth. This term is used interchangeably with the term 'geological hazard'.
- Meteorological: A hazard caused by short-lived, micro- to meso-scale extreme weather and atmospheric conditions that last from minutes to days.
- Hydrological: A hazard caused by the occurrence, movement and distribution of surface and subsurface freshwater and saltwater.
- Climatological: A hazard caused by the long-lived, meso- to macro-scale atmospheric processes ranging from intra-seasonal to multi-decade climate variability.
- Biological: A hazard caused by the exposure to living organisms and their toxic substances (e.g. venom, mould) or vector-borne diseases that they may carry. Examples are venomous wildlife and insects, poisonous plants and mosquitos carrying disease-causing agents such as parasites, bacteria or viruses (e.g. malaria).

- Extra-terrestrial: A hazard caused by asteroids, meteoroids and comets as they pass near-earth, enter the Earth's atmosphere and/or strike the Earth, and by changes in interplanetary conditions that effect the Earth's magnetosphere, ionosphere and thermosphere.

Table 7.2 indicates an extract of 10 of the 43 countries in Africa that experienced some form of natural disaster in the period under consideration.

The coordinates of each location were used to provide the locations for plotting the number of natural disaster events on a map of Africa, as indicated in Figure 7.11. The legend indicates four categories as well as the centre-of-gravity of the mentioned disaster events. The result represents a centralized centre of gravity for all disaster events and is located just west of Kisangani in the DRC. The location is not ideal as accessibility in this remote area is a huge challenge.

However, the locations of all events, as indicated in Figure 7.11, suggest that there could be merit in splitting the continent into three regions or clusters as follows: West, South and East Africa. Such a regional or decentralized network will improve accessibility and

TABLE 7.2 CRED Disaster Centre-of-Gravity Analysis – Africa (Extract)

No	Country name	No of events	Location	Region	Lon°	Lat°
1	Benin	1	Porto Nova	West	2.63	6.50
2	Réunion	1	Saint-Denis	South	55.45	−20.89
3	Togo	1	Lomé	West	1.23	6.17
4	Zambia	1	Lusaka	South	28.32	−15.39
5	Botswana	2	Gaborone	South	25.92	−24.63
39	Kenya	9	Nairobi	East	36.82	−1.29
40	Somalia	10	Mogadishu	East	45.31	2.05
41	Niger	11	Niamey	West	2.12	13.51
42	Tanzania	11	Dodoma	East	35.75	−6.16
43	DRC	12	Kisangani	East	25.20	0.52
	Centre-of-Gravity				21.00	−0.12
	Total number of events	200				

FIGURE 7.11 Centre-of-gravity analysis of all natural disaster events in Africa (January 2014 to June 2017)

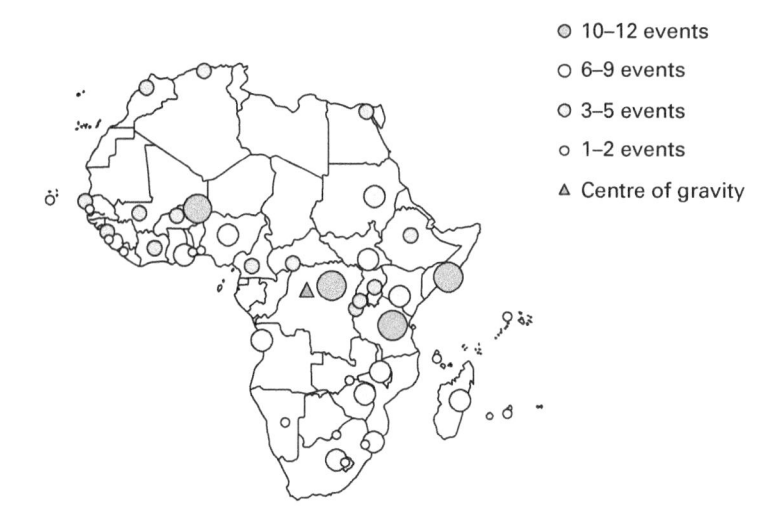

mobility significantly and result in a more responsive network of pre-positioning facilities.

The decentralized network is shown in Figure 7.12.

The centres of gravity for the three regions are indicated in Figure 7.13.

The centre of gravity for West Africa is located at Ouagadougou, the capital of Burkina Faso. The centre of gravity for the South African region is at Bulawayo in Zimbabwe and the third centre of gravity, for East Africa, is located at Nakuru in Kenya.

If the two outlying countries of Morocco (Rabat) and Algeria (Algiers) are ignored, the centre of gravity for West Africa moves closed to Accra in Ghana, where UNHRD currently has a depot. If the outlying countries of Angola (Luanda) and Réunion (Saint-Denis) are ignored, the centre of gravity for the south would probably stay close to Bulawayo, although it might make sense to move it slightly south to Johannesburg in South Africa, due to better global accessibility. If the outlying country of Egypt (Cairo) is ignored, the centre of gravity for East Africa would move closer to Nairobi.

It is important to mention again that the centres of gravity do not necessarily provide the answer to where the facilities should be

FIGURE 7.12 Potential regional clusters of natural disasters in Africa (January 2014 to June 2017)

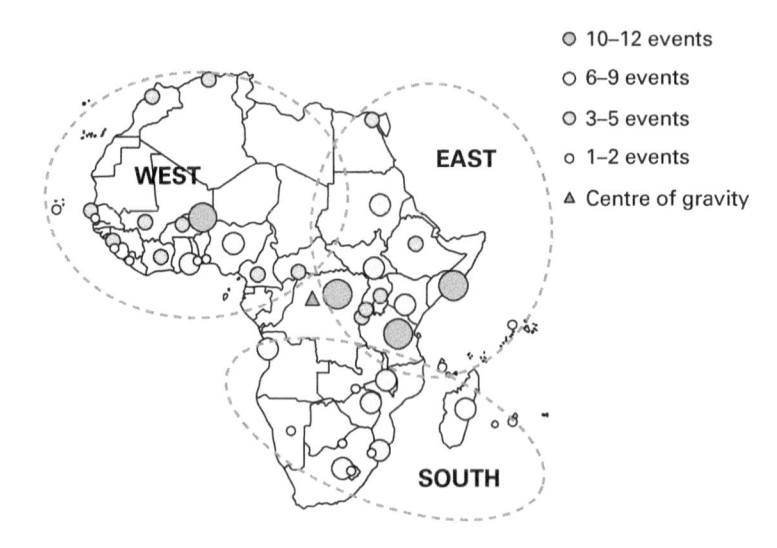

FIGURE 7.13 Centre-of-gravity analysis for potential regional clusters of natural disasters in Africa

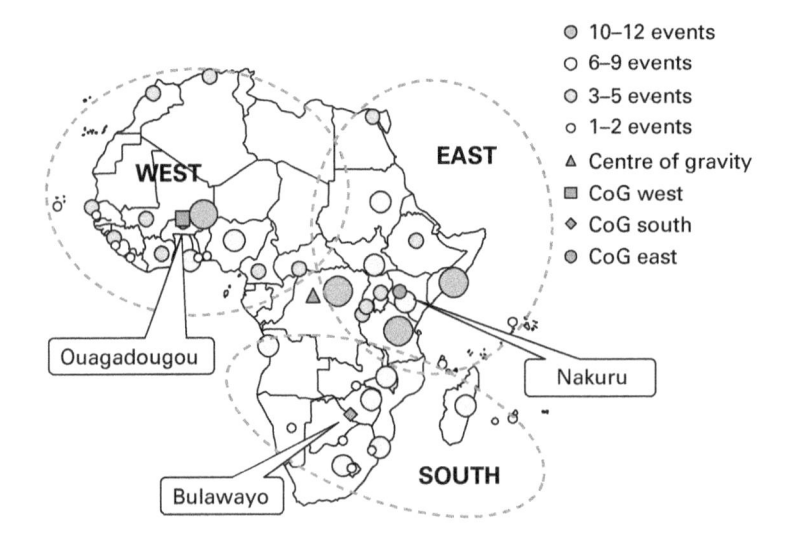

located, but rather the starting point on macro level from where the best location could be found. The calculations are based on optimization of transport costs, and it should be complemented with qualitative indicators before final decisions are made.

The results suggest that the three regions should have pre-positioning facilities located at:

- West: Accra (Ghana)
- South: Johannesburg (South Africa)
- East: Nairobi (Kenya)

This might not be a surprise to those active in emergency response in Africa, although it provides at least comfort that the UNHRD depot at Accra is at the correct location. What is maybe more important, is that if similar facilities could be established at the centres of gravity for the southern and eastern regions, emergency response could be done much more efficiently.

Once the locations have been determined on macro level, further analyses will be required on micro level to ensure that qualitative factors, as mentioned earlier, are taken into account.

Channel strategy[2]

Channel strategy refers to the configuration of supply chains to provide for the required service levels at acceptable costs and the purpose of this last section is to indicate the close relationship between network design and channel strategy.

Although the main focus of this chapter is pre-positioning for emergency response, it is important to mention that humanitarian logistics does not only deal with disasters. Different supply chains exist for different services, such as emergency relief or response, food distribution, distribution of gifts-in-kind and development projects. Long term droughts that might require food and medical supplies can be managed with predictable and known demand. Little safety stock is needed and inventory demand is mostly dependent with sufficient time to negotiate cost-effective logistics services. This is not the case when unpredictable earthquakes happen and independent demand requires sufficient safety stock to be carried in suitable pre-positioning locations.

Gattorna (2010: 51–56) has done much work in the development of supply chain frameworks and he suggests that there are typically four generic supply chains (or strategies):

- *Continuous replenishment supply chain*: Very predictable demand from known customers; easily managed through tight collaboration with these collaborative customers; focus on retention of customer relationships
- *Lean supply chain*: Regular pattern of demand; quite predictable and forecastable although may be seasonable; tend to be mature low risk products/services; focus on efficiency
- *Agile supply chain*: Usually unplanned at least until the last possible moment. May result from promotions; new product launches; fashion marketing; unplanned stock-outs; or unforeseen opportunities. Focus on the service-cost equation
- *Fully flexible supply chain*: Unplanned and unplannable demand due to unknown customers with exceptional, sometimes emergency, requests. Focus on providing creative solutions at a premium price.

This last type is further split between a business event strategy in an entrepreneurial environment and a humanitarian response strategy in an emergency environment. Gattorna (2010: 251–260) recommends that the emergency response or humanitarian supply chain should not only be agile, but fully flexible. In disaster situations, there is usually an initial event (or series of events) that dictates the requirement for a fully flexible supply chain, although it changes from the critical response phase into the ongoing rebuilding phase.

The immediate aim is to quickly provide life-saving essentials to the survivors, who often have no choice of buyer behaviour, but rather 'whatever is provided'. In the next stage, when basic living requirements are restored, and in the rebuilding phase, survivors will exhibit a greater range of buyer behaviours as the situation permits. This situation occurs when an entire complex of supply chains need to be created from scratch because of a major disruption to normal living and business due to situations such as war, terrorist attacks, famine and natural disasters such as earthquakes, tsunamis or cyclones.

TABLE 7.3 Matching humanitarian situations with generic supply chains

Humanitarian Situation	Generic Supply Chain
Emergency relief or response	Fully flexible supply chain
Food distribution	Lean supply chain
Distribution of gifts-in-kind	Continuous replenishment supply chain
Development projects	Agile supply chain

SOURCE Niemann and de Villiers, 2022: 424

Niemann and de Villiers (2022: 423–424) managed to match the four generic supply chains of Gattorna with typical supply chains in humanitarian situations, as indicated in Table 7.3.

Emergency relief or response clearly requires a fully flexible supply chain while food distribution needs a lean supply chain, because of known demand from historic take-off. Gifts-in-kind usually provides for predictable demand due to long term relationships and regular supply of school books or medicines, for example, while development projects often necessitate agility to be able to respond to unplanned or unforeseen demand.

Finally, management of the strategies requires different approaches to activities and challenges in the supply chains in general and network design in particular. Gattorna (2010: 79-86) provides a comparative analysis of the different strategies and his opinion on the differences in channels of distribution (or network design), is indicated in Table 7.4.

The generic supply chain strategies and associated logistics networks all contribute to the objective of developing and implementing a supply chain that will be able to compete effectively on cost, service levels and whatever other performance indicators require. The solution is not one-size-fits-all but a process to develop a customized supply chain with support of modelling techniques such as centre-of-gravity analysis.

TABLE 7.4 Supply chain strategies and network design

Supply Chain Strategy	Channels of Distribution (Network Design)
Continuous replenishment supply chain	Either direct or via trusted outlets
Lean supply chain	Wide distribution through multiple channels for maximum accessibility
Agile supply chain	Fewer, more direct channels to access consumers
Fully flexible supply chain – Business event	Limited and very targeted
Fully flexible supply chain – Humanitarian	As many as needed in a given situation

SOURCE Gattorna, 2010: 79–86

Summary and conclusion

The purpose of this chapter was to discuss network design in pre-positioning emergency relief items and in particular the centre-of-gravity analysis technique. It explained the usefulness of the technique in determining the locations for pre-positioning of emergency relief items. The application of this technique in humanitarian logistics is indeed similar to the application in commercial supply chains and it promises significant improvement in the efficiency of emergency relief response.

Much information is available in databases such as the EM-DAT Emergency Events Database and it is relatively easy to use the available information to start looking for the ideal location for the establishment of global, regional or local pre-positioning facilities.

Notes

1 This section is based on the contribution of the author to Chapter 2 in *Transportation, Land Use and Integration*, edited by Schoeman, I.M., 2017, pp. 40–42.

2 This section is based on the contribution of the author to Chapter 2 in *Transportation, Land Use and Integration*, edited by Schoeman, I.M., 2017, pp. 45–47.

References

Bowersox, D J, and Closs, D J (1996) *Logistical Management: The integrated supply chain process*, McGraw-Hill, Singapore

Coyle, J J, Langley, C J, Gibson, B J, Novack, R A and Bardi, E J (2008) *Supply Chain Management: A logistics perspective*, South-Western Cengage Learning, Mason, USA

De Villiers, G (2008) Supply Chain Management in Humanitarian and Emergency Relief, *SAPICS 2008*, Sun City, South Africa

Duran, S, Gutierrez, M A and Keskinocak, P (2011) Pre-Positioning of Emergency Items Worldwide for CARE International, *Interfaces*, **41**(3), pp 223–237

EM-DAT, (2017a) Map of natural disasters, www.emdat.be/ (archived at https://perma.cc/27N4-6KT5)

EM-DAT, (2017b) The Emergency Events Database - Université catholique de Louvain (UCL) CRED, D Guha-Sapir http://www.emdat.be (archived at https://perma.cc/TS2X-46BR), Brussels, Belgium

EM-DAT, (2021) *Disaster in numbers*, Centre for Research on the Epidemiology of Disasters (CRED), Brussels, Belgium

ESUPS Website (2022) https://esups.org/ (archived at https://perma.cc/DLK3-XU54)

Gattorna, J L (2010) *Dynamic Supply Chains: Delivering value through people*, 2nd edn, Pearson Education Limited, Harlow, UK

Niemann, W and de Villiers, G (2022) *Strategic Logistics Management: A supply chain management approach*, 3rd edn, Van Schaik Publishers, Pretoria, South Africa

Roh, S and Kim, C (2016) Comment: Humanitarian relief logistics: Pre-positioning warehouse strategy, *KMI International Journal of Maritime Affairs and Fisheries*, **8**(2)

TLI White Paper (2016) *Decision Support Preparedness - Requirements for a Network of Emergency Response Facilities in Indonesia*, Volume 16-Jan-HL, National University of Singapore, Singapore

Schoeman, I M (2017) *Transportation, Land Use and Integration: Applications in developing countries*, WIT Press, Southampton, UK

Stock, J R and Lambert, D M (2001) *Strategic Logistics Management*, McGraw-Hill, New York

UNHRD (2009) *United Nations Humanitarian Response Depot (HRD) Network*, UNHRD Customer Service, Rome, Italy

UNHRD (2019) *Supply chain solutions for the humanitarian* community, November 2019, World Food Programme, Rome, Italy

UNHRD (2017) http://unhrd.org/page/about-us (archived at https://perma.cc/9UNU-WZHU)

UNICEF (2021) COVID-19 Impact Assessment on Global Logistics and Supplies, UNICEF Supply Division, September 2021, Copenhagen, Denmark

World Food Programme (2022) *Ukraine Logistics Cluster - Concept of Operations*, Logistics Cluster, Rome, Italy.

World Vision (2010) *HEA Ministry Guiding Philosophy: Principles and Values for Humanitarian Action*, Los Angeles, USA

08

Competing for scarce resources during emergencies: A system dynamics perspective

PAULO GONÇALVES

ABSTRACT

This chapter dissects the complex challenges associated with the competition for scarce resources during humanitarian response to emergencies through a general dynamic complexity lens, adopting system dynamics as methodology to gain insight and depth of understanding of such complex dynamic behaviour. The chapter trails a journey of discovery that characterizes first the intrinsic challenges associated with humanitarian response during emergencies. It proceeds then to present the building blocks of the system dynamics methodology as it applies to complex humanitarian challenges. It finalizes with a mapping of the interplay of those dynamic components in the competition for scarce resources problem. Through this journey, the chapter provides dynamic insights and a system dynamics lens to view complex challenges. In doing so, system dynamics views complex humanitarian challenges not as unpredictable, ambiguous, challenging and invisible, but instead system dynamics sees

> them as evolving, dynamic, adaptive, self-organizing and governed by feedback. Such aspects do not mean that system dynamics can make precise predictions about the future, but it allows us to widen our horizons, and see many more policy alternatives and possibilities before us than previously imaginable, providing a depth of understanding on the behaviour of humanitarian response efforts.

Introduction

Consider the following examples of humanitarian efforts following major disasters: the 1994 Rwandan refugee crisis, the 2010 Haiti earthquake and the 2021 Covid-19 vaccine inequity failure.

The 1994 Rwandan refugee crisis provides a first example of a humanitarian response where the competition among humanitarian organizations (HOs) compromised humanitarian aid. The human suffering associated with the crisis attracted a significant quantity of funds, increasing the pressure for HOs to act and secure profitable contracts (Terry, 2002). With the availability of funds and concerns that failure to respond to the ongoing crisis could compromise future contracts, hundreds of HOs arrived at the theatre. The competitive nature of contracts directed HOs' attention to donors, instead of beneficiaries. Programme evaluations, when done, were incomplete or biased, emphasizing positive aspects and downplaying negative ones (Bare, 2017). HOs competed for current donor funding, media attention and the possibility of future funds. The large number of HOs and competition among them prevented proper coordination, affecting accountability and information-sharing (Eriksson et al., 1996).

The 7.0 magnitude earthquake that struck Port-au-Prince, Haiti's capital, on 12 January, 2010, caused over 230,000 deaths (Britannica, 2022), affected three million people, left one million homeless, and destroyed thousands of official buildings (e.g. government and HO offices) and critical infrastructure (e.g. roads, the only airport, the port, hospitals, water and electricity systems, etc.). The earthquake also completely leveled Léogâne and heavily affected Jacmel. The weeks that followed saw the influx of hundreds of humanitarian organizations to support humanitarian efforts in Haiti. Due to its high population density, concentrated need and media presence, Port-au-Prince became the centre of attention for the humanitarian response. Easy access to a large pool of beneficiaries, availability of local products, global visibility and better logistics infrastructure made Port-au-Prince an attractive location for HOs to deliver aid from operational and cost perspectives (BBC, 2010). As the number of HOs in Port-au-Prince increased, so did the number of conflicts among them. In one case, UN peacekeepers had to intervene in a dispute between two medical HOs that claimed 'ownership' of a hospital. Within weeks, HOs at Port-au-Prince competed for all types of resources: transport (e.g. helicopters, trucks), qualified personnel (e.g. translators), local infrastructure (e.g. accommodation) and data (e.g. on affected population and beneficiaries) (Oloruntoba and Gray, 2006). At the same time, beneficiaries in 'less attractive' affected areas, such as Jacmel and Léogâne, did not receive enough aid (Bhattacharjee and Lossio, 2011).

The Covid-19 pandemic generated devastating global disruptions, leaving almost six million dead (by end of February 2022), driving millions into poverty, significantly increasing food insecurity and reducing years of progress on the fight against malaria. Early vaccines developed by the summer of 2020 brought much hope to control the spread of the disease. With support from COVAX, a coalition supporting the equitable access to Covid-19 tools (e.g. tests, treatments and vaccines), low-income countries aspired to pool their demands and funds to ensure adequate quantities of vaccines at affordable prices (GAVI, 2021). However, countries outside COVAX with greater financial resources secured large quantities of the vaccines by purchasing directly from manufacturers. On April 2021, of the first 700 million Covid-19 doses administered, 87 per cent took place in high- and upper middle- income countries, with only 0.2 per cent administered in low-income countries (WHO, 2021). With over 10 billion Covid-19 vaccine doses administered globally as of February 2022, only 13 per cent of the population in low-income countries (LIC) have been vaccinated, in contrast to 69 per cent of the population in high-income countries (HIC) (UNDP, 2022). The COVAX aspiration 'to ensure an equitable distribution of vaccines regardless of countries' income level... failed' (Kim, 2021). Competition for the scarce supplies of vaccines available early on disrupted the COVAX ambitions, contributing to the emergence of new variants and undermining global economic recovery (Kimball, 2022).

Competition for scarce resources among humanitarian organizations is typical during the early phases of an emergency, when media attention and donor funds are abundant (Stephenson and Schnitzer, 2006; Lindenberg, 2001). Countries' competition for scarce vaccine supplies (e.g. vaccine hoarding) took place in the early stages of vaccine production, when need (demand) was high and availability (supply) low. In both cases, competition for scarce resources led to negative and undesirable effects, such as failure to coordinate (e.g. HOs and countries acted independently and competitively), duplication of efforts (e.g. parallel needs assessments, parallel price-quantity negotiations with manufacturers), inefficiencies (e.g. inadequate allocation of resources), all of which reduced overall performance (e.g. beneficiary well-being, global disease diffusion) (Kent, 1987; Stephenson, 2005).

The examples focusing on humanitarian response during emergencies suggest that they face several complex challenges that span multiple domains (e.g. political, financial, infrastructural, human, etc.), requiring a diverse set of management skills (i.e., ethical and fair resource allocation, ability to assess changing supply and demand conditions, etc.). This chapter seeks first to characterize the intrinsic challenges faced by humanitarian response during emergencies; second to present system dynamics as a methodology capable of modelling such complex challenges; and third to assess the competition for scarce resources during emergencies from a system dynamics lens, to shed light and understanding on the behaviour of humanitarian response efforts.

Challenging aspects of humanitarian response to emergencies

Due to the disaster, operating conditions may be inadequate or hostile. Relief effort takes place in theatres with impaired or lacking basic infrastructure (e.g. roads, railroads, airports, ports). Political unrest and turmoil in affected countries may limit or severely delay the ability of agencies to provide relief. Demand for relief is highly uncertain and changes over time. The need to respond to the disaster requires rapid assembly of teams and allows little time to prepare and structure processes. Humanitarian organizations must evaluate supply conditions on site, requiring them to design and implement their supply chains hastily in short periods and, at times, completely from scratch. HOs resort to rapid on-site assessment to estimate the numbers of beneficiaries and their needs. HOs update those assessments only rarely, if at all, leading to relief efforts driven by inadequate or outdated estimates. With lives at stake and limited time horizons to react to catastrophes, humanitarian professionals face extremely high pressure to perform and compressed project life cycles (Besiou and Van Wassenhove, 2020). High pressure, intense work hours and difficult conditions lead to high turnover among field staff. Field staff turnover limits institutional learning from past emergencies. Attention to short-term pressures and needs in the field lead to a reactive culture that prevents effective long-term organizational planning and capacity building; field and headquarter staff often lack adequate training (Gonçalves, 2011). HOs often manage multiple stakeholders with widely different and sometimes conflicting objectives (Starr and Van Wassenhove, 2014). Unsolicited donations exacerbate the challenges faced in the field often creating bottlenecks in different parts of the supply chain straining use of resources and diverting them from high priority items (Buzogany et al. 2018). Table 8.1 provides a summary of different challenges.

While the challenges discussed above span multiple areas, they also interact with each other, influencing and often exacerbating their effects. Tomasini and van Wassenhove (2004) caution that the complex challenges in humanitarian response can worsen as 'complexity increases and… interact[s] with other vulnerability factors demanding

TABLE 8.1 Challenging factors in humanitarian response

- Operating conditions after a disaster strikes are poor and unpredictable.
- Relief and response efforts take place with inadequate infrastructure as ports, roads, railroads, airports, may not be available.
- Conditions may not allow adequate implementation of information systems and structured logistics processes.
- Political unrest and turmoil in affected countries may limit or severely delay the ability of humanitarian organizations to provide relief.
- Demand for relief is highly uncertain.
- Needs assessment is done on site with little time for preparedness and adjustments.
- Scarce resources and priority to relief means assessment is done quickly, and frequently not updated, inappropriately assessed needs often drive the relief effort.
- Supply planning is typically reactive and involves significant guesswork.
- Supply conditions must be evaluated on site, requiring the whole supply chain to often be designed and implemented in short periods.
- Humanitarian relief operations experience extreme time pressure.
- Pressure to resolve current short-term crisis often precludes attention to long-term capacity building.
- Humanitarian staff often lack adequate training and skills, and field stress leads to high turnover.
- Donors often fund specific relief efforts but not long-term infrastructure building in relief organizations.
- Humanitarian organizations often respond to multiple emergencies at the same time and must manage multiple stakeholders with different and sometimes conflicting goals.
- Unsolicited donations create unpredictable logistics bottlenecks and strain use of limited resources.

NOTE Further details on challenging factors in humanitarian relief and recovery operations can be found in Chomilier et al., 2003; Fenton, 2003; Gonçalves, 2011; Gustavsson 2003; Kaatrud et al., 2003; Kovács and Spens, 2007; Long and Wood, 1995; Tatham and Houghton, 2011; Thomas, 2003; Thomas and Kopczak, 2005; Tomasini and van Wassenhove, 2004; and van Wassenhove, 2006.

a response that challenges the traditional response mechanisms and capabilities'. For instance, China's 2008 Sichuan earthquake caused landslides to form a new lake above the city of Beichuan. Fears that the Tangjianshan quake lake could burst its banks required the additional evacuation of 150,000 people. The evacuation demanded attention beyond and in addition to operations directly addressing earthquake victims (NPR, 2008). As the challenges interact and feed on each other, they intensify the challenges of managing disaster response as the examples below exemplify:

- Rapid early-on needs assessment may serve as the basis for flash appeals for funds. However, when those needs assessments are

flawed, funding requests may be inadequate and may not suffice to meet beneficiary needs.
- Inadequate needs assessment of beneficiary demand may lead to inadequate supply plans as well as unsuitable supply chain designs, insufficient supply rates and procurement of inappropriate items.
- Unsolicited donations requiring processing, sorting, packaging and storing of non-priority items divert human resources from distributing high priority items and provision of relief, reducing overall performance of the humanitarian response, increasing stress levels of relief workers and intensifying donor oversight.
- Pressure to focus on short-term crises at the expense of long-term issues may create a reactive culture where impulsive actions dominate at the expense of proper planning into required information systems, supply and operations plans, staffing needs, etc., all of which reduce organizational capability.

Naturally, it is important for humanitarian practitioners to understand and characterize the complexity that typically exists in humanitarian relief efforts. Understanding existing challenges, possible interactions and their impacts can provide clues on how to address them. But how can humanitarian practitioners manage complex challenges that span multiple areas and not only interact but also intensify each other? We try to answer this question by first focusing on the characteristics of complex systems, their key components and common types of behaviours, and then considering possible managerial tools to support decision-making in such settings.

Characteristics of complex systems

A characteristic of complex systems is their high interactivity. Early on, the strength of such relationships is small. Unless practitioners are aware of their possibility, they may not be visible and can go completely unnoticed. Because the small cause-and-effect relationships are not well understood and distributed across the system, they are also not predictable, leading to high ambiguity. With the compounding effect of other interactions, these relationships incrementally grow over time, escalating their impacts, eventually even coming to dominate the overall dynamics. As humanitarian practitioners focus on the most severe and urgent aspects of a crisis, they often leave other small aspects unchecked, allowing them to interact and grow over time, until they lead to further consequences. This typical focus on urgent and pressing matters, and inability to anticipate which other dispersed aspects may become important generate the emergence of new phenomena, characterizing the unpredictability and challenging nature of complex systems. Adapting the work of Richardson (1994) on disaster management, van Wassenhove (2006), provides a broad characterization of the challenges, described above, associated with anticipating dynamic behaviour of complex systems during disasters (see Table 8.2).

The characterization allows an initial understanding of the different aspects responsible for generating the dynamic complexity in humanitarian response and provides important clues for why such challenges persist. However, the characterization does not provide a playbook to help humanitarian practitioners understand what actions they should take to better manage complex crises. For instance, it does not provide a framework to help them anticipate which aspects may become important, or a set of policies to help them address the emergence of new phenomena and the unpredictability and challenges of complex systems.

System Dynamics provides a useful framework to deal with the dynamic complexity arising from the subtle and delayed cause-and-effect interactions dispersed throughout the system over time (Gonçalves, 2008). Incremental change arises in dynamically complex systems because changes in key system variables (e.g. stocks) take place over time, changing constantly through processes of accumulation. Because such systems are tightly coupled, variables in one part of the system feedback and interact with variables in other parts of these systems, making overall behaviour highly ambiguous and often unpredictable. Because causes and their effects are often not close together in time or space, relationships go unnoticed while they are small, but by the time the undesired effects become evident it may be too late to avoid them. Furthermore, nonlinearity among interactions exacerbates

TABLE 8.2 A disaster management perspective of complexity

Characteristic	Comments
Incremental change	Focus on most severe aspects of a crisis, often leave other factors unchecked, allowing them to interact, grow over time and lead to further consequences
Highly interactive	Interaction among factors accelerates rate at which disasters might escalate
Highly ambiguous	Cause and effect relationships are not clear and dispersed over time
Challenges from new phenomena	Challenges due to unknown effects and impact
Invisibility/ unpredictability	Inability to understand or anticipate which factors are important and may dominate the dynamics

the lack of predictability, with impacts being significantly higher than originating causes. Finally, because complex systems are self-organizing and adaptive, they resist changes and adapt to policy interventions.

> When discussing 'complexity' issues in humanitarian response, it is useful to distinguish between detail and dynamic complexity (Senge, 1990, Sterman, 2000). Detail, or combinatorial, complexity arises when decisions must account for large numbers of components or their possible combinations. For an airline to schedule its fleet of airplanes, it must allocate dozens of planes, across hundreds of routes, manned by thousands of crew members, etc. The large number of possible combinations create the combinatorial complexity to find the scheduling plan that minimizes overall scheduling cost. While detail complexity is a common part of humanitarian relief operations, here we focus on dynamic complexity, which we deem more important.

Because dynamic complexity is constantly changing, tightly coupled, highly adaptive and nonlinear, it is important that decision makers use frameworks, or mental models, that are also dynamic, encompassing and broad to be able to capture their dynamic nature. However,

TABLE 8.3 A dynamic perspective of complexity

Characteristic	Comments
Constantly changing/past-dependent	Changes in variables over time characterize the current state of the system. Past history influences available futures
Tightly coupled/governed by feedback	Variables in the system interact strongly with one another, until they feedback to influence themselves
Self-organizing	The dynamic behaviour of the system arises from the structural interactions in the system
Adaptive/emergent behaviour	Systems are often resistant to changes and adaptive to policies
Nonlinear	Effect is not proportional to cause. Local effects do not apply globally

decision makers tend to adopt 'mental models that are static, narrow, and reductionist' (Sterman, 2001), leading to what is commonly known as misperceptions of feedback (Bendoly et al., 2011). Due to the mismatch between decision makers' mental models and the dynamic nature of complex systems, dynamic complexity poses a far bigger challenge to tackle. In dynamically complex settings, decision makers must account for the nuanced and delayed interactions among parts of the system that are not closely connected causally or temporally. Table 8.3 provides a different characterization of the challenges associated with anticipating the behaviour of complex systems, from the perspective of dynamic complexity (based on Forrester, 1961; Richardson and Pugh, 1981; Sterman, 2000, 2001).

Behaviour in complex systems

Unnoticed and nonlinear cause-and-effect relationships of self-organizing and adaptive systems generate ambiguous, unpredictable, counterintuitive and emergent behaviours. Several system dynamics researchers (Forrester, 1971; Sterman, 2000, 2001) report on different aspects of complex systems. Typically, complex systems behave in unexpected ways, but present some common aspects, such as: unanticipated side-effects, trade-offs between subsystems and the broader system, trade-offs between short-term versus long-term and insensitivity to policy interventions.

In addition to Gonçalves (2008), several Operations Management scholars have discussed the important role that System Dynamics can play in shedding light on the behaviour of complex systems, including considerations of the distribution of therapeutic food items during humanitarian response under different preparedness scenarios (Besiou et al., 2011; Kunz et al., 2014; De Vries and Van Wassenhove, 2020). The description below, following Gonçalves (2008) closely, considers each one of these behaviours in different humanitarian operations settings.

Unanticipated side-effects

It is common for managers to complain about unanticipated 'side-effects' in complex systems. However, as Sterman (2000) puts it: '[T]here is no such thing as side-effects'; there are only effects. There are actions and consequences of those actions. There are causes and their respective effects. However, because cause-and-effect can be far away from each other both in time and in space, managers may easily overlook their connection. 'Side-effects' typically account for unpredicted and often undesired effects resulting from corrective actions. For instance, when the Japanese government established free health clinics to provide aid to over 30,000 people injured in the 1995 Kobe earthquake, private hospitals experienced the 'side-effect' of delayed financial recovery and near-bankruptcy. In another example, the competition for scarce resources during emergency response leads to 'side-effects' such as duplication of efforts, inadequate allocation of resources and lack of operational effectiveness.

Trade-offs between subsystems and the broader system

Another typical behaviour of complex systems includes the trade-offs between subsystems and the overall system. Forrester (1971) states that 'a conflict exists between the goals of a subsystem and welfare of the broader system'. Acting to optimize the subsystem leads to a suboptimal outcome for the broader system. Actors doing their 'best' (e.g. trying to optimize their actions) at a local level (e.g. a subsystem), frequently fail to ensure improved performance in the broader system. Since locally, actors do not observe, understand or control the broader system, their actions rarely translate into broader system optimization. For instance, as HOs concentrated their efforts in Port-au-Prince, they optimized their operational- and cost-effectiveness. At the same time, across Haiti, other affected areas, such as Jacmel and Léogâne, did not receive sufficient humanitarian relief. Furthermore, across the whole humanitarian response HOs' decisions led to duplication of efforts and overall inefficiencies. In another example, in 2007, CARE decided to permanently refuse US$45 million/year from Food Aid, recognizing that the practice of selling large quantities of subsidized farm products in African nations had been driving many local farmers out of business (Dugger, 2007). While Food Aid was beneficial to CARE and local recipients (e.g. subsystems), it drove farmers out of business and increased system wide fragility to future crop shortages (e.g. broader system).

Trade-offs between short-term and long-term

Forrester (1971) acknowledges that 'a policy that produces improvement in the short run is usually one that degrades a system in the long run', recognizing that here is an intrinsic trade-off in complex social systems 'between short-term and long-term consequences of a policy change'. In humanitarian relief, addressing the needs of affected people in the short-term makes local communities more susceptible to the longer-term threat and impact of disasters. Emergency relief often creates recipient dependency, preventing or limiting the ability of local economies to develop sustainably. After the Ethiopian Red Cross Society guaranteed food relief for Ethiopians that were completely destitute, people denied assistance on the grounds that those who were not too poor promised to sell their meagre possessions to qualify

(Keller, 1992). By doing so, they benefited in the short-term of relief aid, but were more vulnerable and exposed in the long-term for letting go of their possessions.

Insensitivity to policy interventions

The self-organizing and adaptive nature of complex systems allow them to adjust to changes and adapt to policy interventions. Forrester (1971) states that 'social systems seem to have a few sensitive influence points through which behaviour can be changed', also highlighting that 'these high-influence points are not where most people expect'. That is, complex systems are typically insensitive to policy interventions, with only a few hard-to-find high-leverage policies. Such high-leverage policies that can effectively influence a complex system are not the conventional ones typically implemented by managers. *Policy resistance* captures a complex system insensitivity to policy change.

Components of complex systems

The macro behaviours of complex systems described above (e.g. unanticipated side-effects, trade-off between short- and long-term, trade-off between subsystem and whole system and policy resistance) arise due to the complex interaction of individual micro-elements (e.g. stocks-and-flows, feedback, delays and nonlinearities) that compose them. Below, we focus on these micro-elements that create the challenging dynamics in complex social systems such as humanitarian response during emergencies.

Stocks and flows

Stocks, or accumulations, capture key attributes of a complex system. In humanitarian response, the number of beneficiaries in need, the number of HOs in the theatre and available supplies represent different types of key stocks in the system. Stocks also provide important sources of information for decisions. To decide how many items to procure, HOs must account for the number of beneficiaries in need and pre-positioned supplies. Stocks give systems inertia and memory. Many beneficiaries in need will require time before HOs can assist them. Flows describe how stocks change over time, through increases or decreases to their original values. Flows are activities that change the system, such as procurement and distribution of items. Flows are enabled, or constrained, by stocks. The stock of trucks enables, or constrains, the distribution of relief items.

Feedback

Complex systems are characterized by feedback. People, however, tend to hold an open-loop, event-oriented view of the world. We learn from an early age that for every action there is a reaction; for every cause an effect. When a child touches a hot surface, she learns that the burning (effect) is due to her touching a hot surface (cause). She also learns that cause and effect are close in time and space. We tend to generalize such lessons with simple systems, frequently searching for a possible cause that is 'close in both time and space to symptoms of the trouble' (Forrester, 1971). Experience with simple systems also suggests that the cause is often outside the control of (i.e. exogenous) the people impacted by the effect. The surface being hot is unrelated with the child's action to touch it. Exogenous perceptions of causes lead to an open loop view of the world, where our reactions (effects) do not shape future actions (causes). Complex systems, however, do not behave like people's open loop view of the world. In complex systems, today's actions often lead to new, unanticipated, reactions. Feedback processes link cause and effect, allowing today's effects to influence tomorrow's causes. Unlike simple systems, such connections may be far removed from each other either in time or space. In the Rwandan refugee crisis, it was the availability of funds (cause) that led hundreds of HOs to arrive at the theatre (effect). However, it was those HOs' focus on donors (effect) that generated increased the availability of funds (cause).

Delays

While in simple systems cause-and-effect may be close in time, delays are all too common in complex systems. Such delays separate causes from their effects, preventing people from recognizing the connection between them. Delays dramatically increase a system's dynamic complexity. In some systems, delays can filter out and attenuate unwanted variability. Because delays can aggravate the original cause in a disaster, humanitarian organizations try to respond as rapidly as possible. Delays in meeting beneficiary needs can exacerbate human suffering and lead to significant loss of life. In other systems, however, delays can cause or amplify instability and oscillation. Quickly committing to a course of action in disaster response may limit valuable alternative options as further understanding of the needs of the disaster are clarified. Committing to supplying goods by ship can be ineffective and block access to scarce supplies if the port infrastructure is inoperative.

Nonlinearity

'Effect is rarely proportional to cause, and what happens locally... often does not apply in distant regions' (Sterman, 2000). Nonlinearity is typical of complex systems. HOs react nonlinearly as they compete for scarce resources. Competition may not be fierce when the perception of scarcity is low. Competition increases sharply as resources run out. Once the resource is exhausted, there is nothing left for which to compete. HOs can distribute items while trucks are available. Distribution stops when stocks run out or when trucks are no longer available. As rivers, lakes and mountains create natural nonlinear barriers to urban sprawl, they may also create barriers or facilitate access for relief operations. Easier road access to Port-au-Prince may influence in a nonlinear way the amount of relief it receives compared to less accessible Jacmel and Léogâne. As disasters damage critical infrastructure (e.g. roads, railroads and airports), they create nonlinear challenges for supply efforts of relief operations.

A system dynamics lens: competing for scarce resources

This section returns to the competition for scarce resources described in the examples in the introduction and analyses it under a system dynamics lens. In the context of humanitarian operations, competition for scarce resources can take the form of hoarding, which has been previously studied in system dynamics in commercial settings (Gonçalves, 2003, 2018; Sterman and Dogan, 2015). As the disaster strikes, it creates an influx to the number of beneficiaries, i.e., the part of the population in need of assistance. When beneficiaries receive assistance from HOs, the number of beneficiaries decreases. Operationally, HOs assist beneficiaries through their relief capacity. Capturing stocks and flows in this system, beneficiaries would represent a stock; people in need due to disaster an inflow; and assistance an outflow. The ability of HOs to assist beneficiaries depends on their available relief capacity. As HO's allocate relief capacity to the region, they can serve more beneficiaries, closing the balancing loop B1 *Beneficiaries served* (Figure 8.1).

The more beneficiaries served, the more the HO shows donors its impact on the field (i.e., its ability to aid beneficiaries), and the more it can attract future funds (as observed in the Rwandan refugee crisis). With increased funds, the HO can further increase its relief capacity, closing the reinforcing loop R1 *HO's Relief capacity* (Figure 8.1). As described, loop R1 operates in a virtuous way. It is also possible for loop R1 to operate in a vicious way. In such a case, a decrease in funds would impose limits on the HO's relief capacity, limiting its ability to provide relief in the future, resulting in a lower overall future impact that would further reduce access to future funds from donors.

Because the HO is not operating alone in the theatre, similar feedback processes are present for competitor HOs, closing reinforcing loop R2 *Competitor HO's relief capacity* (Figure 8.2). Also, because future funding available is itself a limited resource, as more funds are made available to the HO, fewer funds are available to its competitor HOs. With fewer funds, competitor HOs are less able to increase their relief capacity, allowing more beneficiaries to be served by the

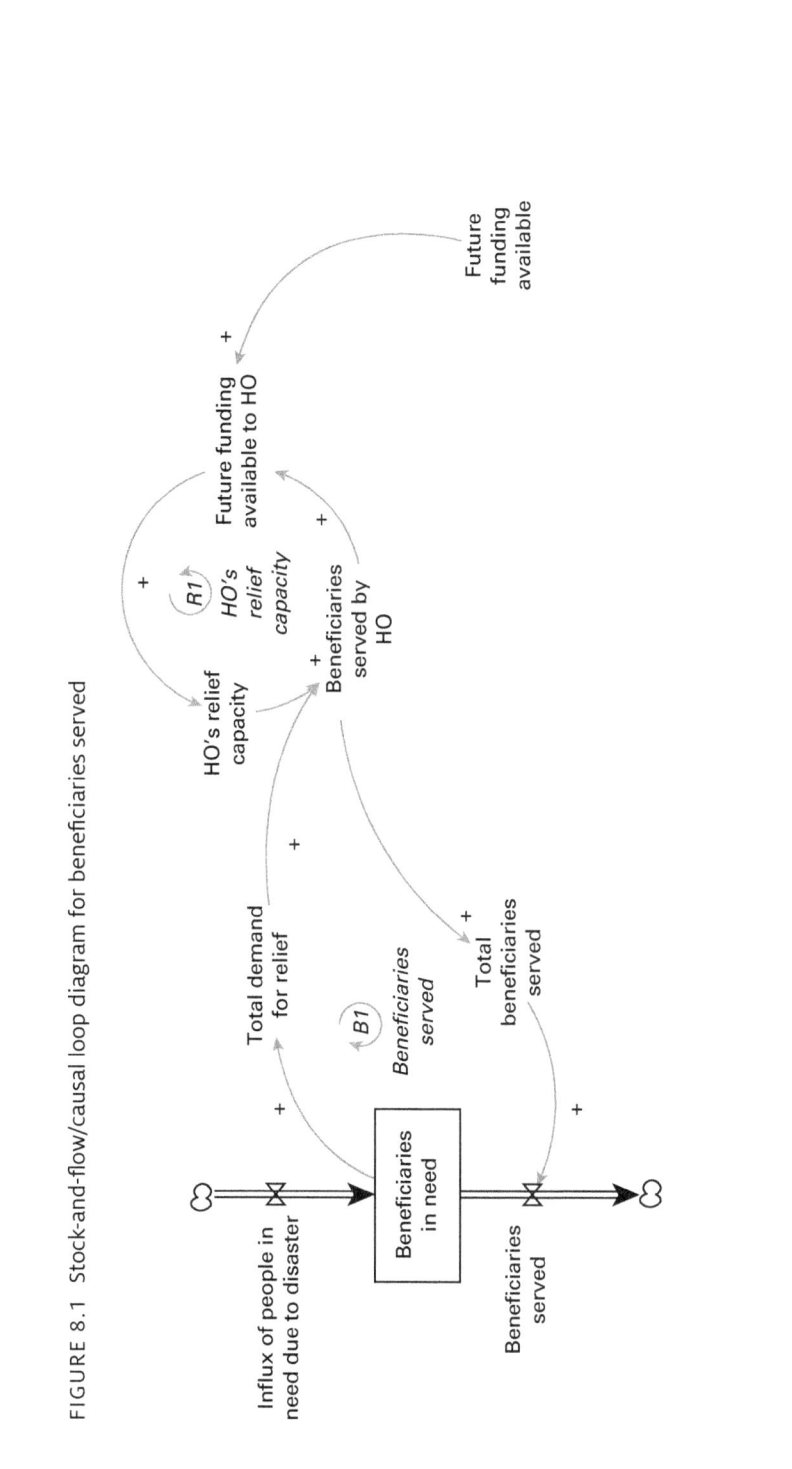

FIGURE 8.1 Stock-and-flow/causal loop diagram for beneficiaries served

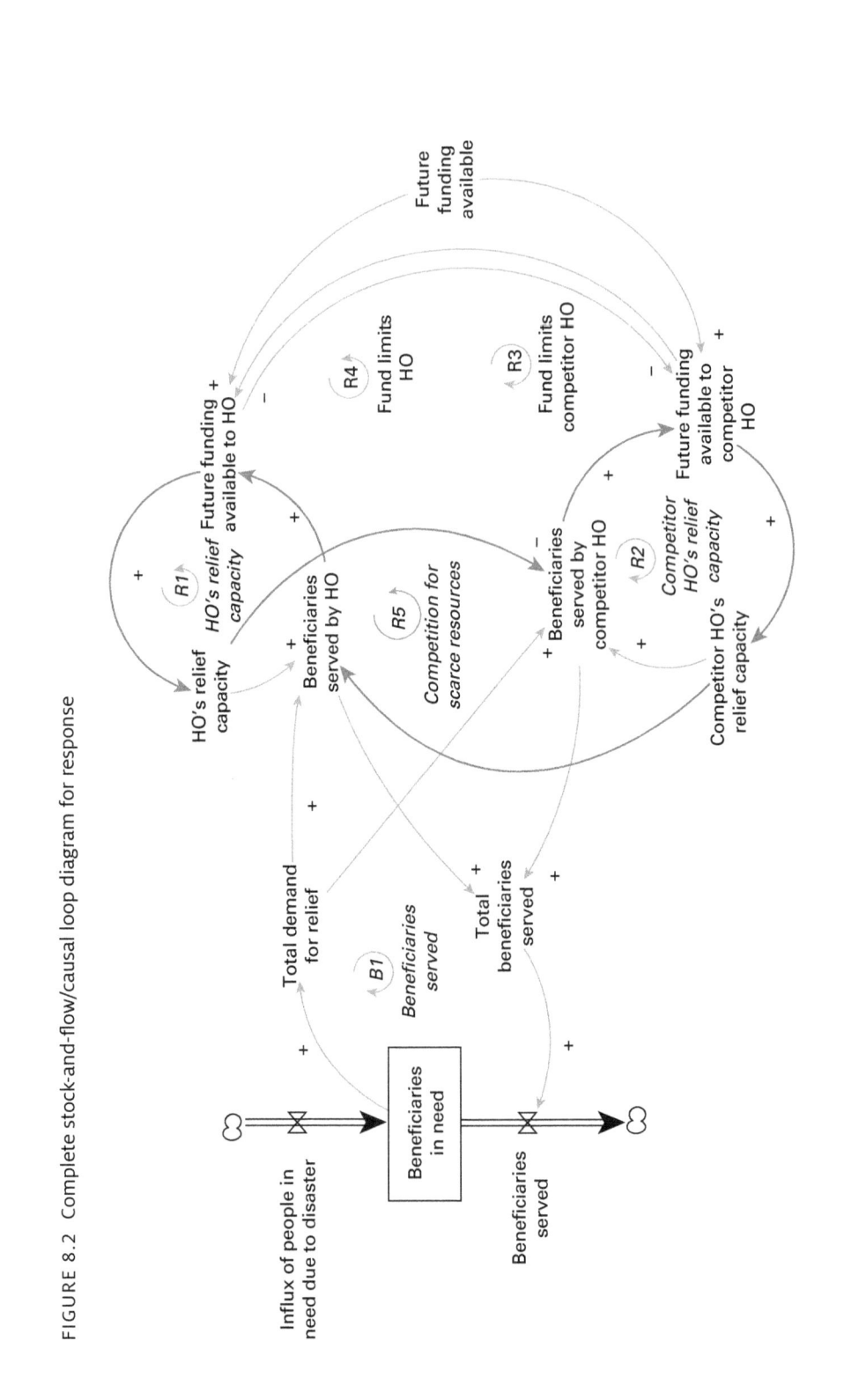

FIGURE 8.2 Complete stock-and-flow/causal loop diagram for response

HO, which attracts even more future funds and further reinforces the funds available to the HOs' competitors, closing reinforcing loop R3 *Fund limits to competitor HOs* (Figure 8.2). An analogous reinforcing loop (R4) exists limiting the HO's funds. Finally, Figure 8.2 captures the *Competition for scarce resources* loop (R5), a reinforcing loop that pits HOs against competitor HOs in terms of their relief capacities and ability to serve beneficiaries.

Loop R5 can either operate in a way that it strengthens the HO's relief capacity (in a virtuous way to the HO), or alternatively in a way that it strengthens the competitor HO's relief capacity (in a vicious way to the HO). Either way, virtuous or vicious, because HOs and competitor HOs compete for scarce resources, the overall response suffers in multiple different ways (e.g. donor funding, media attention, inadequate coordination, duplication of efforts, decreased effectiveness, etc.).

Intended rationality

Since donors, humanitarian organizations, and beneficiaries would benefit from eliminating this competition for scarce resources during humanitarian emergencies, why does it persist? One could hypothesize that decades of awareness about the problem and multiple instances of its costly and detrimental impacts would be sufficient motivation for scholars and practitioners to eliminate the problem. However, this does not yet seem to be the case.

Misperception of feedback can explain the rationality associated with the root cause of the problem and the difficulty in eliminating it. While humanitarian actors make intendedly rational decisions, cognitive limitations prevent them from fully understanding the long-term impacts of their decisions. Humanitarians' mental models typically ignore delays, feedbacks, nonlinearities, and stocks-and-flows, all of which are present in dynamically complex context such as those in humanitarian response. Such boundedly rational decisions in dynamically complex systems lead to typical inadequate behaviour, such as unintended side-effects and policy resistance.

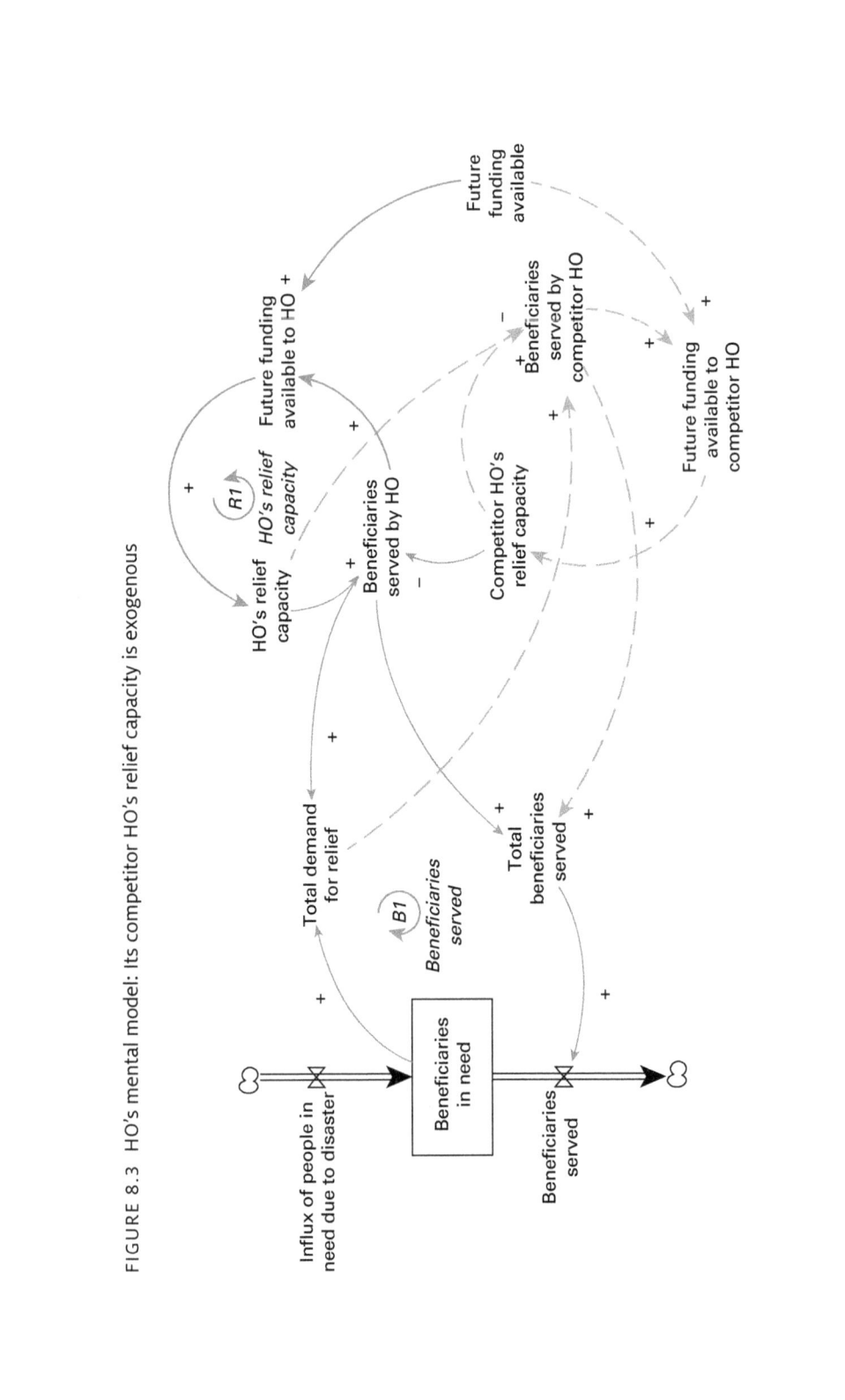

FIGURE 8.3 HO's mental model: Its competitor HO's relief capacity is exogenous

How can the decision to compete for scarce resources during humanitarian emergencies be rational? What makes it *intendedly* rational is that the decision is reasonable in a much simpler environment that matches the humanitarian actor's mental model. Hence, if the humanitarian actor perceives the competitor HO's relief capacity as exogenous, it is reasonable for the HO to increase its own relief capacity to meet beneficiary demand for relief. The HO's implicit assumption is that competitor HOs will not or cannot increase their relief capacity. If the system were as simple as the HO's mental model, then it would be sensible for the HO to invest in its own capacity as much as possible to meet the need.

However, the HO's mental model is incomplete. Competitor HOs are neither unwilling nor unable to increase their relief capacity. In practice, competitor HOs are willing to increase their relief capacity and do so using a similar logic as the HO. As competitors grow their relief capacity, they close the *Competitor HO's relief capacity* reinforcing loop (R2). And, as both HO and competitors expand their capacity, they use their relief capacity to compete for beneficiaries in the current emergency and for future funds, closing several reinforcing loops, such as *Fund limits* (R3 and R4) and *Competition for scarce resources* (R5). By failing to adopt an appropriate (e.g. more complex) mental model, whereby competitor HO's relief capacity is endogenous, the HO's intendedly rational (but short-sighted) decision to expand capacity not only duplicates response efforts, but also creates significant inefficiencies, reducing overall performance for the humanitarian response.

Summary and conclusion

Managers in humanitarian organizations can use system dynamics to understand the complex behaviour associated with humanitarian operations during disaster response and to design and implement high leverage policies that can mitigate possible detrimental effects. A key aspect discussed in the chapter deals with HOs' competition for scarce resources during emergency response. While the notion of humanitarian organizations competing for scarce resources during emergencies seem outrageous and objectionable, conditions in the complex humanitarian system not only allow these types of behaviour to take place, but also foster and perpetuate their occurrence. Because future contracts get awarded to HOs that have proven their ability to perform in the past, there is a strong incentive for HOs today to be first on the ground and ensure that their relief efforts and impacts are visible. Through their impacts today, HOs have access to funds tomorrow. Hence, for HOs it is both rational and critical to endure a 'measured' competition for resources to be able to achieve such impacts.

A possible way to avoid such competition would be for a central authority to coordinate efforts of multiple HOs, appropriately allocating responsibilities and resources. The UN Cluster Approach seeks to avoid overlaps orchestrating 'the assistance delivered by humanitarian organizations' by enhancing 'predictability, accountability and partnership' (OCHA, 2020). While the foundations for the UN Cluster approach were set in the General Assembly resolution 46/182 in December 1991, it was first adopted in response to the 2005 earthquake in Pakistan after the disastrous humanitarian response of the 2004 Indian Tsunami. Despite its aspiration to coordinate the delivery of humanitarian assistance, upon its implementation, the Cluster Approach still faced multiple challenges (Grzelkowski, 2006) such as inadequate deployment, potential conflict of interests, poor planning and coordination, insufficient donor involvement, etc. Addressing those challenges and improving HOs' support of the UN Cluster Approach will be critical to reduce past problems associated with HOs' competition for scarce resources during emergency responses.

References

Bare, F (2017) Competition, Compromises, and Complicity: An Analysis of the Humanitarian Aid Sector (2017), Unpublished CMC Senior Theses. http://scholarship.claremont.edu/cmc_theses/1617 (archived at https://perma.cc/296N-TWFG)

BBC (2010) Haiti quake death toll rises to 230,000, http://news.bbc.co.uk/2/hi/8507531.stm (archived at https://perma.cc/S5QD-MXSZ)

Bendoly, E, Croson, R, Goncalves, P, and Schultz, K (2011) Bodies of knowledge for research in Behavioral Operations. *Production and Operations Management*, **19**(4), pp 434–452

Besiou, M, Stapleton, O, Van Wassenhove, L N (2011) System dynamics for humanitarian operations. *Journal of Humanitarian Logistics and Supply Chain Management,* **1**(1), pp 78–103, https://doi.org/10.1108/20426741111122420 (archived at https://perma.cc/UU8C-HPUC)

Besiou, M, Van Wassenhove, L N, (2020) Humanitarian operations: A world of opportunity for relevant and impactful research, *Manufacturing & Service Operations Management,* **22**(1), pp 135–145

Bhattacharjee A and Lossio, R. (2011) Evaluation of OCHA Response to the Haiti Earthquake OCHA - Office for the Coordination of Humanitarian Affairs, p. 9–14. http://www.alnap.org/resource/6002 (archived at https://perma.cc/HVZ6-ELKN)

Britannica. (2010). Haiti earthquake. Available: www.britannica.com/event/2010-Haiti-earthquake (archived at https://perma.cc/H7KZ-UR9C)

Buzogany, R F, Gonçalves, P, Yoshida, H (2018) Policy Analysis of Material Convergence Challenges During Disasters. Unpublished Università della Svizzera Italiana (USI) Working Paper.

Chomilier B, Samii R, Van Wassenhove L N (2003) The central role of supply chain management at IFRC, *Forced Migration Review*, **18**, pp 15–16

De Vries, H, Van Wassenhove, L N (2020) Do optimization models for humanitarian operations need a paradigm shift? *Production and Operations Management,* **29**(1), pp 55–61

Dugger, C (2007) CARE turns down federal funds for food aid. *The New York Times*, 16 August

Eriksson, J, Adelman, H, Borton, J, Christensen, H, Kumar, K, Suhrke, A, Tardif-Douglin, D, Villumstad, S (1996) The international response to conflict and genocide: Lessons from the Rwanda experience Joint Evaluation of Emergency Assistance to Rwanda, *Steering Committee of the Joint Evaluation of Emergency Assistance to Rwanda*, March, https://www.oecd.org/countries/rwanda/50189764.pdf (archived at https://perma.cc/KPA3-WJ2N)

Fenton G (2003) Coordination in the Great Lakes, *Forced Migration Review*, **18**, 23–24

Forrester, J W (1961) *Industrial Dynamics*, Productivity Press, Cambridge, MA

Forrester, J W (1971) Counterintuitive behaviour of social systems. *Technology Review* **1**

GAVI Alliance (2021) COVAX explained, www.gavi.org/vaccineswork/covax-explained (archived at https://perma.cc/AN9K-C4P6)

Gonçalves, P (2003) Phantom orders and demand bubbles in supply chains, Unpublished PhD Dissertation, MIT Sloan School of Management, Cambridge, MA

Gonçalves, P (2008) System dynamics modeling of humanitarian relief operations. Unpublished MIT Sloan School of Management Working Paper, Cambridge, MA

Gonçalves, P (2011) Balancing provision of relief and recovery with capacity building in humanitarian operations, *Operations Management Research*, 4(1–2), 39–50, doi.org/10.1007/s12063-011-0045-7 (archived at https://perma.cc/33YK-BF4E)

Gonçalves, P (2018) From boom to bust: An operational perspective of demand bubbles, *System Dynamics Review*, 34(3), pp 389–425, doi.org/10.1002/sdr.1604 (archived at https://perma.cc/3WCB-BM47)

Gustavsson, L (2003) Humanitarian Logistics: Context and challenges. *Forced Migration Review* 18, pp 6–8

Grzelkowski, B (2006) Commentary on the implementation and effectiveness of the cluster approach, *Mercy Corps*, September, https://reliefweb.int/sites/reliefweb.int/files/resources/6297BD5CFEFEDACB8525720D006D252D-Mercy%20Corps-clusters-sep2006.pdf (archived at https://perma.cc/RXR2-H5SX)

Kaatrud, D, Samii, R, and Van Wassenhove, L (2003) UN joint logistics centre: A coordinated response to common humanitarian logistics concerns. *Forced Migration Review*, 18, 11–14

Keller, E (1992) Drought, war, and the politics of famine in Ethiopia and Eritrea, *The Journal of Modern African Studies*, 30, pp 609–624

Kent, R C (1987) *Anatomy of disaster relief: The international network in action*, F Pinter Publishers, London

Kim, H (2021) We need people's WHO to solve vaccine inequity, and we need it now, *BMJ Global Health*;6:e006598

Kimball, S (2022) WHO says vaccine inequity undermines economic recovery: It's 'a killer of people and jobs', *CNBC*, 6 January, www.cnbc.com/2022/01/06/who-says-vaccine-inequity-undermines-economic-recovery-its-a-killer-of-people-and-jobs.html (archived at https://perma.cc/S5U4-D88Y)

Kovács, G and Spens, K (2007) Humanitarian logistics in disaster relief operations, *International Journal of Physical Distribution & Logistics Management,* 37, pp 99–114

Kunz, N, Reiner, G, Gold, S (2014) Investing in disaster management capabilities versus pre-positioning inventory: A new approach to disaster preparedness, *International Journal of Production Economics*, 157, pp 261–272

Lindenberg, M (2001) Are we at the cutting edge or the blunt edge? Improving NGO organizational performance with private and public sector strategic management frameworks, *Nonprofit management and leadership*, 11(3), pp 247–270

Long, D and Wood, D (1995) The logistics of famine relief, *Journal of Business Logistics*, **16**, pp 213–229

OCHA (2020) What is the cluster approach? 31 March, www.humanitarianresponse.info/en/coordination/clusters/what-cluster-approach (archived at https://perma.cc/PB98-39MT)

Oloruntoba, R and Gray, R (2006) Humanitarian aid: An agile supply chain? *Supply Chain Management: An international journal*, **11**(2), pp 115–120

NPR (2008) Chinese Troops Work to Prevent Flooding, *National Public Radio, Morning Edition,* 29 May, www.npr.org/templates/story/story.php?storyId=90931426 (archived at https://perma.cc/7448-DUXX)

Richardson, B (1994) Crisis management and management strategy—time to 'loop the loop'? *Disaster Prevention Management*, **3**(3), 59–80

Richardson, G and Pugh, A (1981) *Introduction to System Dynamics Modeling with Dynamo*, Productivity Press, Portland, Oregon

Senge, Peter M (1990) *The Fifth Discipline: The Art and Practice of the Learning Organization*, Doubleday, New York

Starr, M K, Van Wassenhove, L N (2014) Introduction to the special issue on humanitarian operations and crisis management, *Production and Operations Management* **23**(6), pp 925–937

Stephenson Jr, M (2005) Making humanitarian relief networks more effective: Operational coordination, trust and sense making, *Disasters*, **29**(4), pp 337–350

Stephenson Jr, M, and Schnitzer, M H (2006) Interorganizational trust, boundary spanning, and humanitarian relief coordination, *Nonprofit Management and Leadership*, **17**(2), pp 211–232

Sterman, J D (2000) *Business Dynamics: Systems Thinking and Modeling for a Complex World,* McGraw-Hill, New York

Sterman, J D (2001) System dynamics Modeling: Tools for Learning in a Complex World. *California Management Review*, **43**(4), pp 8–24.

Sterman, J D, Dogan, G (2015) 'I'm not hoarding, I'm just stocking up before the hoarders get here.': Behavioral causes of phantom ordering in supply chains. *Journal of Operations Management,* **39**, pp 6–22

Tatham, P and Houghton, L (2011) The wicked problem of humanitarian logistics and disaster relief aid, *Journal of Humanitarian Logistics and Supply Chain Management* **1**(1), pp 15–31, doi.org/10.1108/20426741111122394 (archived at https://perma.cc/KR3J-DAAL)

Terry, F (2002) *Condemned to Repeat?: The paradox of humanitarian action.* Cornell University Press, Ithaca, NY

Thomas, A (2003) Humanitarian Logistics: Enabling Disaster Response, *Fritz Institute*,, ftp.idu.ac.id/wp-content/uploads/ebook/ip/BUKU%20TENTANG%20LOGISTIK%20MILITER/RISET%20LM/enablingdisasterresponse.pdf (archived at https://perma.cc/NS7U-SHQ9)

Thomas, A and Kopczak, L (2005) From logistics to supply chain management: The path forward in the humanitarian sector. *Fritz Institute*, 1 January

Tomasini, R M and van Wassenhove, L (2004) A framework to unravel, prioritize and coordinate vulnerability and complexity factors affecting a humanitarian response operation, Working Paper, *INSEAD*, Fontainebleau, France

Van Wassenhove, L (2006) Blackett memorial lecture: Humanitarian aid logistics: Supply chain management in high gear, *Journal of the Operations Research Society*, 57, 475–489

UNDP (2022) Global dashboard for vaccine equity, data.undp.org/vaccine-equity/ (archived at https://perma.cc/5RF6-9XTM)

WHO (2021) Director-General's opening remarks at the media briefing on COVID-19, 9 April, www.who.int/director-general/speeches/detail/director-general-s-opening-remarks-at-the-media-briefing-on-covid-19-9-april-2021 (archived at https://perma.cc/Y9PB-GAKV)

09

Preparing for cash and voucher assistance: Developing capabilities and building capacities

RUSSELL HARPRING

ABSTRACT

The growth of cash and voucher assistance in humanitarian settings has led to a recognition of the importance of developing preparedness measures to ensure programmes are designed and implemented effectively. An organization's ability to deliver cash and voucher assistance (CVA) is dependent upon available capacities and acquired capabilities, emphasizing the importance of information-sharing networks and continual development of skills and knowledge. This chapter discusses how humanitarian organizations can improve their level of preparedness and presents a practical case example for analysis.

Introduction

Over the past decade, cash and voucher assistance (CVA) has changed how humanitarian organizations prepare for and respond to crises. What was once a limited modality has become a core component of humanitarian responses, now accounting for approximately 17.9 per cent of total international relief provided by humanitarian organizations, more than double the amount in 2016 (Jodar et al., 2020). As the evidence has mounted about the benefits of CVA in relation to efficiency and effectiveness, international humanitarian organizations have shifted their approach to institutionalize and routinely consider CVA in their response planning. Global initiatives, such as Workstream 3 of the Grand Bargain, advocate for the systematic use of CVA across the humanitarian sector to ensure better coordination, delivery and monitoring of cash assistance (Metcalfe-Hough et al., 2021). These movements have gathered support from United Nations (UN) agencies, Red Cross and Red Crescent Movement (RCRCM) societies and non-governmental organizations (NGOs), aiming to create a large impact across the humanitarian sector.

The integration of CVA as a main-stay mode of providing aid has led organizations to recognize the importance of developing preparedness measures for CVA programming. These measures must be formed in the preliminary stages of planning, well before implementation takes place to ensure continuity of programming throughout the intended operational cycle (Castillo, 2021). Thus, there has been a global push for organizations to be 'cash ready' (Spencer et al., 2016), or 'cash proficient' (UNHCR, 2016), to provide efficient, effective and timely responses. However, preparing for CVA programming is a challenging task that requires logistical competencies, such as forecasting, planning and procurement (Heaslip et al., 2019), developed capabilities (e.g., operational knowledge and skillsets) and adequate capacity to deliver the assistance (e.g. physical assets, human resources and technology) (Jodar et al., 2020). Not having these in place limits the effectiveness of CVA and may hinder the performance of operations.

This was made evident during the Covid-19 pandemic as supply chain disruptions, lockdowns and social distancing measures interfered with 'traditional' in-kind assistance, leaving many humanitarian organizations rushing to set up or scale up CVA programmes. While some organizations were able to respond rapidly and provide additional cash disbursements to offset the effects of the disruptions, other organizations faced severe operational delays (Beazley et al. 2021). Merely having plans or CVA programmes in place was not a failsafe against disruptions during the pandemic though. Operations that relied on manual or analogue cash transfers experienced more interruptions from the pandemic compared to operations that utilized technology and digital transfers (Beazley et al. 2021), mainly due to social distancing restrictions. Thus, the digital divide between various countries and communities was once again brought to light, putting already vulnerable populations at greater risk through a lack of digital inclusion.

For humanitarian organizations, adapting responses to face these challenges simultaneously demonstrated the potential effectiveness of CVA programming for rapid responses as well as the need for preparedness measures and contingency planning to allow for operational flexibility. Since the start of the pandemic, over 1.3 billion individuals have received Covid-19-related cash transfers from governments or international organizations, with more expected to follow (Gentilini et al., 2021). As the quantity of CVA programmes continues to expand across diverse contexts and throughout different phases of crises, the need to retain quality in programming will also continue to increase. This infers strategic planning for the development, implementation and monitoring of programmes throughout the entire project cycle.

This chapter explores the concept of preparedness in relation to CVA programming and provides insights from leading organizations in the field of CVA about the different interpretations of preparedness as well as actions that can lead to a greater state of response readiness. A framework is presented to illustrate how knowledge, skills and resources affect an organization's ability to deliver CVA, as well as how information-sharing networks and joint initiatives can affect performance. These concepts are then illustrated through a case study of a cash working group facing critical delivery challenges amid the Covid-19 outbreak.

A brief introduction to cash and voucher assistance

In the humanitarian context, 'cash and voucher assistance' is defined as the provision of money or vouchers, either as emergency relief to address basic needs or as support for recovery activities to help re-establish livelihoods (CALP, 2018). It is also synonymous with several other terms used by humanitarian practitioners, such as cash-based interventions, cash-based relief, and cash transfer programming (Maghsoudi et al., 2021). While the terms all refer to the same concept, 'CVA' is adopted for this chapter for clarity and consistency, as it explicitly includes both cash transfers and voucher programmes. It must be noted that the term does not include remittances and microfinance programmes.

In relief operations, CVA may be used as a standalone form of assistance or delivered alongside goods or services in-kind. Delivery of CVA may be conducted either physically or digitally, depending on beneficiary preference, operational objectives and context of implementation. Table 9.1 presents the different mechanisms used to deliver cash and vouchers, as well as examples of each.

CVA is provided to individuals or households and may be given conditionally (i.e., to specific qualifying recipients, such as mothers

TABLE 9.1 Types of CVA

Modality	Means of delivery	Types
Cash transfer	Physical cash distribution	- Direct cash - Check
	Electronic transfer	- Pre-paid card - Stored-value card - Bank account - Debit/ATM Card - Smart card - Mobile money
Voucher	Physical voucher distribution	- Paper voucher - Scratch card
	Electronic transfer	- Pre-paid card - Stored-value card - Smart card - Mobile voucher

SOURCE Adapted from the CALP Network (2018)

with school-aged children) or unconditionally with no qualifying requirements. Similarly, there can be restrictions on how the assistance is used. Figure 9.1 illustrates how these conditions and restrictions relate to spending flexibility on the side of the beneficiary. Vouchers are more restrictive as they are limited by where and how the recipient can use them, whereas cash assistance is flexible and allows the recipient to choose how to spend the cash. CVA may have specifically designed purposes (e.g., for food, water or shelter) or be used to address multiple needs across various sectors simultaneously. In general, the more needs a programme is designed to cover, the less restrictive it will be in terms of usage by the recipients.

Providing financial assistance through cash and vouchers differs from the traditional form of aid, which provides physical goods to beneficiaries by allowing recipients to purchase what is needed locally. As noted by Heaslip et al. (2015), this has a profound effect on the

FIGURE 9.1 CVA conditionality and restrictions matrix

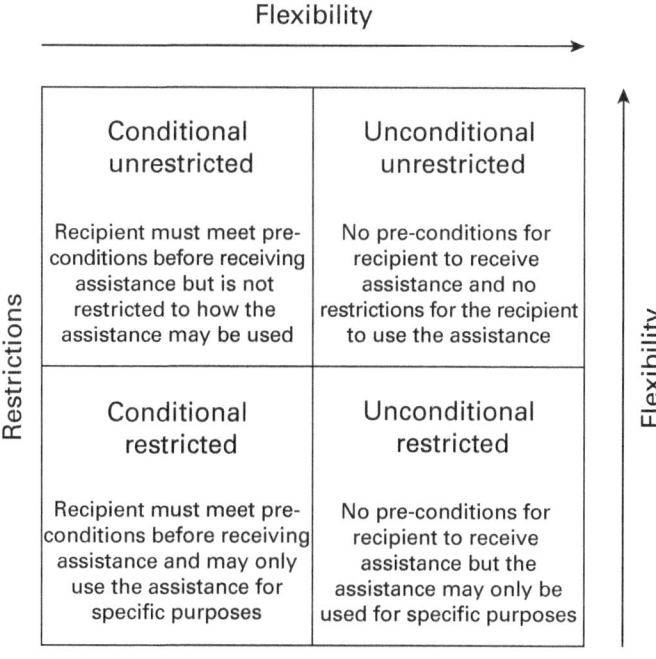

	Conditional unrestricted	Unconditional unrestricted
	Recipient must meet pre-conditions before receiving assistance but is not restricted to how the assistance may be used	No pre-conditions for recipient to receive assistance and no restrictions for the recipient to use the assistance
	Conditional restricted	Unconditional restricted
	Recipient must meet pre-conditions before receiving assistance and may only use the assistance for specific purposes	No pre-conditions for recipient to receive assistance but the assistance may only be used for specific purposes

Axes: Flexibility (horizontal), Restrictions (vertical left), Flexibility (vertical right), Conditionality (horizontal bottom)

movement of goods and finances throughout the supply chain as well as the actors in the supply network. Figure 2 shows the impact this has on material and financial flows, creating a 'pull' effect as beneficiaries choose what is needed, rather than goods being 'pushed' to them from organizations. This change not only supports local suppliers but also provides empowerment and dignity of choice to the beneficiaries (Harvey, 2007). Humanitarian organizations benefit through the reduction in logistical activities (e.g., purchasing, warehousing, transportation and customs clearance) for goods already available in local marketplaces and can then focus on procuring materials that are not readily available for beneficiaries (Tatham et al., 2018). This strategic usage of both in-kind assistance and CVA can provide operational flexibility by switching between delivery modalities depending on the situation. Furthermore, these measures are important metrics for donors when assessing the relative quality of a humanitarian operation's work (Wakolbinger and Toyasaki, 2018), which has an impact on financial flows for relief aid.

These noted benefits have led to increased use of CVA globally in humanitarian operations, rising to a record amount of US$6.3 billion in 2020 (Development Initiatives, 2021). United Nations (UN) agencies

FIGURE 9.2 Changes typical humanitarian supply chain flows undergo through CVA programming

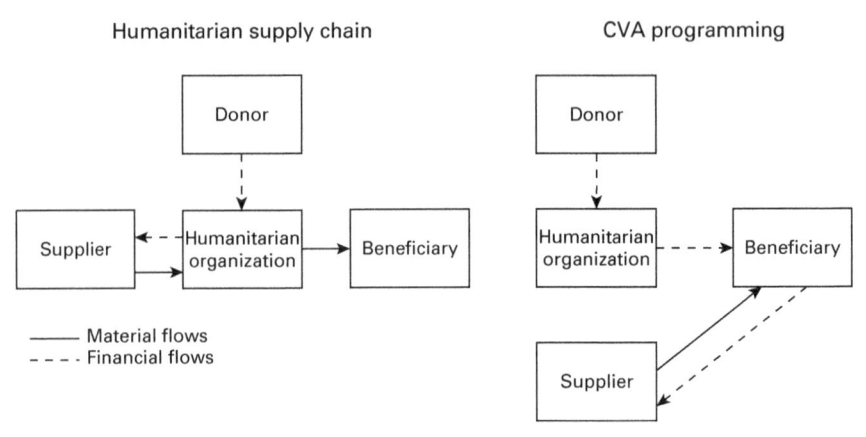

SOURCE Adapted from Heaslip et al. (2015)

and the International Red Cross and Red Crescent Movement (RCRCM) accounted for 75 per cent of the transfer amount, with the largest actors being the World Food Programme (WFP), RCRCM and the United Nations High Commissioner for Refugees (UNHCR) (Development Initiatives, 2021). This trend is likely to continue further upward as humanitarian organizations continue to expand existing operations and implement new projects (Spencer et al., 2016).

Despite the growth of CVA, further research is needed to better understand how various delivery mechanisms can be used effectively in different contexts (Maghsoudi et al., 2021). In each situation, financial constraints may change depending on the disaster phase, which also influences the effectiveness of the response (Wakolbinger and Toyasaki, 2018). Furthermore, across cultures the concept of 'cash' may carry different social meanings, creating different perceptions between actors (e.g., donors, beneficiaries and host community) of which type of aid is most suitable and how it should be distributed (Vogel et al, 2021). Therefore, CVA programmes must not only be feasible and measurable but also contextually appropriate. For this, numerous tools and guides exist to help practitioners make informed decisions regarding CVA programming.

These guides and checklists vary by complexity and purpose, and often reflect the objectives of the organization that created them. For instance, UNHCR created a CVA feasibility and analysis toolkit that uses a series of nestled assessments and decision trees across various levels to determine an appropriate *protection* response, in line with their mandate (UNHCR, 2017). The assessments first address whether CVA is feasible and safe to use in each context, then help determine which delivery mechanism should be selected, and finally how to implement the chosen delivery mechanism (UNHCR, 2017). While the objective is to deliver the correct response according to each context, many of the CVA guides require in-depth knowledge of markets, financial service providers and the targeted beneficiary population. According to Castillo (2021), although these decision trees and assessments are comprehensive, their practicality is often limited by the resources available to complete all the tasks accurately and in a timely fashion, which may ultimately affect the efficiency of the response. This is especially true in disaster situations when time is critical, and resources are already limited. Thus, there exists the need for long-term and continuous investment in preparedness activities to enhance an organization's state of readiness.

What does preparedness mean for cash and voucher assistance?

In the humanitarian context, *preparedness* is broadly defined as 'the knowledge and capacities developed by governments, response and recovery organizations, communities and individuals to effectively anticipate, respond to and recover from the impacts of likely, imminent or current disasters' (UNDRR, 2009). Preparedness is one of the main phases of disaster management and includes five major areas, which should be continuously developed (Tomasini and Van Wassenhove, 2009):

i. human resources,
ii. knowledge management,
iii. logistics,
iv. financial resources,
v. and community.

Investment in preparedness activities that fall under these five areas can positively impact responsiveness during a disaster and may reduce the overall cost of the response. In research related to preparedness activities, the development of physical resources (e.g., inventory, infrastructure, technology) often receives more focus than intangible resources (e.g., capacity building and training) (Kunz et al., 2014). However, the proper transfer of knowledge and skills, especially at the local level, is imperative for the effective and timely management of disasters. In other words, responsiveness depends upon contingency planning and ensuring the right systems and tools are in place when and where an emergency occurs.

For CVA, the concept of preparedness differs slightly. Although it still relates to all the five major areas, there are distinct elements that differentiate it from other forms of assistance. This means that if organizations wish to provide CVA alongside in-kind aid, additional resources, knowledge and logistical considerations are needed. For instance, disaster preparedness plans may emphasize pre-positioning of supplies and optimal transportation, whereas CVA preparedness activities are more concerned with identifying financial suppliers, mapping potential cash distribution points in a specific geographic region and conducting market assessments. In this manner, while similarities exist between preparedness measures using in-kind distribution, local procurement and CVA, there are important differences that must be recognized for aid to be delivered effectively.

Unfortunately, literature related to preparedness for CVA is fragmented and limited (Maghsoudi et al., 2021), despite calls to expand the body of evidence (Spencer et al., 2016). This has led to multiple interpretations of what 'CVA preparedness' entails. Of the two most common interpretations, one considers 'CVA preparedness' to be the minimum requirements necessary to implement and operate CVA programmes (IFRC, 2021), whereas the other interpretation refers to the inclusion of CVA in disaster preparedness and contingency plans so that cash may always be considered alongside other forms of relief (UNHCR, 2016; WFP, 2021). The first interpretation can be applied to both disaster relief and development aid, but the second is directed towards emergency response. By integrating both definitions, a third interpretation emerges: if humanitarian organizations choose to institutionalize CVA into their response plans, then they must understand what activities need to be carried out, who is responsible for each activity, what the minimum skill requirements are to carry out each activity and the capacities needed to scale up in times of a disaster.

At present, no standardized minimum requirements exist across the humanitarian sector for short-term operability and long-term integration of CVA. This is partly because there are no recognized standard operating procedures across the humanitarian community for using CVA. Initiatives have been made, including the integration of CVA as a core modality into the *Sphere Handbook* and the development of the *Minimum Standard for Market Analysis (MISMA)*. These handbooks are designed to ensure quality and accountability during humanitarian responses through a consistent and principled approach. Additionally, the Tracking Cash and Voucher Assistance (CVA) Working Group, part of the Grand Bargain Cash Workstream, has undertaken the challenge of finding a systematic approach to better track and trace global volumes of CVA. Despite the progress being made, there is still no benchmark to determine whether an organization is 'cash-ready' prior to implementation or not. A lack of preparedness and operational knowledge can lead to reduced cost-efficiency and effectiveness of the programme. Well-designed projects may utilize more resources initially, but often reduce long-term logistics and supply-related costs. Therefore, investment in capacities and capabilities is necessary to execute each individual step, which can improve the overall state of preparedness for CVA programming. The next section provides an overview of the steps in the cash and voucher assistance operations cycle, along with a framework for building capacities and capabilities.

Preparedness throughout the cash and voucher assistance operations cycle

Even though no single set of standard operating procedures exists across the entire humanitarian sector for using CVA, there are specific activities that must be completed to adequately evaluate, implement and manage cash projects. The cycle in Figure 9.3 presents the high-level phases necessary for cash and voucher assistance programming, based on the steps outlined by the CALP Network. The CALP Network is a diverse group of more than 90 members, including donors (e.g., USAID, SIDA SDC), UN agencies (e.g., WFP, UNHCR, UNICEF), and NGOs (e.g., IFRC, Oxfam and Care International), who collectively provide the majority of all humanitarian CVA delivered globally. Their aim is to enhance the skills, knowledge and policy of CVA, both within and between organizations in the humanitarian ecosystem. Thus, the cycle shown in the figure represents a general consensus of the major steps involved in CVA programming, though each individual organization's process may vary to meet its specific objectives and mandates.

FIGURE 9.3 CVA operational cycle

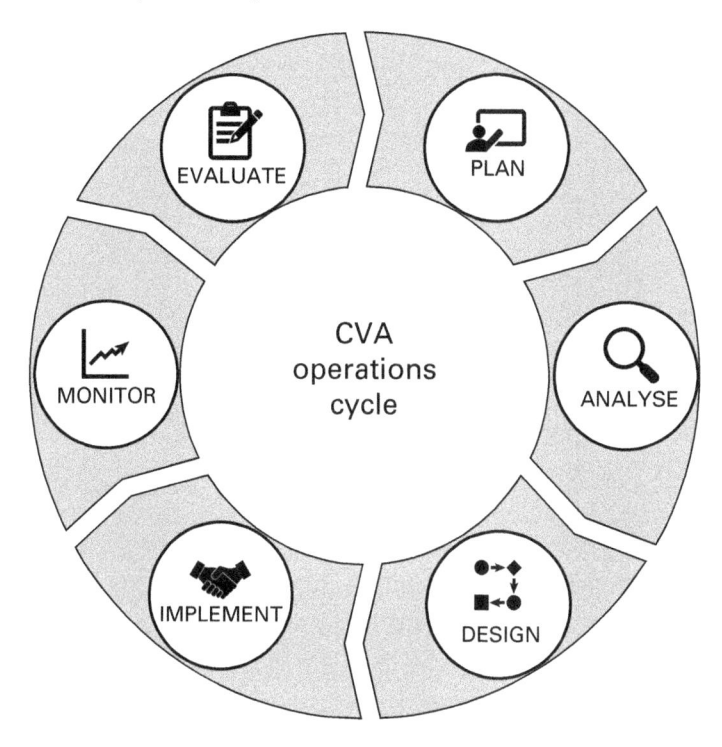

CVA operations cycle: EVALUATE, PLAN, ANALYSE, DESIGN, IMPLEMENT, MONITOR

In the *planning* phase, the primary objective is to create a basis for the CVA project and develop a flexible strategy to implement and operate the programme, both in the current state and in potential crisis situations. During the *analysis* phase, assessments are conducted to ensure that CVA is feasible, addressing the adequacy of the local infrastructure, market functionality, security risks, access to financial services and supply chain constraints. Once the operational context has been assessed, the different options for delivering CVA can be explored in the *design* phase. This step involves establishing contact with financial and payment service providers, vendors, suppliers and conducting procurement activities to determine which delivery mechanisms are feasible and implementable at scale.

When the *design* phase has been carried out and an action plan is in place, *implementation* can begin. Contracts and standard operating procedures must be established with suppliers, partners and other relevant actors involved in the distribution, monitoring, and reconciliation of CVA. Once the programme has been set up, periodic *monitoring* of the distribution cycles, market prices, inflation and contract ceilings must be conducted throughout the project so that adjustments can be made as needed. When the project is completed or terminated, a thorough *evaluation* should be made to retain lessons learnt, improve institutional knowledge and keep records to have a thorough audit trail.

Annex 1 provides a more detailed list of activities to be performed at each phase of the operational cycle. The activities listed for each phase are derived from operational procedures and toolkits for CVA from leading humanitarian agencies such as RCRCM, UNHCR, WFP and UNICEF, as well as consortiums of NGOs, such as the Collaborative Cash Delivery Network, the Cash Learning Partnership Network, and EMMA Partners. It must be noted that each activity consists of its own set of steps and sub-activities which vary between organizations. The organizations and consortiums listed provide valuable guidance for each of these activities. Thus, the set of activities may be revised according to each organization's needs, operating procedures and objectives. The initial list shown in Annex 1 has been compiled to provide a starting point based on common approaches.

Capacities, capabilities and competencies

The activities to be completed for each phase require knowledge of finance, supply chain, security, and information technology. According to Maghsoudi et al. (2021), this requires a capable workforce and available resources for each task. By developing capabilities (i.e., skills and knowledge) and investing in capacities (i.e., human capital, allocated budget, time and assets), an organization increases its ability to perform certain tasks. The level of performance for each task is dependent upon how competent the organization is, thus there is a need to continually develop capabilities and build capacities to ensure successful implementation and operation.

This follows what is known as 'the resource-based view' (Barney, 1991), in which an organization must know what its competencies

FIGURE 9.4 Hierarchy of capacities, capabilities, and competencies in an organization

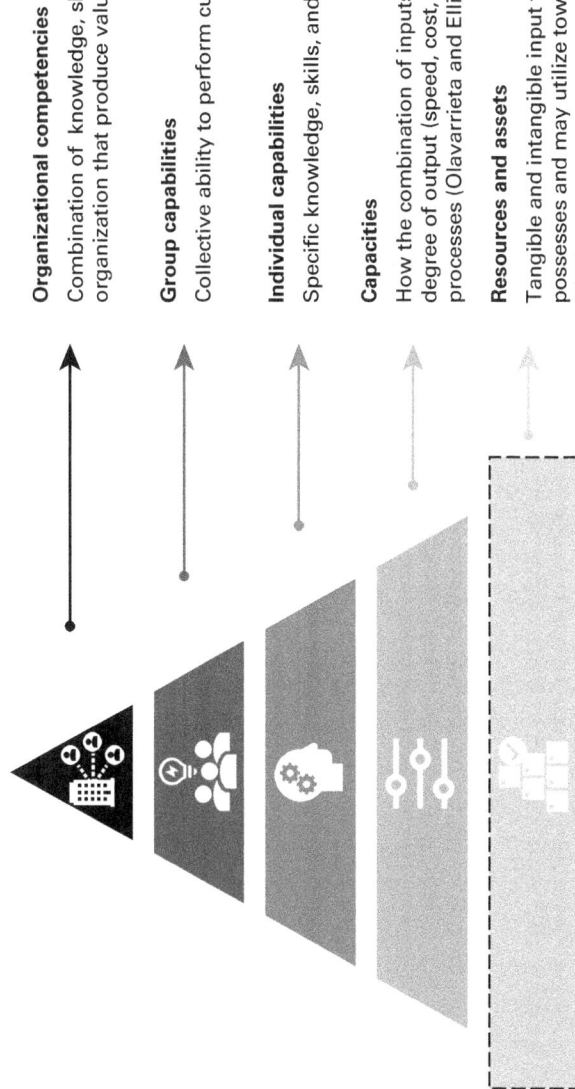

Organizational competencies
Combination of knowledge, skills, and resources within an organization that produce value (Prahalad and Hamel, 1990)

Group capabilities
Collective ability to perform current and future tasks

Individual capabilities
Specific knowledge, skills, and abilities to perform task

Capacities
How the combination of inputs (resources, assets, etc.) influences degree of output (speed, cost, quality) through organizational processes (Olavarrieta and Ellinger, 1997)

Resources and assets
Tangible and intangible input factors which an organization possesses and may utilize toward a task (Olavarrieta and Ellinger, 1997)

SOURCE Adapted from Javidan (1998) and Zoiopoulos et al. (2008)

and capabilities are before it can fully utilize its own capacities. Inversely, investing in the development of knowledge, skills and tools increases capacity, which impacts an organization's capabilities. Figure 9.4 presents a framework adapted from Javidan (1998) and Zoiopoulos et al. (2008) which represents the hierarchical relationship between capacities, capabilities and competencies. From this perspective, an organization works from the inside out by inventorying its own resources, identifying its abilities and recognizing strengths that can be leveraged to its advantage to respond to the needs of the external environment (Prahalad and Hamel, 1990). Applying this concept to the humanitarian sector, organizations that understand their own competencies can develop responses based on their known strengths and weaknesses.

An organization's capacity is made up of a pool of both tangible and intangible resources and assets available for utilization toward any specific task (Sanchez et al. 1996). The number of resources dedicated to a task will influence the outcome. In CVA programmes, outcomes must be measured to gauge performance related to time, cost and quality. While it is essential that organizations track performance, there are inherent challenges in quantifying data that has both tangible and intangible dimensions. For instance, one key metric used by leading donors is cost-efficiency, which compares a project's input costs related to the outputs achieved. While tangible resources such as money, time, and assets are relatively static and measurable, other resources related such as human capital are dynamic and can only be partially measured. In measuring human capital, quantifiable elements such as the number of employees assigned to the task, including their salary and billable hours to the project can be tracked, while intangible elements related to each individual's ability to complete a task cannot be objectively calculated. As an individual's knowledge and skills increase, so does the organization's overall ability to perform tasks more efficiently, including complex or unanticipated tasks (Heaslip et al., 2019). In this sense, continual investment in skills and training should be viewed as a form of preparedness, as it affects a humanitarian organization's available capacity, which in turn may affect metrics such as performance and funding.

Developing skills for CVA, however, is not an easy task and requires an investment of resources. In addition to general knowledge related to a task or activity, the humanitarian sector requires a set of context-specific skills to operate in dynamic environments (Kovács et al., 2012). This creates a paradoxical challenge for decision-makers in humanitarian organizations; in a resource-constrained environment, how can a manager balance allocating time and money to training when it may reduce the time and money available toward relief operations? At the same time, investment in training has a long-term effect that has a positive impact on the efficiency of relief provided (Gonçalves, 2011). This challenge has been especially prevalent for humanitarian organizations beginning to use CVA, as well as those seeking to expand existing cash programmes. In a survey conducted by the CALP Network (2019), humanitarian organizations responded that one of the biggest challenges faced in CVA programming was the risk of sacrificing quality, timeliness or appropriateness when scaling up coverage. As CVA is a cross-sectoral tool that requires skills and knowledge from sectors such as finance, supply chain, security and information technology, additional training and close collaboration are often required so that a team can capably perform the tasks required by the project.

To address this challenge, several leading humanitarian organizations have begun to 'institutionalize' CVA into their relief operations, adopting strategies to regularly consider CVA in their responses and contingency plans (Jodar et al. 2020). This strategic vision is a phased approach that requires comprehensive planning and continuous investment in capacities and capabilities, which may take place over several years, to develop competencies for delivering cash. According to Stephenson's (1998) theory of developing capabilities through learning in dynamic environments, individuals and organizations will become more capable of performing certain tasks over time, even in unfamiliar environments or against new challenges (Figure 9.5), such are the conditions immediately following a disaster. By institutionalizing CVA, humanitarian agencies are fostering an environment of continuous development of capabilities and competencies, which influence their ability to respond to disasters.

FIGURE 9.5 Complex task competency framework with learning feedback loops

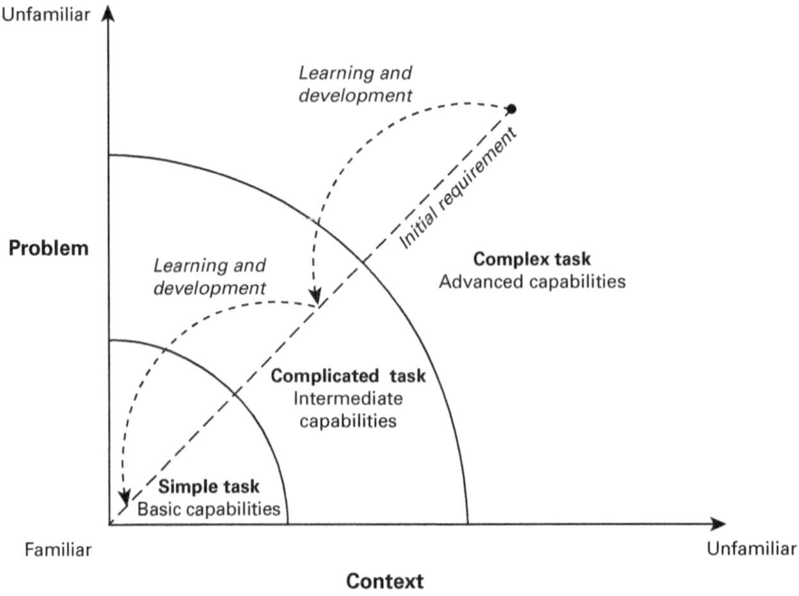

SOURCE Adapted from Stephenson (1998) and Stacey (1996)

As shown in Figure 9.5, the more familiar an operator or team is with the context and problem, the simpler the task is perceived to be. More complicated tasks often require refined skills and knowledge to carry out. When a task is considered 'complex', a novel approach must be taken as no known solution yet exists. In complex crises, external factors such as integrity of infrastructure and political stability are often beyond the control of any single organization, meaning that humanitarian organizations must adapt their strategies for the operational context (Harpring et al., 2021). This level of difficulty requires competency in the area to develop a new solution or approach. However, this creates a state of learning and development as information is gathered about a problem, then an action is performed, followed by a result being produced, which then creates a feedback loop by providing additional information about the original problem, thus developing competence. Over time, as experience is gained and the context becomes more familiar, an organization can perform a task more capably. The feedback loop continues to be present even after an organization becomes competent at performing a task, as the organization will constantly review and evaluate the performed tasks, creating a continuous state of development (Black, 2015).

Mapping an organization's baseline capacities and capabilities then is the first step in being able to know which areas need development. The following tables provide a basic assessment template for documenting capacities and capabilities at the individual level (Table 9.2) and team level (Table 9.3). The information gathered from these tables should then be used to complete the operational assessment (Annex 1). This assessment is designed to help organizations understand their level of preparedness at each stage of the operations cycle.

Working from the bottom up, all individual operators should provide feedback for the tasks they are assigned to. Then, the same exercise can be conducted at the team level, as tasks may be shared or tightly coupled with other tasks. At the operational level (Annex 1), the assessment gives insights into the level of support available from the organization and whether activities can be executed in the current state. Identifying deficiencies in capabilities and inadequate capacities can help an organization know which areas need additional investment, both at the individual level and group level. The completed assessment provides a snapshot of an organization's current state of preparedness to conduct CVA activities, and may also provide insights toward performance in a potential disaster situation. As exemplified in Figure 9.5, the results of the assessment may shift over time depending on the degree of familiarity with the problem. Therefore, the assessment may need to be updated periodically to provide relevant results.

Beyond the organizational level, intra-organizational networks may enable higher performance through information sharing and knowledge access. The ability to share information effectively amid an ongoing disaster is a dynamic capability, which can significantly impact the speed and efficiency of relief operations (Suifan et al., 2020). Particularly, the concept of solidarity networks applies to the humanitarian sector, which posits that organizations in the network

TABLE 9.2 Individual-level CVA capacity and capability assessment

		Individual capacities		Individual capabilities	
Phase	Activity	My tasks for this activity include:	I have the adequate resources available to carry out this task	If no, what additional resources do I need to carry out the task?	I have the adequate skills and knowledge to carry out this task	If no, what additional skills and knowledge do I need to carry out the task?
Analyse	Financial service provider mapping					

TABLE 9.3 Team-level CVA capacity and capability assessment

				Team capacities		Team capabilities	
Phase	Activity	Our Team's tasks for this activity include:	Which team member(s) will carry out this task?	Do we have the adequate resources available to carry out this task?	If no, what additional resources do we need to carry out the task?	Do we have the adequate skills and knowledge to carry out this task?	If no, what additional skills and knowledge do we need to carry out the task?
Analyze	Financial Service Provider Mapping						

value mutual interest over self-interest without focusing on the return of capital (Smith, 2009). In these network formations, organizations form a collective entity to coordinate their efforts under a set of common guiding principles. The humanitarian cluster system is an example of such a network, which can improve information diffusion across a network of actors (Altay and Pal, 2014).

This concept has expanded to include cash operations as well, leading to the formation of the Global Cash Working Group, which acts as a focal point for the humanitarian community. Information is gathered from field operations, which then help formulate good practices that can be disseminated to national, regional and local working groups. In this way, smaller organizations with limited capacities and capabilities can leverage information provided by partner organizations. Other initiatives in the humanitarian community are also underway to address the need for more effective coordination mechanisms in CVA operations, including the UN Common Cash Statement for UN agencies and the Cash Coordination Caucus to adopt system-wide approaches to CVA. Both initiatives share the same objective: to ensure accountability and predictability of CVA programming within the humanitarian ecosystem. These initiatives not only aim to enable continuity of operations against limited resources but also to allow inexperienced organizations to gain knowledge from more experienced partners. This is exemplified through a case study of the cash working group in Mexico in the following section.

CASE STUDY

Preparing for cash assistance in Mexico

During the Covid-19 outbreak, many organizations sought to expand their use of CVA to help overcome procurement issues caused by global supply disruptions of personal protective equipment and hygiene items. However, the organizations faced several challenges as they scaled up existing CVA programmes or implemented new projects, largely due to constraints related to organizational capacities and capabilities. To address these challenges amid the crisis, five humanitarian organizations with varying degrees of experience formed a cash working group (CWG), which is analysed in this case study.

This case primarily focuses on the *planning, analysis,* and *design* phases for CVA through a joint procurement exercise. Procurement of financial services for beneficiaries can be a challenging task that requires knowledge of finance, logistics and local laws, customs and regulations. The process results in the establishment of a contract or legal framework agreement with a financial service provider (FSP) or payment service provider (PSP) for the distribution of cash or vouchers to beneficiaries using bank accounts, debit cards, mobile money, etc. Following the procurement phase, organizations must be able to conduct monitoring and evaluation activities throughout the CVA project cycle, adjusting distributions when necessary to maintain effectiveness. These tasks generally require more advanced skills and knowledge of CVA programming, which creates barriers for smaller or less experienced organizations.

A note on methodology

To understand how members of the CWG overcame the challenges related to capacity and capability constraints, semi-structured interviews were conducted with nine participants representing all five organizations of the CWG, as shown in Table 9.4. Secondary data, such as external reports, meeting minutes, studies and publications, were used for contextualization and triangulation of information (Seuring, 2008). In collecting and storing the data, a strict protocol was followed to ensure the privacy of the organizations and participants involved. To protect the identities of the

TABLE 9.4 Cash Working Group members and interviews conducted

Cash Working Group Member	Dedicated CVA Staff?	Years Using CVA in Mexico*	Interviews Conducted
IHO 1	No	3	1
IHO 2	Yes	5	4
IHO 3	Yes	2	1
IHO 4	No	0	2
IHO 5	No	0	1

interviewees and the integrity of the research, the names and organizations have been anonymized.

The data gathered from each interview was recorded and transcribed using Nvivo software, and later reviewed and cleansed. During the analysis, key insights emerged in two ways: within each organization and between organizations. When complemented with the hierarchy presented in Figure 9.4 and the framework in Figure 9.5, the data presents a story of how each organization contributed their individual capabilities and capacities in the formation of a new group, which transcended typical organizational level boundaries.

Contextual background

Mexico has a history of providing cash assistance to its citizens to improve livelihoods, education and healthcare access. In 1997, the government of Mexico founded a social assistance programme, *Progresa* (rebranded as *Oportunidades* in 2002, and later as *Prospera* in 2014), which used conditional cash transfers to counter the effects of the economic crisis from 1994 to 1995. The programme sought to alleviate poverty by providing financial assistance to families below the food poverty line to cover the costs of healthcare, education, and nutrition. By the end of 2015, the programme had provided support to over 25 per cent of Mexico's population and created both short-term and long-term impacts on school enrolment rates and children's nutritional intake (Masino and Niño-Zarazúa, 2020). The programme acted as a model for national social protection and was replicated in 52 other countries.

Nevertheless, this social protection programme was not designed as a response option for disasters or humanitarian crises, nor could the registries of beneficiaries always be used to identify populations vulnerable to natural hazards. Since 2014, Mexico has been at the forefront of an influx of immigrants fleeing from violence and poverty in the Northern Triangle of Central America (Guatemala, Honduras and El Salvador). From 2014 to 2020, the number of asylum seekers and refugees fleeing those countries increased by over 1,400 per cent from 66,588 to 1,030,898 (UNHCR, 2021). Many of these migrants were women and unaccompanied minors seeking refuge in Mexico, the United States and Canada.

Although refugees have the right to permanent residence in Mexico, they have been largely excluded from *Prospera* due to issues regarding documentation and targeting. UN agencies sought to close this gap and began operating more closely with the Mexican Commission for Refugee Assistance (COMAR) and the National Migration Institute (INM). These partnerships enabled more robust support to be provided to asylum seekers and refugees, including determination of status and temporary cash and voucher assistance. The objective was to create a transitory pathway for refugees and migrants into Mexico's social safety net system. Although much of the preparatory work had to be redone and redesigned to meet the needs of the refugees and asylum seekers, the success of *Progresa/Oportunidades/Prospera* did have an indirect impact on subsequent CVA provided by humanitarian organizations in Mexico. Most notably, banking infrastructure had expanded and grown throughout the country to include additional rural locations that had not previously had access to formal financial services. Humanitarian organizations were able to utilize the matured national financial network in the planning, implementation and management of their own CVA programmes.

Joint Initiatives for Preparedness

As more humanitarian organizations began to use CVA throughout the country, the need arose to better track *who does what, where, and when* (the 4Ws) to avoid duplication of aid and to ensure no gaps exist in coverage. In early 2020, shortly before the onset of the pandemic, the CWG in Mexico was formed to improve collaboration and communication among agencies that use CVA. As the group became operational, the Covid-19 pandemic arrived and brought about several new sets of challenges.

Mexico was hit particularly hard during the first year of the pandemic. Over 1.4 million cases and 123,845 deaths (WHO, 2020) were confirmed by the end of 2020, which was the third-highest mortality rate in the world at the time. Humanitarian organizations recognized that vulnerable migrant and refugee populations in rural, urban and camp settings that were not included under social protection programmes like *Prospera* were at a greater risk than other groups. The humanitarian organizations worked quickly to procure personal protective equipment and hygiene items, but faced shortages and severe delays due to global supply chain issues. Rather than compete with the commercial sector for items, organizations adapted their strategies to rely on cash assistance and in-kind to meet the additional needs. However, not all organizations had the capacities and capabilities in place to readily scale up their CVA programmes.

During the initial onset of the pandemic, the CWG's activities slowed down as inter-organizational duties were prioritized over intra-organizational matters.

Although all five organizations had begun institutionalizing CVA at the organizational level, only three had conducted in-country CVA activities prior to the pandemic. Even for those three organizations, CVA was still a relatively new concept for their team in Mexico. In carrying out the tasks, only two of the five organizations had dedicated CVA positions, while the other three organizations integrated CVA-related tasks into an existing function. This presented an acute capacity constraint during the pandemic, especially for the three organizations with cross-sectoral functions. Even the two more experienced organizations noted that the lack of capacity caused delays in the distribution of aid, especially during the period of the pandemic when it was unclear in which ways the virus could be transmitted. One organization reported that they considered the possibility of temporarily providing only aid through digital cash transfers to reduce the risk of spreading the virus to beneficiaries but realized it would not be feasible to do so in a short timeframe.

The pandemic pressed the members of the CWG to evaluate how to strengthen their individual in-country operations. Three of the five members had existing CVA operations, but all mentioned that they would be needing to establish a new contract with an FSP or PSP in the next year or two. Therefore, the members agreed to initiate a joint procurement exercise. The task would be led by the member of the CWG with the most experience in CVA, but all members would contribute to the tasks.

One of the first objectives was to create a database of nationwide FSPs and PSPs operating in Mexico. Each organization contributed to the database by sharing information related to past experiences with FSPs and PSPs. This included the services and delivery mechanisms offered by each FSP/PSP, their network of operation, fixed and variable costs, transfer time and prerequisite, such as 'know-your-customer' (KYC) requirements. Similar to an organization's internal baseline assessment in the *planning* phase, the compilation of this database acted as an intra-organizational exercise to document the group's shared resources. Contact information for each organization was provided where possible, to allow each member of the working group to reach out to any FSP/PSP in the database.

To build upon this database, an extensive FSP/PSP mapping exercise was conducted to identify additional financial institutions with nationwide coverage. As the members of the CWG operated in different Mexican states, FSPs/PSPs with national presence were targeted first as they could mutually benefit all organizations. Once a list of organizations had been generated, an Expression of Interest was launched targeting those FSPs/PSPs with nationwide coverage to gather more information. The response rate remained low as several of the FSPs mentioned that they did not have the capacity to take on additional projects during the pandemic. If the organization did respond, a meeting was set up to orientate the FSPs/PSPs with CVA and inform them of the upcoming joint tender. In these meetings, the FSPs/PSPs also advised the CWG of their requirements (e.g. KYC) and possible delivery mechanisms. Regardless of their interest, each FSP/PSP identified in the exercise was recorded in the database for future reference.

As the database was expanding, the CWG also began activities in the *design* phase by developing a charter document for the group. This initial document described information about the group as a whole, including its purpose, scope and structure, as well as descriptions of each individual member, including the background of the organization, goals and populations they worked with. Not only did this exercise help the group to form a common goal, but also saved time during the tendering phase, as this information was used to help shape the solicitation documents (e.g. *the terms of reference*).

Procurement

With the information gathered from the FSP/PSP meetings, and the initial charter established, the tender documents could start to be assembled. Four key documents needed contributions and agreement from all parties before the tender could be launched:

i. Terms of Reference,
ii. Technical Requirements and Offer Form,
iii Financial Offer Form,
iv. Data Protection Policy.

Creating, editing and finalizing these documents was a rigorous task as each organization of the CWG needed approvals and reviews from other departments within their organizations (e.g., finance, legal, and programme). Other documents, such as the general terms and conditions of the contract and code of conduct, aligned with the leading organization's principles – but were accepted by all members of the CWG.

The *technical requirements* and *financial offer form* would provide the CWG with the necessary information for awarding a contract, thus those documents particularly needed to be representative of each individual organization and the group as a whole. To allow the procurement activity to serve all organizations equally, bidders were requested to denote their available delivery mechanisms as well as areas of operation. While preference was given to nationwide operators, this also increased competition in the tendering process by allowing bidders with only regional or local coverage, despite the desire for nationwide coverage.

The solicitation process resulted in the leading organization of the CWG establishing a contract with the FSP which offered the best value for money and nationwide coverage for CVA. The winning bidder was also the most flexible service provider, by offering cash or voucher packages, with physical and digital means for both. Each of the members of the CWG could either 'piggyback' onto the established contract or establish their own contract with the winning bidder, or another bidder, depending on their procurement rules and individual programming needs.

Lessons learnt and future activities

Following the closure of the joint procurement activity, the CWG held a meeting to discuss lessons learnt and plan for the next steps. Two organizations commented that they gained significant knowledge during the procurement process and would be able to adopt a more assertive role in the next procurement process. The lead organization also commented that through the CWG, more FSPs were able to be identified on the market, leading to a contract with one that offered the best value for money. All members agreed that they were able to achieve a greater outcome through this collaboration, as opposed to conducting the process individually, especially considering the strain caused by the pandemic.

Moving forward, the group addressed future activities related to *implementation, monitoring,* and *evaluation* phases. Even though the individual members worked in different contexts and with different populations of interest, they agreed that sharing information for monitoring and evaluation activities could reduce redundant work and increase efficiency at the network level. For instance, one member advertised that they were conducting financial literacy training through an NGO and would be willing to share their findings and information if other members were interested. Another member said they were re-evaluating their CVA exit strategies and would be willing to receive any inputs from other members. The information exchange continued and created a snowball-like effect which resulted in the creation of a 'good practices' document to record how CWG members carried out CVA activities and challenges experienced. The document would act as a reference guide that would be continuously updated by the group based on their learnings. This would also provide orientation and a useful tool for future organizations that were interested in joining the group.

In addition to bringing other organizations into the CWG, the members also discussed how they could link their information to the government of Mexico's various social safety net programmes, such as *Prospera*, to ensure sustainable social protection. For the group, their CVA programme needed to have thorough exit strategies so that the vulnerable populations would remain covered as they transitioned into society, or otherwise changed situations. This common protection aspect adheres to UN Sustainable Development Goal number 17 (Partnerships for the Goals) and further strengthens the link between humanitarian relief aid and development work.

The creation of the CWG enhanced the level of preparedness for all members. It provided an opportunity for each member to develop their capabilities and share resources, which created a larger capacity than any one single organization possessed. Thus, the group as a whole was able to perform higher-level tasks, saving time and resources across all organizations.

Summary and conclusion

Cash and voucher assistance is becoming an increasingly vital component of humanitarian aid, with organizations expanding their operations and piloting projects in new contexts. As interest continues to grow, the demand for skilled CVA operators will also grow. The sector currently faces a shortage of experts, largely due to the specialized training needed to manage programmes. Consortiums of NGOs and working groups have been working to build the global capacity for CVA programming to address this problem. For instance, the CALP Network has begun piloting the *Building Individual Expertise Programme*, to enhance CVA knowledge at the individual level, so that personnel may then join a humanitarian organization with qualifying skills. This too will take time to build a developed workforce in various regions though. Against this constraint, organizations must be willing to also invest time and resources to develop the necessary skills and knowledge within their organization to deliver cash and vouchers effectively. Utilizing extant CVA toolkits and conducting assessments, such as the ones provided in this chapter, are good practices to help identify which areas should be invested in and how resources can be allocated.

Preparedness for CVA not only means having contingency plans built into an organization's framework but also ensuring that the organization has the means to carry out the plans. Despite the evidence linking collaboration and information sharing to preparedness, organizations still often operate in silos, focusing on their own mandates and objectives. Though, this is not because of an unwillingness to form partnerships, but rather due to the fact that the formation of working groups and coordination of CVA activities is still done on an *ad hoc* basis. Initiatives at the global level are underway, such as the UN Common Cash Statement for UN agencies, the Collaborative Cash Delivery Network for NGOs and Cash Coordination Caucus, although these initiatives face several hurdles due to the lack of standardization across the sector. As changes may be slow to take effect across the sector, it remains the responsibility of each organization to actively seek out opportunities for joint initiatives in the face of shared challenges.

Acknowledgements

The author would like to thank all the members of the humanitarian organizations who contributed their time to this study. Furthermore, this work would not have been possible without the generous support of the Hans Bang Foundation, the Hanken Support Foundation, and the Foundation for Economic Education (Liikesivistysrahasto).

References

Altay, N, and Pal, R (2014) Information diffusion among agents: Implications for humanitarian operations, *Production and Operations Management*, **23**(6), pp 1015–1027

Barney, J (1991) Firm resources and sustained competitive advantage, *Journal of management*, **17**(1), pp 99–120

Beazley, R, Marzi, M, Steller, R (2021) Drivers of Timely and Large-Scale Cash Responses to COVID19: What does the data say?, in *Social Protection Approaches to COVID-19 Expert Advice Service (SPACE)*, DAI Global UK Ltd, United Kingdom

Black, S A (2015) Qualities of effective leadership in higher education, *Open Journal of Leadership*, 4(2), pp 54–66

CALP (2018) Glossary of Terminology for Cash and Voucher Assistance, *CALP Network*, www.calpnetwork.org/wp-content/uploads/2020/03/calp-glossary-english.pdf (archived at https://perma.cc/45ZN-QQ76)

CALP (2019) Consolidated Feedback from Field Consultations on the Draft Cash Coordination Guidance for Cluster Coordinators and Draft Standard ToRs for Cash Working Groups, *CALP*, Oxford

Castillo, J G (2021) Deciding between cash-based and in-kind distributions during humanitarian emergencies, *Journal of Humanitarian Logistics and Supply Chain Management*, 11(2), pp 272–295

Development Initiatives (2021) Global Humanitarian Assistance Report 2021, devinit.org/documents/1008/Global-Humanitarian-Assistance-Report-2021.pdf (archived at https://perma.cc/HKL3-CXWB)

Gentilini, U, Almenfi, M, Orton, I, and Dale, P (2021) 'Social Protection and Jobs Responses to COVID-19: A Real-Time Review of Country Measures', World Bank, Washington, DC

Gonçalves, P (2011) Balancing provision of relief and recovery with capacity building in humanitarian operations, *Operations Management Research*, 4(1), pp 39–50

Harvey, P (2007) 'Cash-Based Responses in Emergencies', HPG Report 24, Overseas Development Institute (ODI), London

Harpring, R, Maghsoudi, A, Fikar, C, Piotrowicz, W D and Heaslip, G (2021) An analysis of compounding factors of epidemics in complex emergencies: a system dynamics approach, *Journal of Humanitarian Logistics and Supply Chain Management*, 11(2), pp 198–226

Heaslip, G, Haavisto, I, and Kovács, G (2015) Cash as a form of relief, in *Advances in Managing Humanitarian Operations*, Zobel, C, Altay, N, and Haselkorn, M, (eds) pp 59–78, Springer, Cham

Heaslip, G, Vaillancourt, A, Tatham, P, Kovács, G, Blackman, D, and Henry, M C (2019) Supply chain and logistics competencies in humanitarian aid, *Disasters*, 43(3), pp 686–708

IFRC (2021) 'Guidance for Mainstreaming Cash and Voucher Assistance Cash Preparedness for Effective Response, /cash-hub.org/wp-content/uploads/sites/3/2021/06/CVAPreparedness-Guidance_-Chapter-1_CVAP-Areas_v2-Jun21.pdf (archived at https://perma.cc/L3VP-E9DD)

Jodar, J, Kondakchyan, A, McGormack, R, Peachley, K, Phelps, L and Smith, G, (2020) 'The State of the World's Cash 2020', www.calpnetwork.org/publication/the-state-of-the-worlds-cash-2020-full-report/ (archived at https://perma.cc/UEK2-4CF2)

Javidan, M (1998) Core competence: what does it mean in practice?, *Long range planning*, **31**(1), pp 60–71

Kovács, G, Tatham, P, and Larson, P D (2012) What skills are needed to be a humanitarian logistician?, *Journal of Business Logistics*, **33**(3), pp 245–258

Kunz, N, Reiner, G, and Gold, S (2014) Investing in disaster management capabilities versus pre-positioning inventory: A new approach to disaster preparedness, *International Journal of Production Economics*, **157**, pp 261–272

Maghsoudi, A, Harpring, R, Piotrowicz, W and Heaslip, G (2021) Cash and voucher assistance along humanitarian supply chains: A literature review and directions for future research, *Disasters*, https://doi.org/10.1111/disa.12520 (archived at https://perma.cc/QC9C-LUJH)

Masino, S, and Niño-Zarazúa, M (2020) Improving financial inclusion through the delivery of cash transfer programmes: The case of Mexico's progresa-oportunidades-prospera programme, *The Journal of Development Studies*, **56**(1), pp 151–168

Metcalfe-Hough, V, Fenton, W, Willitts-King, B, and Spencer, A (2021) 'The Grand Bargain at five years: an independent review', HPG commissioned report, ODI, London

Olavarrieta, S, and Ellinger, A E (1997) Resource-based theory and strategic logistics research, *International Journal of Physical Distribution and Logistics Management*, **27**(9/10), pp 559–587

Prahalad, C K, and Hamel, G (1990) The core competencies of the corporation, *Harvard Business Review*, **68**, pp 79–91

Sanchez, R, Heene, A, and Thomas, H (1996) *Dynamics of Competence-Based Competition*, Elsevier Pergamon, Oxford

Seuring, S (2008) Assessing the rigor of case study research in supply chain management, *Supply Chain Management*, **13**(2), pp 128–137

Smith, J (2009) Solidarity networks: what are they? And why should we care?, *The Learning Organization*, **16**(6), pp 460–468

Spencer, A, Parrish, C, and Lattimer C (2016) 'Counting Cash: Tracking Humanitarian Expenditure on Cash and Vouchers', ODI and Development Initiatives, London

Stacey, R D (1996) *Complexity and Creativity in Organizations*, Berrett-Koehler Publishers, San Francisco

Stephenson J (1998) The Concept of Capability and its Importance in Higher Education, in *Capability and Quality in Higher Education,* Stephenson, J and Yorke, M (eds), pp 1–14, Routledge, New York

Suifan, T, Saada, R, Alazab, M, Sweis, R, Abdallah, A, and Alhyari, S (2020) Quality of information sharing, agility, and sustainability of humanitarian aid supply chains: An empirical investigation, *International Journal of Supply Chain Management*, **9**(5), pp 1–13

Tatham, P, Heaslip, G and Spens, K (2018) Technology Meets Humanitarian Logistics: A View on Benefits and Challenges, in *Humanitarian logistics: Meeting the challenge of preparing for and responding to disasters,* 3rd ed, Christopher M and Tatham P (eds), pp 76–97, Kogan Page, London

Tomasini, R M and Van Wassenhove, L N (2009) *Humanitarian Logistics*, Palgrave Macmillan, Houndmills, Basingstoke, UK

UNHCR (2016) 'Policy on Cash-Based Interventions' www.unhcr.org/protection/operations/581363414/policy-on-cash-based-interventions.html (archived at https://perma.cc/SCF9-ZDRZ)

UNHCR (2017) 'Cash Feasibility and Response Analysis Toolkit' www.unhcr.org/5a8429317 (archived at https://perma.cc/63E3-VJAB)

UNHCR (2021) 'Asylum-seekers, refugees and returned refugees from Guatemala, Honduras and El Salvador' www.unhcr.org/refugee-statistics/insights/forcibly-displaced-and-stateless-persons/visualisation-americas.html?situation=3 (archived at https://perma.cc/B8GB-82H2)

UNDRR (2009) Terminology, UNISDR: Geneva

Vogel, B, Tschunkert, K, and Schläpfer (2021) The social meaning of money: Multidimensional implications of humanitarian cash and voucher assistance, *Disasters*, doi.org/10.1111/disa.12478 (archived at https://perma.cc/4UCV-477Y)

Wakolbinger, T, and Toyasaki, F (2018) Impacts of funding systems on humanitarian operations, in *Humanitarian logistics: Meeting the challenge of preparing for and responding to disasters,* 3rd edn, Christopher M and Tatham P (eds), pp 41–57, Kogan Page, London

WFP (2021) 'WFP Strategic Vision (2022-2026)', World Food Programme, Rome

WHO (2020) 'WHO COVID-19 Dashboard' covid19.who.int/ (archived at https://perma.cc/7BCY-34S6)

Zoiopoulos, I I, Morris, P W G and Smyth, H J (2008) Identifying organizational competencies in project oriented companies: an evolutionary approach, in *Procs 24th Annual ARCOM Conference,* Dainty, A (ed) pp 547–555, Association of Researchers in Construction Management, Cardiff

10

Pandemic response and humanitarian logistics

GYÖNGYI KOVÁCS, TINA COMES AND IOANNA FALAGARA SIGALA

ABSTRACT

The effects of the Covid-19 pandemic have been exacerbated by unpredictable human behaviour, a lack of coordination and disrupted health supply chains. As we will be confronted with more epidemics and pandemics in the future, in this chapter we develop an overview of the most critical lessons learnt from humanitarian (health) logistics that can be applied to pandemic response to prevent disruptions and ensure an efficient and fair delivery of health services. Starting from an overview of lessons from past pandemics, we investigate two specific supply chain problems in the Covid-19 response: (i) the personal protective equipment (PPE) supply chain at the onset of the pandemic as a case for rapid scale-up, and (ii) the vaccine supply chains that combine volatile supply and demand with the complexity of large batch-sizes and demanding cold chain requirements. For each part, we develop concrete lessons learnt and make recommendations for tools from humanitarian logistics that can be applied to the specific health logistics challenge.

Introduction

Epidemic and pandemic response has always been an essential part of humanitarian operations, and thereby humanitarian logistics. Whenever one looks at the disaster alert map of ReliefWeb or any compilation of ongoing large scale humanitarian efforts, one will spot several operations related to infectious disease outbreaks. Typical efforts include responses to cholera, measles, polio, yellow fever or even Ebola outbreaks, but the list also extends to dengue, zika, Lassa fever, Marburg, SARS and MERS, to name but a few. Some of these have also overlapped with the Covid-19 pandemic. While in the case of Covid-19, the health emergency is at the origin of the humanitarian disaster, often, man-made or natural disasters lead to an outbreak of infectious diseases (Charnley et al., 2021). Accordingly, humanitarian healthcare extends much beyond infectious diseases to healthcare provision in conflict zones, in the aftermaths of so-called natural disasters, in internally displaced people's (IDP) and refugee camps or even as healthcare provision to people who would otherwise not have access to healthcare services. Epidemic and pandemic response is therefore just a small part of humanitarian health care overall.

While the world struggles to respond to the Covid-19 pandemic, there is this vast area of humanitarian healthcare we can all learn from. This also goes for humanitarian healthcare logistics and healthcare supply chains. Overall, there are some similarities between humanitarian logistics and public healthcare logistics. First and foremost, both areas subscribe to the Hippocratic oath of 'first, do no harm', and both assist people in need. There are similarities in:

- processes, with both starting with needs assessment and prioritization or if need be, triage, under capacity constraints
- performance objectives, especially the explicit inclusion of equity as an objective, while neither type of organization is seeking profit
- data and information management, as both beneficiary and patient data are highly sensitive and of high importance for the coordination of the response and the decision making.

This chapter focuses on the overlap of humanitarian logistics and public health, and more specifically, on the area of pandemic response. We will therefore revisit the particularities of humanitarian logistics in pandemic response first, before turning to the case of the Covid-19 pandemic. We will present our lessons learnt in the first wave of the pandemic concerning personal protective equipment (PPE) supply chains, and then turn the focus to the supply chain tools that are at hand in the humanitarian logistics domain to set up and accelerate Covid-19 vaccination programmes.

Humanitarian healthcare logistics

Public health and medical logistics are highly regulated areas. These regulations also apply to medical humanitarian logistics and supply chains. They come in many different shapes and forms.

Regulations extend to national *drug lists* that are set by ministries of health and the WHO Model List of Essential Medicines (Chomilier and Chomilier, 2019) that set what can be imported to any given country. These challenge procurement and import at the same time as they facilitate customs clearance in healthcare humanitarian supply chains. Also outside of any emergencies, the procurement of medicines is an area that humanitarian organizations are much involved with. For example, UNICEF procures medicines including vaccines on behalf of many ministries of health around the world. And yet, import and customs in humanitarian logistics remain a difficult and vastly non-standardized area. Initiatives such as the current Importation and Customs Clearance Together (IMPACCT) project and its various working groups under the United Nations Office for the Coordination of Humanitarian Affairs (OCHA) tries to address this challenge. Not surprisingly, an urgent working group under the project focuses on the import of medicines.

In this regard it is important to note that medical supply chains are inherently global. This is not only due to the few places where, for example, pharmaceuticals are manufactured. Also upstream the supply chain, ingredients for these pharmaceutical products may not exist or may not be produced everywhere in the world. Safety stocks are thus essential for bridging any potential variations in demand or in lead times, thereby guaranteeing security of supply. On the one hand, such safety stocks can be local or part of a national preparedness stock, on the other, humanitarian organizations include medical items also in their international *preparedness and global pre-positioning*.

Healthcare preparedness extends to other activities as well, such as training, the seeking of interoperability, and the development of standard modules. This is an area in which humanitarian organizations have come together to develop so-called 'interagency emergency health kits' (Vaillancourt, 2016). The sizes of these kits vary. Some have been developed for specific epidemics (e.g. cholera kit, measles kit) or health care units (trauma and emergency surgery kit), others include or supply entire field hospitals. While when putting together kits more upstream the supply chain may be more complicated, they reduce outbound picking times and eliminate the problem of missing items on site. What is more, with many agencies using the same kits, they can also support and deliver to one another, with the receiving end correctly identifying in which box to find which item, and which to treat differently in terms of e.g., temperature control.

Temperature control is an important material-specific aspect to consider in medical humanitarian logistics. There are several different temperature ranges to consider; with special considerations going to vaccines especially with a live ingredient (that needs to be kept alive) and to countries with extreme temperatures but unstable energy sources for cooling systems. The latter problem is particularly challenging for humanitarian cold chains (Comes et al., 2018). But even outside the humanitarian context, few logistics service providers meet the requirements of the pharmaceutical industry's Good Distribution Practices certification.

Temperature control is not the only item-related issue. The quality of medical items and the sensitivity of medical equipment posit further restrictions on humidity, vibration, and dust. In addition, healthcare regulations specify who is permitted to handle specific types of drugs and equipment, which extend not only to healthcare centres but also to anyone receiving, storing and moving items at ports, airports and warehouses.

Insights from past pandemic response

Disease outbreaks, epidemics and pandemics come with even more challenges. Again, humanitarian logistics has dealt with them before, and there is substantial research on pandemic response from the humanitarian context. The Covid-19 pandemic has resulted in numerous special issues in disaster management, healthcare, and logistics and supply chain management journals alike.[1] The first special issue summarizes past pandemic response research in humanitarian logistics, and the second comprises first results from humanitarian logistics research focusing on the Covid-19 pandemic.

Disease outbreaks, epidemics and pandemics typically go beyond the usual health care provision that humanitarian organizations can also be involved with. Many humanitarian organizations are specialized in healthcare provision, after all, they often run healthcare centres, clinics and hospitals in various areas including conflict zones around the world. What distinguishes disease outbreaks is that they result in a sudden increase of demand for very specialized healthcare services. This surge results in a sudden increase in volumes in the supply chain and many decisions that need to be taken in a time-compressed fashion to meet the urgency requirements of dealing with the outbreak. In other words, we are dealing with a health emergency.

As with other types of disasters, health emergencies can also be reoccurring or seasonal, as is the case with seasonal cholera. Others are the cascading effects of other disasters. For example, disasters resulting in unstable sanitation may become the breeding ground for waterborne diseases. In other words, many disease outbreaks are not as unpredictable as they may seem at the first glance. This is good news for both mitigation and preparedness. Yet others are prevalent under specific climate conditions (e.g., malaria). In their case, climate change sadly enables the migration of their disease vectors (in this case certain types of mosquitoes) and a potential further spread of such diseases as well over time.

Both large-scale and especially difficult outbreaks lead to the mobilization of the wider humanitarian healthcare community. Depending on the disease, in the outbreak area, all sorts of activities can be disrupted by the outbreak. Infected people are not only parts of families and social networks but are also a part of the workforce. Companies and critical infrastructures are impacted by the epidemic through their employees, both if employees get sick, or in terms of taking protective measures for their employees. During the Covid-19 epidemic, many companies shut their operations at different points, either due to their workforce being directly affected, or due to lockdown policies. Healthcare providers and emergency services had to limit operations. Lockdowns and closures affected shops, manufacturing facilities and transportation routes. They caused supply chain disruptions both for items available in quarantine zones and lockdown areas as well as globally through cascading effects in the supply chain. Already by February 10, 2020, at the very beginning of the Covid-19 outbreak, shipping lines alone had been estimated to have lost US$350 million per week in lost volumes (Paris, 2020) and multiple airlines suspended their flights to China. Every aspect of life of people in affected regions was impacted, with a stark reduction to only essential services.

At the same time, medical supply chains are impacted not just by a lack of remaining transportation capacity, and a lack of access to the affected regions, but also interruptions of the cold chain (Comes et al., 2018), and by a sheer lack of critical materials (ranging from face masks to IV fluids to gloves and PPE) both due to a lack of pre-positioned quantities but also irregular purchasing behaviour that includes bullwhipping, panic buying, as well as speculative pricing. Similar effects of supply chain disruptions have been observed in other pandemics (Kumar and Chandra, 2010).

As for medical supply chains to outbreak and especially quarantine locations, one key fear of freight forwarders and couriers themselves is to risk contagion. During the 2015 Ebola crisis, people delivering aid risked being quarantined as well. This included flight crews and lorry drivers as well as grounded flights and lorries restricted from exiting quarantine zones again. This calls for different approaches for medical deliveries such as the use of unmanned aerial vehicles (UAVs, drones) that do not expose any person to contagion.

UAVs have been proposed and used in other areas of humanitarian logistics as well, including for medical humanitarian logistics (Tatham et al., 2017). UAVs are in place already in medical deliveries, for example in Rwanda, including complex hub and spoke systems to deliver rare, critical items such as snake antivenom, vital blood supplies and other medication.

The *key lessons from past pandemics* can be summarized as follows:

- Healthcare emergencies lead to cascading effects that ripple across many sectors and domains, affecting critical infrastructures, public sector and businesses alike by a shortage of workforce and lack of transportation capacity.
- Direct and indirect supply chain disruptions hamper the efforts to effectively manage the pandemic, exacerbated even more by panic buying and speculative behaviour.
- New technologies such as UAVs are promising alternative avenues to re-organize logistics processes in epidemics.

Lessons learnt from the Covid-19 pandemic: The case of personal protective equipment

Many global material shortages caused supply chain disruptions during the first wave of the Covid-19 pandemic. Not surprisingly, the first outcry concerned adequate PPE. Those countries that subsequently included them in their preparedness and pre-positioning have bought themselves some breathing time during the Covid-19 outbreak (Falagara Sigala et al., 2020). Pre-positioning is a common risk mitigation strategy in disaster relief that accelerates disaster response through enabling first deliveries in parallel to activating procurement for next ones. It significantly reduces delivery lead times yet enables regular order lead times. Pre-positioning is most effective if included in coordination mechanisms and focusing on commonly needed items across organizations, regions, and disasters. Across these, 'white' stock that is labelled only upon activation for a specific purpose minimizes obsolescence while maximizing inventory turnover (Mochizuki et al., 2015; Sabbaghtorkan et al., 2020). But while common in the humanitarian space, even the EU's coordination mechanism RescEU was activated late in the first wave of this pandemic.

Pre-positioning can be done for individual items, or by kitting packages of interdependent items. As mentioned earlier, across medical humanitarian organizations, inter-agency health kits have been developed for this purpose (Vaillancourt, 2016). These kits reduce picking and packing times, i.e. accelerate deliveries (Kovács and Falagara Sigala, 2021), and eliminate the problem of missing items on site; after all, most vaccines also need syringes, needles and gloves for their administration. Sadly, our findings indicate a lack of recognition of interdependencies in the first wave of the Covid-19 pandemic, thus supply chain disruptions followed the critical path across them (findings here are presented from the HERoS project).[2] For example, in the testing supply chain, first, test swabs were not available in adequate quantities, then once that bottleneck was resolved, reagents were missing for labs (Falagara Sigala et al., 2020). From a supply chain perspective, bottleneck analysis techniques such as *critical path analysis* are useful for spotting interdependencies and turning one's focus to the removal of the next bottlenecks in time.

Coordination mechanisms also apply to procurement (Kovács and Falagara Sigala, 2021), but little joint procurement was done in the first wave of the pandemic. Instead, different healthcare providers, countries and governmental and non-governmental organizations all tried to procure the same PPE in parallel. This led to supply chain disruptions due to price wars for scarce items (e.g. facemasks) and additional bullwhipping and panic buying when facing shortages (Falagara Sigala et al., 2020). Such behaviours often surface when organizations face sudden surges in demand, yet they only exacerbate the problem (Patrinley et al., 2020). Some countries employed desperate measures such as export bans, ever-changing export regulations, and the seizure of already customs-cleared items. Together, these measures led to extreme price fluctuations and a lack of oversight in quality control. Yet production was diminished by the effects of the same pandemic, thus none of these behaviours facilitated access to PPE (Falagara Sigala et al., 2020). Later, more coordinated procurement by June/July 2020 helped calm the effects of these irrational behaviours.

Production changeover, the means of refitting existing production lines to manufacture related items, was the only measure that alleviated shortages at the onset of the pandemic. This way the automotive industry could manufacture respirators, while pulp and paper and textile industries switched to PPE. Further additions to global capacity, and thereby the availability of critical items, were achieved by setting up new production lines or adding shifts in manufacturing. Any such measures require additional certified personnel, machinery and/or quality control. Remarkably, qualified personnel became the largest capacity constraint in the first wave, both in PPE and respirator production and with regards to ICU capacity at hospitals (HERoS Covid-19 map, 2021; Uimonen et al., 2020). Other supply chains were constrained by upstream raw material scarcities, such as cotton for test swabs. Notably, reshoring moved final assembly lines closer to consumption but did not overcome upstream component and raw material shortages.

Further bottlenecks exist in the downstream supply chain. In a pandemic, many of these are caused by export and travel restrictions. Prior research highlighted the ineffectiveness of travel bans for reducing effective contact tracing as people travel in hiding (Nuzzo et al., 2014; Pruyt et al., 2015). From a supply chain perspective, travel bans complicate freight transportation if truck drivers and pilots alongside their vehicles get trapped in outbreak zones. During the Covid-19 pandemic, truck drivers have indeed refused deliveries to outbreak zones (e.g. New York), or have temporarily been barred from exiting them (e.g. crossing from the UK to France due to the B1.1.7 variant). Also, the grounding of passenger planes has disrupted the movement of medical supplies that are usually transported as belly cargo of passenger planes. Higher prices did not resolve the fact that certain routes were completely unavailable, hampering especially PPE transportation out of China (Falagara Sigala et al., 2020). Finally, applying production changeover principles, passenger planes were converted to cargo planes. Alternatively, as highlighted earlier but not yet used in this pandemic, autonomous vehicles such as driverless trucks and unmanned aerial vehicles (Comes et al., 2018) could remove the risk of contagion for drivers and pilots.

The key lessons learnt from PPE supply chains in the Covid-19 pandemic are:

- *Pre-positioning* of critical goods provides the much-needed buffer capacity especially at the onset of a health emergency.
- *Kitting* along with an analysis of interdependencies of different items along with a *critical-path-analysis* can help overcome bottlenecks and increase robustness of supply chains.
- *Coordination* in procurement and an alignment of regulations together are effective in preventing supply chain disruptions due to price wars and bullwhipping or price fluctuations.
- *Production capacity* is key to rapidly make critical items available locally. *Production changeover* was the only measure to alleviate shortages in the beginning of the pandemic.

Supply chain tools to accelerate vaccination programmes

Rarely is a vaccination programme under so much scrutiny as Covid-19 vaccinations worldwide. In record time, an increasing number of Covid-19 vaccines have been granted market authorizations, but problems with vaccine production and distribution have shifted the attention to vaccine supply chains. The sheer size of a global vaccination endeavour is astonishing. At the same time, as vaccination programmes continue worldwide, and new Covid-19 variants are developing, there is an ever-increasing demand for more doses of vaccines. In addition, while the vaccination programmes in the European Union (EU) and the United States (US) are quite advanced, the same cannot be said for many other countries. There are numerous factors that contribute to the roll-out of vaccination programmes, ranging from the operational and supply chain ones to public health policy and even global politics. Vaccine supply chains have an important role to play in this mix.

Joint preparedness and pre-positioning are common for inter-agency health kits in the humanitarian space when it comes to vaccination programmes as well. However, these do not extend to vaccines in their trial phases, or those waiting for licences and market authorizations. COVAX is a welcome exception in this regard. The vaccine sharing part of the COVAX scheme was very promising until the political appearance of wanting to share did not deliver in actual quantities or volumes. This is partly due to prioritizing certain populations in the world, but also to the sheer problem of global production capacity lagging behind as well as of the stakeholders and the governance of COVAX. The prioritization issue is at the heart of vaccine nationalism versus vaccine diplomacy. Vaccine nationalism prioritizes one's own population, and is safeguarded by export stops and non-deliveries, for example. These have serious repercussions for vaccination programmes that have expected but not received those vaccines. A first such non-delivery has made it to a Belgian court, ordering AstraZeneca to deliver vaccines to the EU. Similarly, the COVAX scheme has suffered from empty promises of vaccine sharing and of exports, to the extent that the *Lancet* stated on their June 18, 2021 cover page that 'COVAX "was a beautiful idea, born out of solidarity. Unfortunately, it didn't happen… Rich countries behaved worse than anyone's worst nightmares."' (*Lancet*, 2021)

Conversely, some countries have preferred to export large quantities of vaccines, even prioritizing those exports over vaccinating their own populations. Such behaviour has also drawn criticism for vaccine diplomacy.

Specific for the Covid-19 pandemic was the question of accelerating vaccination programmes even prior to the market authorization of a vaccine. Even though *pre-positioning of vaccines without a market authorization* is not common, legally, in most cases (e.g. in the EU and the UK) it is possible. This was done for the Pfizer BioNTech and Moderna vaccines that were delivered to warehouses close to vaccination centres throughout the United States, and then activated from there once the US Food and Drug Administration granted them their emergency use authorizations (EUAs). This speculation strategy of pre-positioning combined with parallel processing of the authorization significantly reduced delivery lead times and accelerated vaccine distribution. While quality control may complicate matters in terms of a qualified person assuring the quality of any such pre-positioned batch, batchwise quality assurance was also done in the UK that granted batch-specific EUAs.

The joint procurement of vaccines has commenced early in the EU as well as across the COVAX scheme. Nonetheless, panic buying and bullwhipping had already been observed at the very outset of these in early 2021. These are exacerbated by production shortages and delays in vaccine production, the arrival of new variants of the SARS-CoV-2 virus adding to the urgency of vaccination efforts, but also procedural changes in vaccination programmes. For example, the US changing their vaccination programme from warehousing second doses of initially two-dose vaccine regimes to direct distribution without informing key partners has caused them to first decline a next delivery, but then lack exactly those quantities. Consequently, new orders were placed, resulting in a global price hike plus non-deliveries of already promised quantities to other countries (Norway, EU, etc.). Politically, additional orders of vaccines, and additional EUAs, create the appearance of caring for one's citizens, regardless of the fact that global production capacities are constrained, and such behaviour only adds to higher prices and further supply chain disruptions, as we have seen with PPE. In late 2021/early 2022, some countries adding third, or even fourth, booster doses to their vaccination schemes further exacerbated the sheer lack of vaccines in other parts of the world, as pharmaceutical companies are scrambling to adapt their vaccines to the new variant.

Recognizing interdependencies through a kitting strategy and a critical path analysis would also facilitate an early recognition of next possible bottlenecks. Importantly, such analyses must account for a dynamically changing system, in which available vaccines, production capacities, procurement strategies, but also restrictions on transportation (border closures or customs procedures) are in flux. As such, planning needs to account for the tremendous uncertainties inherent to the system, and favour robust approaches that succeed under a broad range of scenarios. While traditionally, robust and flexible vaccine supply chains have been advocated at a regional level, here, robustness considerations need to consider a rapidly changing global network. During the various waves of the Covid-19 pandemic, it has become evident that syringes and needles, not only vaccines, are in short supply globally.

Throughout all these waves and developments, *production capacity* being the largest bottleneck, adding capacity is the obvious choice for coping with global demand. Unlike the redistribution of intensive care unit (ICU) patients across countries for operational smoothing (and the coping of some health care systems) earlier, no geographical cross-distribution of vaccines addresses global shortages. In addition, there are the many challenges related to cold chains that favour regional production. Some vaccine producers have from the beginning reached out to facilities outside their own network of factories to add production capacity, with limited success. Some agreements have since been reached, for example between AstraZeneca and Serum Institute of India, as well as Siam Bioscene, and Pfizer BioNTech with Sanofi and Novartis. Production changeover is smoothest at pharmaceutical companies with the same technologies and skills available that do not (yet) produce their own Covid-19 vaccines. Yet even those facilities need months to set up production lines for this purpose. In addition to incentivizing vaccine research and development, it is time to incentivize the manufacturing capacity of Covid-19 vaccines (Sun et al., forthcoming), e.g. through advance purchase agreements, specific loans or grants and funding for companies to scale up and refit their factories for this purpose.

In summary, the factors that drove supply chain performance for PPE also hold for the vaccine supply chains in Covid-19. These points are mainly focusing on creating the much-needed capacity to source and produce the vaccines:

- *Coordination* and *joint procurement* as well as procedural consistency are needed for the rapid set-up of fair vaccination programmes.
- *Kitting* along with *critical-path-analyses* can help detect and overcome critical bottlenecks.

- *Production capacity* is key and can be supported by early speculative investments, advance purchase agreements, specific grants and refitting of factories.

Yet, there are also problems and tools that are specific to the vaccine supply chains, namely tools to manage the volatility of the delivery schedule and demand in combination with the strict cold chain requirements. Therefore, vaccination programmes need careful planning for the distribution and administration of vaccines, not the least due to cold chain requirements in combination with large package sizes and the two-dose regimen of most Covid-19 vaccines, not counting booster schemes.

The combination of *time and temperature control* is critical in cold chains. Three main temperature ranges exist for Covid-19 vaccines, either requiring an 'ultra-cold' chain (−80 to −60°C), a 'frozen' chain (−25 to −15°C) or a 'refrigerated' cold chain (+2 − +8°C). (Fourth, −40°C is used upstream the supply chain for vaccine raw materials.) Of these, refrigerated cold chains are most typical for vaccines (Comes et al., 2018; Duijzer et al., 2018), but also food and nutrition products require similar refrigeration. Frozen chains fall in the same category as frozen foods. There is warehousing and transportation capacity for both, albeit with higher dependence on steady cooling systems the lower the required temperature. However, while violating temperature constraints may limit shelf-lives in food supply chains, cold chain violations render vaccines unsafe and unusable. Typical breaking points of temperature control supply chains are the loading and unloading of vehicles, intermodal operations, movements at hubs, customs clearance and quality control. The complexity of international vaccine cold chains is aggravated by the absence of standardized cooling technologies (Comes et al., 2018), and different infrastructure systems. In particular, unreliable electricity or delays in last-mile transportation can be barriers in low- and middle-income countries (de Boeck et al., 2020).

Ultra-cold chains exacerbate vaccine supply chain complexity, as ultra-cold freezer units for vaccination centres are scarce and ultra-cold transportation requires active plus passive cooling, i.e. reefer containers with power cells plus dry ice. Dry ice, consuming oxygen, is heavily regulated in air transportation. Yet that is the preferred transportation mode for such vaccines, to comply with their movement time restrictions while covering the maximum distance from production facilities. Each Covid-19 vaccine comes with additional limits on different temperature ranges and their time limits during production versus movement versus materials handling and final use. Strict time compression requirements from production to vaccination delimit the numbers of utilizable nodes and links in the supply chain. Every step needs detailed planning; with a good benchmark being the ultra-cold chain of the STRIVE Ebola vaccine in the most challenging environments (Jusu et al., 2018). Even package sizes matter, for example, packages of 1,000 doses with ultra-cold chain requirements need large vaccination centres for safe administration.

To date, there are three main strategies to deal with unforeseen fluctuations despite the large batch sizes and taxing cold-chain requirements:

- *Direct distribution* that maximizes the number of first dose inoculations especially to the most vulnerable populations, but delays the second
- The accumulation of *safety stock* and establishing large-scale infrastructure and vaccination centres, partially along with export bans of vaccines manufactured in one's own country/region
- Waiting for *easier administrable vaccines* to be distributed later within the current infrastructure and public health system. The latter is used especially for rural areas and also developing countries, not always by choice.

All these strategies slow down vaccination efforts. Better predictability of deliveries could reduce the need for safety stock, maintain required time intervals between doses, and avoid frequent rescheduling of planned deliveries, which has created frustration and even unrest in many different countries. This can be achieved with advance notifications, better information flows, and improved *supply chain visibility*. For instance, after the announcement of the French 'passe sanitaire', a record-breaking one million appointments were booked within one day. As such, announcements and regulations should be aligned with the vaccine supply chain. On the other hand, many developing countries are yet to receive vaccines they were promised and that were procured a long time ago.

Summary and conclusion

While there are already some Covid-19 vaccines with market authorizations in specific parts of the world, and more are in the making, it is a false hope for any healthcare system to rely on any specific one. With the globe's population needing to be vaccinated, and the current lack of production capacity for any of these vaccines, this is a race against time. No system has the luxury to wait for specific vaccines that would not require certain temperature ranges, or second or more doses; rather, it becomes a question of preparing for all alternatives in parallel.

Summa summarum, as long as vaccine production capacity is the biggest impediment for vaccination programmes, the main focus should be on (a) increasing production capacity; (b) the use of co-ordination mechanisms in procurement, kitting, pre-positioning and regulations; and (c) increasing supply chain visibility and ensuring predictable and steady deliveries.

Acknowledgements

This research could not have been achieved without the kind support of the H2020-SC1-PHE-CORONAVIRUS-2020 project No. 101003606 HERoS (Health Emergency Response in Interconnected Systems).

Notes

1 The *Journal of Humanitarian Logistics and Supply Chain Management* (JHLSCM): www.emeraldgrouppublishing.com/journal/jhlscm/logistics-and-supply-chain-management-pandemic-response (archived at https://perma.cc/A2A7-2A42), and www.emeraldgrouppublishing.com/journal/jhlscm/preparing-humanitarian-supply-chain-epidemics-and-pandemics-response-0 (archived at https://perma.cc/UMT6-XSQQ)

2 For further details of the HERoS project, see 'D3.1 Gap analysis and recommendations for securing medical supplies for the COVID-19 response' at www.heros-project.eu/output/deliverables/ (archived at https://perma.cc/2NW2-9JNV)

References

Charnley, G E, Kelman, I, Gaythorpe, K A, and Murray, K A (2021) Traits and risk factors of post-disaster infectious disease outbreaks: A systematic review. *Scientific reports*, **11**(1), 1–14

Chomilier, B, and Chomilier, C (2019) Medical logistics in humanitarian settings, in Kravitz, AS and van Tulleken, A (eds) *Oxford Handbook of Humanitarian Medicine*, Oxford, UK, pp 229–250

Comes, T, Bergtora Sandvik, K and Van de Walle, B (2018). Cold chains, interrupted: The use of technology and information for decisions that keep humanitarian vaccines cool, *Journal of Humanitarian Logistics and Supply Chain Management*, **8**(1), pp 49–69

Comes, T, Van de Walle, B, and Van Wassenhove, L (2020) The coordination-information bubble in humanitarian response: theoretical foundations and empirical investigations, *Production and Operations Management*, **29**(11), 2484–2507

De Boeck, K, Decouttere, C and Vandaele, N (2020). Vaccine distribution chains in low- and middle-income countries: A literature review, *Omega*, **97**, 102097

Duijzer, LE, van Jaarsveld, W and Dekker, R (2018) Literature review: The vaccine supply chain, *European Journal of Operational Research*, **268**(1), pp 174–192

Falagara Sigala, I, Kovács, G, Alani, H, Smith, B, Xu, J, Wu, G, Rollo, A, Cicchetta, G, Boersma, K, Grant, D, Riipi, T and Wan, K-M (2020) D3.1 – Gap analysis and recommendations for securing medical supplies for the COVID-19 response, Health Emergency Response in Interconnected Systems, www.heros-project.eu/wp-content/uploads/HERoS_D3.1_Final.pdf (archived at https://perma.cc/LFE4-XGF2)

HERoS COVID-19 map (2021), Overview of the ICU occupancy rates in Europe, Health Emergency Response in Interconnected Systems, at https://nhg.fi/en/covid19map/

Jusu, M O, Glauser, G, Seward, J F, Bawoh, M, Tempel, J, Friend, M, Littlefield, D, Lahai, M, Jalloh, H M, Sesay, A B and Caulker, A F (2018). Rapid Establishment of a Cold Chain Capacity of –60 C or Colder for the STRIVE Ebola Vaccine Trial During the Ebola Outbreak in Sierra Leone. *The Journal of Infectious Diseases*, 18;217(suppl_1):S48–55.

Kovács, G and Falagara Sigala, I (2021) Lessons learned from humanitarian logistics to manage supply chain disruptions, *Journal of Supply Chain Management*, 57(1), pp 41–49

Kumar, S and Chandra, C (2010). Supply chain disruption by avian flu pandemic for US companies: A case study, *Transportation Journal*, 49(4), pp 61–73

Mochizuki, J, Toyasaki, F and Sigala, I F (2015) Toward resilient humanitarian cooperation: Examining the performance of horizontal cooperation among humanitarian organizations using an agent-based modeling (ABM) approach. *Journal of Natural Disaster Science*, 36(2), pp 35–52

Nuzzo, J B, Cicero, A J, Waldhorn R and Inglesby, T V (2014) Travel bans will increase the damage wrought by Ebola, *Biosecurity and bioterrorism: Biodefense strategy, practice, and science*, 12(6), pp 306–309

Patrinley, J R, Berkowitz, S T, Zakria, D, Totten, D J, Kurtulus, M and Drolet, B C (2020). Lessons from operations management to combat the COVID-19 pandemic, *Journal of Medical Systems,* 44(7) p 129

Paris, Costas, Logistics report: Coronavirus toll on shipping reaches $350 million a week, *Wall Street Journal,* 10 February, www.wsj.com/articles/coronavirus-toll-on-shipping-reaches-350-million-a-week-11581366671 (archived at https://perma.cc/TE8Y-4HM7)

Pruyt E, Auping W L and Kwakkel J H (2015) Ebola in West Africa: Model-Based Exploration of Social Psychological Effects and Interventions, *Systems Research and Behavioral Science*, 32(1), pp 2–14

Sabbaghtorkan M, Batta R and He, Q (2020) Prepositioning of assets and supplies in disaster operations management: Review and research gap identification, *European Journal of Operational Research*, 284(1), pp 1–19

Sun, H, Toyasaki, F, Falagara Sigala, I (forthcoming) Incentivizing at-risk production capacity building for COVID-19 vaccines, *Production and Operations Management*, doi.org/10.1111/poms.13652 (archived at https://perma.cc/Q6QQ-N4NV)

Tatham, P, Stadler, F, Murray, A and Shaban, R Z (2017) Flying maggots: A smart logistic solution to an enduring medical challenge, *Journal of Humanitarian Logistics and Supply Chain Management*, 7(2), pp 172–193

Uimonen T, Mulari M, Niemi A, Rissanen A, Nuutinen M, Riipi T, Väljä A, Karvinen S, Haavisto I, Leskelä R-L, Boersma K and de Vries M (2020) D2.2 Healthcare system analysis, *Health Emergency Response in Interconnected Systems*, www.heros-project.eu/output/deliverables/ (archived at https://perma.cc/2WAN-SA6W) (2021-02-04)

Vaillancourt A (2016) Kit management in humanitarian supply chains, *International Journal of Disaster Risk Reduction*, 18, pp 64–71

11

Helping people and planet: Making the humanitarian supply chain more sustainable

MARIA BESIOU, SARAH JOSEPH, SOPHIE
T'SERSTEVENS AND JONAS STUMPF

ABSTRACT

The humanitarian sector both contributes to and suffers from climate change consequences. Disasters requiring emergency assistance rarely leave time for the long-term perspective, while a focus on short-term interventions and immediate relief may further damage local communities and the planet. In this chapter we investigate the concept of sustainability in the uncertain context of humanitarian operations and humanitarian supply chains, and define it based on three major objectives: people, planet and profit (Elkington, 1998). We then present the link between humanitarian operations and the United Nations Sustainable Development Goals (SDGs), climate change and the challenges of the Covid-19 pandemic. The contribution of the humanitarian sector to environmental challenges is also discussed, especially in regards to carbon emissions, waste and resource use. Lastly, we identify current local and global initiatives by humanitarian organizations and present opportunities to shift towards a more sustainable system, especially in regards to transparency, developing metrics to measure sustainability in supply chains and evidence-based decision-making.

Introduction

Disasters requiring emergency aid often leave little room for decisions that focus on the long-term perspective – decisions need to be made swiftly in a complex and dynamic environment, with limited information and time for good planning. The prime objective of humanitarian organizations (HOs) is to mitigate the suffering of beneficiaries but focusing solely on immediate relief may result in further damage to local communities and the planet. This also rings true for the broader process of development, recovery and mitigation, in which many of the regions in need of the greatest humanitarian aid are – and will continue to be – most at risk to climate change consequences of the future.

Despite the increased attention on sustainability in organizational operations beginning in the late 1990s (Elkington, 1998), implementation is still in the early stages in many humanitarian organizations and several challenges stand in the way of improving the sustainability of supply chains. This is especially true in the context of humanitarian supply chains (HSCs) (Kovács & Spens, 2009). Uncertainty regarding the timing, location, type, size of demand (as well as lead times and sudden changes in demand), high stakes associated with the timeliness of deliverables and a lack of resources differentiate HSCs from commercial supply chains (CSCs) (Kovács & Spens, 2009). Thus, a unique perspective is required to develop interventions to improve sustainability and reduce the environmental impact of operations while still providing efficient and effective disaster response.

In this chapter we investigate the current and future role sustainability plays in HSCs. First, we provide an overview of sustainability in the uncertain context of HSCs, introduce the link between the United Nations (UN) sustainable development goals (SDGs), our changing planet, and the Covid-19 pandemic. Next, we describe environmental sustainability-related challenges facing the sector, identify current initiatives and discuss potential options to continue to shift perspectives towards more sustainable operations in the future. Finally, we provide concluding remarks and recommendations for future work.

Sustainability in the context of uncertainty

The aspects that differentiate HSCs from CSCs – such as logistics, material convergence and coordination (Corbett, et al., 2021) – also make the integration of sustainability in operations more challenging and unclear. The need to provide immediate relief and save lives can harm communities in the long-term, increasing risk of disasters in the future. Additionally, those countries that experience disasters more often tend to stay or get poorer, with the number of disasters growing over time (Besiou, et al., 2021). This implies an increasing number of people that are vulnerable to future natural, man-made, and complex disasters. Difficulties in development through recovery and mitigation operations are also at risk of being unsustainable (lack of sufficient community investment) or inequitable (lack of resources) (Besiou, et al., 2021). Identifying opportunities to increase sustainability and enhance success of humanitarian operations requires a broad understanding of both the macro (global) and micro (local) context. This includes cooperation between regional and global stakeholders, as well as a focus on evidence-based decision-making for preparedness and planning. These tools can also be used to measure sustainability performance, identify target areas, improve transparency and enhance information flows.

What is sustainability?

A broad, yet widely quoted definition of sustainability comes from (UN, 1987): 'Development that meets the need of the present without compromising the ability of future generations to meet their own needs'. Here, sustainability can be defined as social, economic or environmental sustainability. In the organizational context, sustainability is often measured by the 'Triple Bottom Line' framework (Elkington, 1998): people, profit and planet, in which success can be measured by the adherence to a certain standard of all three. Social metrics (people) include relative poverty and health-adjusted life expectancy (Corbett, et al., 2021). Economic metrics (profit) can be income and expenditures (Corbett, et al., 2021). Environmental metrics (planet) may refer to pollution, soil quality, water and carbon footprint and waste management. Environmental sustainability is also a critical point for social and economic sustainability, as more frequent and strong weather events can make improvements in the social and economic contexts even more challenging.

SDGs as a complex system

In 2015, the United Nations (UN) called for specific actions in adopting seventeen sustainable development goals (SDGs) aimed at improving people's lives and protecting the planet. These goals should be achieved by 2030. In 2020, the UN published a report measuring progress towards the goals of each individual SDG. In some areas, such as those related to peace, the trend is negative, and conflict is often at the core of humanitarian crises (Besiou, et al., 2021). Additionally, more people need urgent food aid worldwide in 2021 compared to 2020, and the Covid-19 pandemic has placed further barriers in front of progress and the 2030 goals.

Several studies (Besiou, et al., 2021; Sumner, et al., 2020) and organizations (World Humanitarian Summit, 2017) have indicated the link between humanitarian operations and the SDGs, noted at the 2016 World Humanitarian Summit, organized by the UN. Specifically, the summit proposes that a 'substantial reduction in humanitarian needs also requires increased investment in national, local, and regional [disaster] preparedness and establishing predictable response arrangements, such as shock responsive social protection and safety nets' (World Humanitarian Summit, 2017). As well, it identifies an increased need for 'investing in data, analysis and early warning, and developing evidence-based decision-making processes that result in early action' (World Humanitarian Summit, 2017). Viewing the link between the longer-term 2030 SDGs and shorter-term humanitarian operations as a complex system can provide valuable insights to move closer towards accomplishing those goals.

Countries requiring the largest volume of humanitarian aid also tend to be the countries which face the greatest difficulties in accomplishing the SDGs. The poorest people often suffer from extreme poverty (SDG1), hunger (SDG2), limited or zero access to health services (SDG3), and a lack of water and sanitation (SDG6). (Besiou, et al., 2021) proposes these four be called *survival SDGs*. The closer countries are to achieving these goals, the better opportunity they have in reaching the *individual SDGs*: promoting gender equality (SDG5), decent work and economic growth (SDG8), access to quality education (SDG4), and affordable and clean energy (SDG7) (Besiou, et al., 2021). Individuals in areas that reach these goals are more apt to accomplish the *communal SDGs*: to invest in industry, innovation and infrastructure (SDG9), sustainable cities and communities (SDG11), responsible consumption and production (SDG12) and reducing inequalities (SDG10). Investing in communal goals is crucial to achieving the *global SDGs:* climate (SDG13), life below water (SDG14), life on land (SDG15) and peace and justice (SDG16). Building partnerships (SDG17) between national governments, non-governmental organizations (NGOs), the private sector and civil society is the capstone objective to reaching the SDGs, according to the UN (Besiou, et al., 2021). Fulfilling survival SDGs increases the opportunity to accomplish individual SDGs, with the same logic following for communal SDGs, global SDGs, and the capstone SDG. Furthermore, the more SDGs a country fulfills, the further it matures past survival SDGs, which has implications for the humanitarian sector.

The Covid-19 humanitarian crisis

Not only has the Covid-19 pandemic added difficulties and uncertainty to an already overstretched humanitarian system, but it has also become a humanitarian crisis in its own right, further challenging social, economic and environmental sustainability. As the pandemic continues to spread, crises within the sector have become increasingly visible, such as inequality, poverty, hunger, gender-based violence, climate-related disasters and armed conflict. In 2021, there were 235 million people in need of humanitarian aid, 40 per cent more than in 2020 (Onabanjo & Julitta, 2021). Vulnerable health systems became overstretched, putting crucial health services such as sexual and reproductive health to the side, affecting women and girls. Based on past global crises, non-financial metrics such as infant and maternal mortality, undernutrition and malnourishment, and lack of educational development will rise (Sumner, et al., 2020).

The crisis also had a major impact on agriculture and food supply chains, especially in low- and middle-income countries (LMICs). In Africa, for example, Covid-19 challenges such as trade blockages, smaller labour force, restriction of movement, and higher food prices added to food insecurity, especially in areas highly reliant on imports (Nkamleu, 2020). The pandemic overall had a major impact on the global economy, but certain regions may face more disastrous consequences than others, especially as resources may be diverted for Covid-19 efforts. Research suggests that global poverty may increase for the first time since 1990, with the number of people living in poverty increasing by up to 420–580 million (Sumner, et al., 2020).

The pandemic has, however, triggered an increased interest in strengthening local and regional systems, especially in response to unpredictable supply chains and to reduce reliance on imports. This is expected to have significant implications for HSCs, especially as donors are pushing for localization (Besiou, et al., 2021). This push to improve the capacity of local communities also has impacts for social, economic, and environmental sustainability, as well as for increasing transparency and efficiency in humanitarian operations.

The environmental impact of HSCS and current initiatives

HSCs' contribution to climate change

As a result of climate change, HOs need to react to more frequent and severe natural disasters. The International Federation of Red Cross and Red Crescent Societies (IFRC) estimated that the number of people needing humanitarian assistance as a direct or indirect consequence of climate change might double from more than 100 million people yearly today to 200 million people yearly by 2050. This forecast reflects *the cost of doing nothing* (which is how the IFRC named its report). Climate change has also been identified as one of the largest drivers of medium to long-term risks in global development, predicted to have severe consequences across a number of sectors including ecosystems, agriculture, industry, commerce, residences and transportation (Laguna-Salvadó, et al., 2019). Not only does climate change pose a risk to the success of humanitarian operations – the humanitarian sector, like many others, also contributes to climate change and can negatively impact the environment.

One widely discussed impact, especially in the HSCs, is global warming. Greenhouse gases (GHG) are emitted all along the supply chain of relief items: for the extraction and processing of the raw materials used as inputs for relief items, during the manufacturing of the relief items, throughout the transportation of the relief items from the suppliers down the supply chain to the distribution to beneficiaries, and in some cases in the use of the item itself (such as food preparation). Transport and distribution is often identified as an area to reduce emissions to mitigate the environmental impact (e.g., by relying on sea rather than air transportation or by optimizing truck utilization rates), since HOs often have the most control or leverage over these steps, rather than upstream activities (such as where suppliers source from). However, upstream activities should certainly not be neglected as these typically represent a proportionally larger carbon footprint. Médecins Sans Frontières (MSF) estimated that the GHG emissions resulting from the extraction and manufacturing processes (also referred to as upstream Scope 3 emissions) needed for their relief items represented around 70 per cent of its total supply chain carbon footprint (The New Humanitarian, 2021).

Waste, and the subsequent environmental degradation and pollution is another consequence of HOs' operations. The distribution of relief items often comes with waste, mainly consisting of packaging materials, used disposables, damaged and expired stock, as well as unsolicited in-kind donations. In some cases, donations are requested, but due to delays the item may no longer be urgent/useful when it arrives at the site (i.e. what is urgent now may not be urgent in a few days). Disposal of unneeded items can be challenging at the site of disasters, and when not managed well, waste pollutes land and/or water. It can also be dangerous for humans and animals in local communities. This is especially true for the case of medical waste. Expired medical items, as well as medical waste, can cause a significant risk to the health and safety of the local community when not properly disposed of.

However, waste is also not only related to the downstream effects. It is also tied to the depletion and deterioration of natural resources at the early stages of the supply chain. Unused goods still required inputs such as natural resources, energy and other materials for production. Challenges in transparency and communication make it difficult to predict demand. This is also relevant for procurement in general, and the activities of suppliers should also be considered here: what impact do they have on their surrounding environment? For example, in 2016, the ICRC realized it purchased a significant amount of palm oil (a total of 4,000 tons in 2016) and were thus contributing to deforestation in Indonesia and Malaysia (IFRC/ICRC, 2019). They are now purchasing 20 per cent of their palm oil from sustainable sources (ICRC, 2021) and are further planning to identify and act against environmental risks embedded in their product portfolio. As well, for several reasons (such as cost restrictions) many food products are procured through suppliers who practice resource-intensive industrialized agriculture. These products are typically produced on a large-scale to maximize efficiency without consideration of the externalities associated with production. *Externalities* refer to the costs or benefits that are externalized to other parties besides the user. Negative externalities include pollution of water, damage to ecosystems, effects on human health and soil degradation. These are not typically included in the price of a conventional food product, but rather paid by society later (through a variety of mediums such as the cost to clean up waterways).

Current initiatives

Today, many HOs are committed to helping the communities they support become climate resilient. This is reflected in the first commitment (out of six) of the Climate and Environment Charter for

Humanitarian Organizations: *Commitment 1 – Step up our response to growing humanitarian needs and help people adapt to the impacts of the climate and environmental crises.* The Climate and Environment Charter for Humanitarian Organizations was co-developed by the International Committee of the Red Cross (ICRC) and the IFRC with the input from local and international NGOs, as well as from UN agencies. The Charter was created with the objective of rallying HOs around shared environmental principles, and subsequently enabling a sector-aligned response to the climate and environmental crises (Grayson, 2021). The Charter was published online in May 2021 and has now been signed by more than 200 HOs (ICRC / IFRC, 2021). The importance for HOs to help the communities they support become climate-resilient is also reflected in ALNAP's report *Adapting humanitarian action to the effects of climate change,* published end of 2021 (ALNAP, 2021).

Many HOs also recognize the environmental load of their operations and are committed to mitigate it. Again, this is reflected in the second commitment of the Climate and Environment Charter for Humanitarian Organizations: *Commitment 2 – Maximize the environmental sustainability of our work and rapidly reduce our greenhouse gas emissions.* Note that this commitment is not just about greenhouse gas emissions, it also considers how humanitarian programmes at field level might inadvertently degrade their local environment, for example by failing to manage waste or by depleting surrounding natural resources (ICRC / IFRC, 2021).

The humanitarian sector thus presents a dual objective in reaction to the climate and environmental crises: (i) to help communities at risk become climate-resilient and (ii) to operate in an environmentally sustainable manner. Generally known to be short-term oriented, the sector seems to have adopted a longer-term view in recent years, which will further feed the humanitarian-development nexus. This is evidenced by the many initiatives in the humanitarian sector that aim at limiting the environmental burden of humanitarian activities. Initiatives can be locally launched and managed, or they can be led globally, at organizational or cross-organizational level.

Examples of local initiatives:

- The UN Refugee Agency (UNHCR) supported natural regeneration in Chad. Since 2004, because of violence in Darfur, about 300,000 Sudanese people have sought refuge and settled in camps in eastern Chad. With time, these camps led to local deforestation, forcing refugees to sometimes walk up to 20 kilometres to collect wood. In response to this environmental issue, UNHCR launched an assisted natural regeneration (ANR) project in 2017. With ANR, degraded environments regenerate mostly by themselves; human interaction is limited to protection and surveillance of the ANR areas and a little bit of planting to accelerate the natural regeneration process. The ANR project has now transitioned from UNHCR to GIZ, a German development organization (ECHO, 2021).

- Solar panel installation in South Sudan by the Norwegian Refugee Council (NRC). NRC South Sudan has nine offices across the country, and until 2020 all had been running on old diesel generators (for which fuel must sometimes be brought in by helicopter). In 2020, NRC South Sudan replaced the diesel generator in one country office with solar panels and is now further planning to do the same with other offices. Not only are solar panels less polluting, but they also offer a more stable energy source for hard-to-reach locations and at a cheaper cost (NRC, 2020).

Examples of global organizational initiatives:

- Climate-Smart at MSF. In 2019, MSF's Operational Centre in Geneva (MSF OCG) and MSF Canada piloted their so-called Environmental Impact Toolkit in five countries. This toolkit helps MSF projects and entities measure their carbon footprint (including direct and some indirect emissions, like freight) as well as waste volumes. Following the success of the pilot phase, the tool was further implemented in many other countries, ensuring consistent carbon and waste measurement across the movement. Besides the Environmental Impact Toolkit, the Climate-Smart initiative also aims to identify and disseminate environmentally sustainable best practices and to build an organizational roadmap for carbon-neutrality (MSF, 2020).

- Sustainable Supply Chain Alliance at the ICRC. The ICRC launched the Sustainable Supply Chain Alliance in 2020. The objective of the Alliance is to make the supply chain activities of the Red Cross and Red Crescent Movement (ICRC, IFRC and national societies) more sustainable (considering all three pillars of sustainability). It includes initiatives around sustainable procurement, sustainable fleet management and waste management (ICRC, 2021).

Examples of global cross-organizational initiatives:

- Joint initiative for sustainable humanitarian packaging waste management led by USAID. The joint initiative was launched in 2020 by USAID with the purpose of reducing the negative impact that packaging waste can have on local communities. It addresses this objective through different angles: strengthening of collaboration within the humanitarian sector and with the private sector, definition of best practices and standards, mapping of existing waste management infrastructure, development of innovative recycling procedures, and more. The initiative includes the following organizations: the World Food Programme (WFP), the UN Refugee Agency (UNHCR), the UN Environment Programme (UNEP), the UNEP/OCHA Joint Environment Unit (JEU), the Global Logistics and Shelter Clusters, the International Organization for Migration (IOM), as well as the IFRC and the ICRC (USAID, 2020).
- WREC led by the Global Logistics Cluster (GLC). WREC stands for Waste management and measuring, reverse logistics, environmentally sustainable procurement and transport, and circular economy. This project started in 2021 and aims at supporting the sector in mitigating the negative impact that humanitarian logistics can have on the environment. It specifically considers two environmental dimensions: greenhouse gas emissions and waste. Practically, WREC is about raising awareness through the GLC network, enabling collaboration within the sector and

ensuring collective response at scale and providing guidance and support as needed. Besides the Global Logistics Cluster, this project includes the following HOs: the Danish Refugee Council (DRC), Save the Children, WFP and the IFRC (Logistics Cluster, 2021).

Approaching the challenge of climate change from both local and global initiatives provides the potential to make the greatest impact. Local initiatives understand the specific context of local communities, and can have a direct impact on the environment and beneficiaries' livelihoods. They also provide guidance and vision of HOs with global initiatives. Global cross-organizational ones are essential to enable a sector-wide response at scale and form more collaborations between organizations and/or the local communities.

Shifting perspectives: Opportunities for HSCS

The environmental load of humanitarian operations is highly concentrated in supply chain activities. When considered end-to-end, HSCs are the main source not only of costs during disaster response (Van Wassenhove, 2006) but also of greenhouse gas emissions and environmental degradation in the sector. On the positive side, this also means that the most important levers that will help HOs improve their environmental performance relate to supply chain management. Here, we identify some opportunities to shift perspective for more sustainable humanitarian operations. These are focused on environmental sustainability, but can also have impacts on social and/or economic sustainability.

Sustainable sourcing strategies

The activities performed by the suppliers of HOs (and the suppliers of these suppliers) typically account for a significantly large carbon footprint and could also present various environmental risks (on top of social ones, depending on the procurement process and region). Yet, today the supplier selection processes in the sector mostly rely on price, lead time and quality compliance; the environmental performance of the suppliers might be considered as a bonus or simply not considered at all. There is, however, increased attention in the sector on sustainable procurement and initiatives to analyse and understand the impact that sourcing decisions (supplier and/or product selection) can have on the environment are growing. (Procuring locally or regionally, while generally encouraged in the sector for obvious socio-economic reasons, might appear as more environmentally sustainable than global procurement alternatives, but this is not always the case.) The Climate-Smart initiative at MSF and the Sustainable Supply Chain Alliance at the ICRC, mentioned above, are for example both looking into how to include sustainability in their procurement processes. Other organizations that are similarly evolving in that direction include Save the Children, Action Contre la Faim (ACF), the United Nations Population Fund (UNFPA) and more.

Supply chain network redesign could also help decrease the carbon footprint of humanitarian supply chains, but while HOs might be willing to adapt their suppliers' landscapes for environmental reasons, they might be less willing to do so when it comes to their own supply chain network. Regional hubs or the relocation of kit assembly plants at supplier–beneficiary cross-roads do however represent potential carbon reduction opportunities and are certainly worth the thought. Zarei et al. (2019) carried out a case study with one HO on this particular topic. They studied how the introduction of two regional hubs in the organization's supply chain network would impact its total transportation emissions (from suppliers to in-country point of entry): the two-hub scenario was estimated to present more than five times less greenhouse gas emissions than the as-is situation with no hubs (Zarei, et al., 2019).

HOs are increasingly replacing the distribution of relief items with cash-based transfers. Cash-based transfers, like relief items, also have an environmental impact: the items purchased by cash-based transfers can both contribute to global warming and environmental degradation, the extent of this depends on their source. Cash-based transfers do however present the potential to reduce waste resulting from stock excesses or mismatches because beneficiaries would only purchase what they need. Brangeon and Léon also argue that cash-based transfers can represent an opportunity for environmental sustainability: HOs could define conditions to these cash-based transfers to lead beneficiaries to more sustainable sources (Brangeon & Léon, 2020).

Collaborate with other HOs

As a consequence of an increasing funding gap, humanitarian actors have been looking into how to operate more cost-efficiently. Particular attention has been given to supply chain activities because these are generally assumed to represent as much as 60 to 80 per cent of the total costs incurred by HOs (Van Wassenhove, 2006). Pooled logistics represent one way to decrease the financial burden of humanitarian supply chain activities. With pooled logistics, HOs outsource their logistics activities to common service providers, either humanitarian actors or commercial third parties, depending on the context, and thanks to economies of scale, the common service providers can fulfil the requested logistics activities at a total cost that is lower than if each HO were to organize these separately. On top of reducing total costs, pooled logistics also mitigate the total carbon footprint of the HOs participating in the initiative because of higher truck and warehouse capacity utilization averages. One well-known example of common service providers is Atlas Logistique. Atlas Logistique is a sub-branch of Humanity & Inclusion (HI) and can thus be labelled as a humanitarian common service provider. It offers both transport and storage services to the humanitarian community, as well as engineering and mechanical expertise. (Réseau Logistique Humanitaire, 2019).

Some HOs are going one step further by looking into opportunities for pooled procurement: by partnering up, they wish to increase their purchasing power and negotiate better long-term agreements with suppliers (Réseau Logistique Humanitaire, 2019). From an environmental point of view, this approach also represents an opportunity: HOs could share the effort and/or cost to assess suppliers against environmental standards, and/or take advantage of their increased purchasing power to consider sustainable procurement sources that might have been too pricy with lower purchasing volumes.

Plan and re-plan

An often-cited study released in 2019 estimated that over half of all humanitarian crises are at least somewhat predictable (Start Network, 2019). Many humanitarian crises are protracted (according to the United Nations Office for the Coordination of Humanitarian Affairs (UN OCHA), the average length of a humanitarian crisis requiring international humanitarian assistance has increased from five years in 2014 to nine years in 2018) and thus represent a relatively stable and predictable demand. Some tropical diseases, like malaria, are seasonal and demand peaks can therefore be anticipated. The same holds true for tropical cyclones and other extreme weather events. Hence, humanitarian needs can to a certain extent be planned. Yet, most HOs are biased towards sudden-onset disasters, they invest in pre-positioning strategies but neglect broader demand and supply planning processes. It is a missed opportunity; supply chain planning has the potential to improve both the cost-efficiency of humanitarian supply chains as well as their environmental sustainability. By anticipating humanitarian needs and organizing the supply chain network to meet these in due time, supply chain planning could lead to less reliance on air transportation (items can be shipped upfront) as well as less waste (only what is needed is distributed). Note that to meet their purpose, supply chain plans should be updated regularly; many organizations still rely on demand plans prepared yearly for budgeting purposes.

Supply chain planning, just as sustainable procurement and pooled logistics, has however received increased attention in the sector. Some HOs have or are working on designing and implementing new planning processes. For example, over recent years, the ICRC has implemented new collaborative planning processes which have proved successful so far. WFP is another planning champion with a unit dedicated to end-to-end supply chain planning which not only supports country and regional offices in their planning processes but also provides services to external NGOs.

In the commercial sector, a small but increasing number of organizations convert their supply chain plans from volume to CO_2e (based on predefined emissions factors). Through supply chain planning, these organizations thus estimate the carbon footprint of their supply chain activities for the next one to two years and can work on mitigating their footprint through informed supply chain decision-making. Would a similar approach be possible for the humanitarian sector? Laguna-Salvadó et al. (2019) paved the way for this. They developed, together with the IFRC, a supply chain network optimizer that considers economic, environmental and social dimensions. The model has multiple objective functions: maximize the service level (economic dimension), minimize the total costs (economic dimension), minimize the total CO_2e volume (environmental dimension) and maximize the local sourcing rate (social dimension). These functions are considered one after the other, the results of one constraining the next one – the sequence and constraint deviation tolerance are provided by the user (Laguna-Salvadó, et al., 2019).

Increased transparency along the supply chain

Next to these three supply chain trends, it is important to also note that the humanitarian sector is improving on reporting and understanding their environmental performance and potential risks. One key element to this is data collection and transparency along the supply chain for evidence-based decision-making. This implies collecting data along the supply chain to be used for measuring the environmental performance of the activities. Examples of collected data can include transport distances, load capacity, transport mode, fuel type, inputs for the production of items, waste scenarios, etc.

Many HOs are now reporting their carbon footprint yearly following the GHG protocol. At this date, and to the authors' knowledge, two large carbon accounting initiatives exist in the sector. The ICRC, in consultation with various HOs, is currently developing a carbon accounting tool that would be offered pro bono for any HO wishing to have it. A similar initiative is ongoing with a consortium of 10 French NGOs. These initiatives do not just enable cost-efficient carbon accounting tool development (because each humanitarian does not need to develop their own) but also ensure consistent carbon accounting principles in the sector, and the possibility to compare carbon accounting reports. At local level, humanitarian actors can use NEAT+, a tool provided by the UNEP/OCHA Joint Environment Unit. NEAT+ helps in the process of doing a screening on the environmental surroundings from a given programme – it highlights potential environmental risks and proposes mitigation actions.

One method that has been used in research, but less in the organizational context is life cycle analysis (LCA). LCA is a methodology used to quantify the environmental performance of a product during its complete life cycle. Life cycle impact assessment (LCIA) is the further step that quantifies the impact of the life cycle of the product across several impact measures, such as global warming potential (GWP), particulate matter, freshwater eutrophication and land use. The ICRC has performed life cycle assessments for some of their most distributed products and have encouraged the sector to do the same in an online training which more than 14,000 people have enrolled on (IFRC / ICRC, 2019). There are other examples of life cycle assessments in the humanitarian sector: van Kempen, et al. (2017) describe one for kitchen sets with UNHRC and Oberhofer, et al. (2015) for woven blankets, plastic sheeting and jerrycans with the French Red Cross.

Summary and conclusion

Improving sustainability in humanitarian operations is – and will continue to be – an important topic for HOs. Economic, social and environmental sustainability targets (such as the SDGs) are tightly incorporated in not only the activities of HOs today, but also will play a role in how operations develop in the future. Increasing climate change poses a further risk, especially in regions already in need of aid. Although the uncertain context of the humanitarian sector (e.g.

knowing when and/or where the next disaster will strike) will likely remain unchanged, there are many shifts that can be made to improve sustainability.

In this chapter, we presented an overview of sustainability in the context of the humanitarian sector, identified specific ways in which humanitarian operations contribute to environmental challenges as well as already-existing initiatives by several HOs. We also propose potential opportunities to continue to shift towards a more sustainable humanitarian sector, specifically through enhanced planning and collaboration, but perhaps the key lies in data collection. Data is needed to improve planning, understand demand, select sustainable sourcing and reduce waste. Methods for improving transparency and developing metrics to track HO's activities are opportunities for further research.

References

AIF (2021) *Local Sourcing*, https://africaimprovedfoods.com/local-sourcing-2/ (archived at https://perma.cc/BYH5-E7GY)

ALNAP, 2021. Adapting humanitarian action to the effects of climate change, s.l.: s.n. Besiou, M., Pedraza-Martinez, A. J. P.-M. & Van Wassenhove, L. N., 2021. Humanitarian Operations and the UN Sustainable Development Goals. *Production and Operations Management*, 30(12), pp. 4343–4355.

Besiou, M., Pedraza-Martinez, A. J. P.-M. & Van Wassenhove, L. N., 2021. Humanitarian Operations and the UN Sustainable Development Goals. *Production and Operations Management*, 30(12), pp. 4343–4355.

Brangeon, S. & Léon, V., 2020. *The environmental impact of cash and voucher assistance*, s.l.: Groupe URD.

Corbett, C. J., Guide Jr., V. D. R., Pedraza-Martinez, A. J. & Van Wassenhove, L. N., 2021. *Sustainable Humanitarian Operations: A Forward-Looking Perspective*, s.l.: s.n.

ECHO, 2021. *Compendium of good practices for a greener humanitarian response*. Available at: https://reliefweb.int/report/bangladesh/compendium-good-practices-greener-humanitarian-response (archived at https://perma.cc/8PP3-N2DL)

Elkington, J., 1998. Partnerships from cannibals with forks: The triple bottom line of 21st-century business. *Environmental Quality Management*, 8(1), pp. 37–51.

FAO, 2015. *Soils are endangered, but the degradation can be rolled back.* Available at: https://www.fao.org/news/story/en/item/357059/icode/ (archived at https://perma.cc/N54Z-NLQT)

FAO, 2017. *Chapter 2: Energy for Agriculture.* Available at: https://www.fao.org/3/x8054e/x8054e05.htm (archived at https://perma.cc/F3TB-KH97)

GRAIN and IATP, 2018. *Emissions impossible: How big meat and dairy are heating up the planet,* s.l.: s.n.

Grayson, C.-L., 2021. *Interview with Catherine-Lune Grayson (ICRC) on 'the Climate and Environment Charter for Humanitarian Organizations'.* Available at: https://www.urd.org/en/review-hem/interview-with-catherine-lune-grayson-icrc-on-the-climate-and-environment-charter-for-humanitarian-organizations/ (archived at https://perma.cc/EZ4W-LDS4)

ICRC / IFRC, 2021. *Climate Charter.* Available at: https://www.climate-charter.org/ (archived at https://perma.cc/K2TU-F8XR)

ICRC, 2021. *Introduction to ICRC's approach to green Logistics.* Available at: https://www.icrc.org/en/document/introduction-icrc-green-logistics (archived at https://perma.cc/69Y8-Q7XA)

ICRC, 2021. *Introduction to ICRC's approach to green logistics.* Available at: https://www.icrc.org/en/document/introduction-icrc-green-logistics (archived at https://perma.cc/69Y8-Q7XA)

IFRC, 2019. *The cost of doing nothing,* s.l.: s.n.

Joseph, S., Peters, I. & Friedrich, H., 2019. Can Regional Organic Agriculture Feed the Regional Community? A Case Study for Hamburg and North Germany. *Ecological Economics,* Volume 164.

Kovács, G. & Spens, K., 2009. Identifying challenges in humanitarian logistics. *International Journal of Physical Distribution & Logistics Management,* 39(6), pp. 506–528.

Laguna-Salvadó, L., Lauras, M., Okongwu, U. & Comes, T., 2019. A multicriteria Master Planning DSS for a sustainable humanitarian supply chain. *Annals of Operations Research,* Volume 283, pp. 1303–1343.

Laguna-Salvadó, L., Lauras, M., Okongwu, U. & Comes, T., 2019. A multicriteria Master Planning DSS for a sustainable humanitarian supply chain. *Annals of Operations Research,* Volume 1343, pp. 1303–1343.

Logistics Cluster, 2021. *The WREC Project: Environmental Sustainability in Humanitarian Logistics.* Available at: https://logcluster.org/blog/wrec-project (archived at https://perma.cc/25GQ-SXT3)

MSF, 2020. *Climate-Smart MSF.* Available at: https://msf-transformation.org/news/climate-smart-msf/ (archived at https://perma.cc/4KET-NCR6)

Nkamleu, G. B., 2020. African agriculture in the context of COVID-19: Finding salvation in the devil. *African Journal of Agricultural and Resource Economics,* 15(4), pp. 302–310.

NRC, 2020. *Greening humanitarian operations – one solar panel at the time.* Available at: https://www.nrc.no/expert-deployment/2016/2020/greening-humanitarian-operations--one-solar-panel-at-the-time/ (archived at https://perma.cc/2JMR-MCTF)

Onabanjo & Julitta, D., 2021. *Conflict, Climate and COVID: Tackling humanitarian crises on multiple fronts.* Available at: https://reliefweb.int/report/world/conflict-climate-and-covid-tackling-humanitarian-crises-multiple-fronts (archived at https://perma.cc/TQ9A-3SBT)

Réseau Logistique Humanitaire, 2019. *Strength in numbers: towards a more efficient humanitarian aid, pooling logistics resources,* s.l.: s.n.

Start Network, 2019. *Analysing gaps in the humanitarian and disaster risk financing landscape,* s.l.: s.n.

Sumner, A., Hoy, C. & Ortiz-Juarez, E., 2020. *Estimates of the impact of COVID-19 on global poverty,* s.l.: United Nations University.

The New Humanitarian, 2021. *What's the aid sector's carbon footprint?.* Available at: https://www.thenewhumanitarian.org/investigations/2021/10/27/aid-sector-carbon-footprint-environmental-impact (archived at https://perma.cc/C597-RGCR)

UN, 1987. Our Common Future. Available at: https://sustainabledevelopment.un.org/content/documents/5987our-common-future.pdf (archived at https://perma.cc/NB96-NV5H)

UNEP, 2021. *A Practical Guide to Climate-resilient Buildings & Communities,* s.l.: s.n.

USAID, 2020. *Joint Initiative for Sustainable Humanitarian Packaging Waste Management.* Available at: https://eecentre.org/wp-content/uploads/2021/03/Fact-sheet-Joint-Initiative-On-Sustainable-Humanitarian-Packaging-Waste-Management.pdf (archived at https://perma.cc/N3B8-VV7B)

van Kempen, E. A., Spiliotopoulou, E., Stojanovski, G. & de Leeuw, S., 2017. Using life cycle sustainability assessment to trade off sourcing strategies for humanitarian relief items. *International Journal of Life Cycle Assessment,* 22(11), pp. 1718–1730.

Van Wassenhove, L., 2006. Humanitarian aid logistics: supply chain in high gear. *Journal of the Operational Research Society,* Volume 57, p. 475–489.

World Humanitarian Summit, 2017. *Natural Disasters and Climate Change: Managing Risks and Crises Differently,* s.l.: s.n.

Zarei, M. H., Ronchi, S. & Carrasco-Gallego, R., 2019. On the role of regional hubs in the environmental sustainability of humanitarian supply chains. *Sustainable Development,* Volume 27, p. 846–859.

12

What next for humanitarian logistics?

GEORGE FENTON AND TIKHWI JANE MUYUNDO

ABSTRACT

This chapter considers the current state of humanitarian logistics (during a pandemic) from the personal perspectives of a number of practitioners. It considers developments within the last decade, with emphasis on 2020 and 2021, revisits identifiable challenges, changes in managing supply chains, and discusses the possibly less obvious challenges and means to address these.

The chapter goes on to discuss the potential influence of several emerging issues, such as the limitations of using only internal means to measure logistics performance and value for money (VfM), arguing in favour of a sector-wide and comparative process.

The prominence of cash transfers as a means of humanitarian assistance is highlighted, together with the adoption of the latest AI (artificial intelligence) technology and the need to update the profile and capabilities of humanitarian logisticians (and logistics) in order to accommodate new methods of assistance.

Introduction

In 2020, 389 natural disasters were reported, affecting 98.4 million people and costing US$171.3 billion. The year rivalled 2016 as the world's hottest recorded year despite the absence of a strong El Niño effect. Apart from the Covid-19 pandemic, the year was dominated by climate-related disasters. The impacts of the events were not equally shared: Asia experienced 41 per cent of disaster events and 64 per cent of total people affected. Heatwaves in Europe accounted for 42 per cent of total reported deaths. In a year of record-breaking storms and wildfires, the Americas suffered 53 per cent of total economic losses, largely in the USA which experienced the bulk of the year's most costly climate-related disasters (CRED, 2021).

In 2021, the UN estimated that 235 million people would need humanitarian assistance and protection. This represented a 40 per cent increase from 2020. Conflict, economic instability, disasters and climate related events and the Covid-19 pandemic displaced tens of millions of people from their homes. Some agencies estimate that a quarter of a billion people could be displaced worldwide by 2030.

Over the past two decades, and especially during the pandemic, humanitarian logistics, and supply chain management more broadly, has become a key component of health and humanitarian assistance. Now better recognized as an essential stakeholder in crisis response, logisticians face many challenges, due in part to the frequency, intensity and impact of emergencies, but also because investment in building capacity, recognition and cross-sector connections for the profession is still limited, in particular at the local level. There is increasing opportunity to build on synergies between commercial, public sector and humanitarian logistics activities as was evident during the pandemic response and to build homegrown solutions for local supply chain challenges. However, the quality, capability and effectiveness of any humanitarian intervention will be directly proportional to the capacity, competence and preparedness of its logistics teams. 'Last mile' operational assistance must be adaptable and flexible if the aid community and other responders – internationally, regionally and locally – are to be effective in their actions (Fenton, 2013).

Painting the picture of humanitarian logistics requires a broad canvas but it is still relevant to focus on the 'rights' of supply chain planning: the right product, at the right cost, at the right place, and at the right time. There are of course differing opinions as to the appropriate number of rights (the rights model could also have a negative impact when considered for non-emergency operations as it may imply responsiveness rather than essential collaborative planning among partners), but these four are the most commonly used. A fifth, the right quantity, was critical during the pandemic with the World Health Organisation (WHO) on 3 March 2020 calling on industry and manufacturing organizations to significantly increase production of PPE (WHO, 2020).

This chapter explores the personal reflections of humanitarian logisticians with respect to important cross-cutting developments, challenges, ideas and opportunities for practitioners and raises a number of important questions regarding the evolution of humanitarian logistics over the next five to ten years, both as a critical function within humanitarian response and as a developing sector in its own right. It also takes into consideration the resilience of humanitarian supply chains during crises.

The right product

Ensuring that the right product or service reaches those for whom it is intended has been challenging. Incorrect or limited assessments often do not take into consideration cultural beliefs, gender, diversity and inclusion, and the absence of common standards or inadequate specifications provided by those requesting an item creates significant challenges. The result not only risks wasting resources by delivering the right thing at the wrong time (or the wrong thing at the right time) but also a worsening of relationships between those requesting and those supplying and, more critically, failing to adequately serve those in need.

Obtaining the right product is further complicated when operations receive unsolicited bilateral donations. These donations are given by individuals and organizations in good faith, but often they are not usable or are culturally inappropriate (ARC, 2020). Using common humanitarian standards for emergency items – principally those developed by the International Federation of the Red Cross (IFRC) and now made accessible in the humanitarian space (ULS, 2021) – enables the requester to see options and specifications for any commodity. Some catalogues even provide background information (such as fumigation recommendations for certain food stuffs) or suggestions for linked items ('Would you like a kitchen set to accompany that stove?'). Simple online catalogues for relief items have worked well for logisticians. Having prior knowledge of the standard packaging size, quantity, dimensions and weight of an item allows for far more efficient supply chain planning. Standardization works well for suppliers too, as knowledge of item specifications agreed by aid organizations, disaster management authorities, or even within a large organization, will encourage investment in stock holdings or potentially local manufacturing.

Indeed, one step further is the option to tender to establish long-term supply contracts that introduce confidence that inventory accurately meets commodity specifications, packaging and labelling and customs documentation requirements. Thereafter, a disaster will trigger the rapid mobilization of 'standard' relief items, with all the time-consuming procurement and administration processes having been concluded in advance. Unfortunately, the Covid-19 pandemic shattered our normal reality. This was a disaster that was not foreseen, nor was the world prepared for it. Supply chains all over the world were completely disrupted as the virus spread rapidly through the continents. Organizations all over the world have over the last decade focused on supply optimization, which meant reducing costs through reduced inventory holding and this meant that humanitarians had little or nothing to fall back on when the pandemic hit. Factories closed and countries had to make do with whatever was available, and it was not always the right product, but it helped to minimize the health risks and exposure to the virus.

Innovative approaches that had worked before the pandemic were no longer viable. Organizations established long-term framework agreements to procure as and when required. Manufacturers would also produce as and when a need was communicated. This meant only planned requests could be met and therefore an absence of stocks to fall back on in the event of an emergency. Capacity to produce the right product in a short time frame became a problem.

Standard product specifications and global contracts for, say, pre-positioning stock or for vendor managed inventory, should be well matched with preparedness activities for disasters of a magnitude (and media profile) that generate the funds required for large-scale mobilization. Humanitarian organizations should in turn strategically position themselves to be part of 'first responders' with minimal quantities that cater for the initial intervention. The Philippines typhoon Haiyan emergency (November 2013), where the UK Government and other charities accessed pre-positioned stocks that were immediately deployed, and the 2017 Hurricane Irma in the Caribbean (unofficially, the fourth-costliest hurricane on record) are examples of this. The small to medium disasters, with little to no profile and funding, are less suited to the deployment of these globally 'standard' items. Pre-positioned stock can still be mobilized from global locations, but the costs involved in moving smaller consignments over large distances are often prohibitive.

The solution is increasingly to invest in more numerous, but smaller, pre-positioned stock-holdings, with shorter and less costly supply chains, for situations when funds are at a premium. This in turn opens up the possibility to support local capacities and markets for disaster response and to source from regional, national and local suppliers with the added benefit to support the local economy, which also ensures that supplies are familiar with the intended beneficiaries. However, there are several challenges: what if the preferred local specification, familiar to the local populace, doesn't meet the global specifications? Do logisticians refuse to purchase locally and instead import items that meet global specifications and which may be of a higher quality and often at a higher cost? Consider the spun aluminium cooking sets, used throughout many disaster-prone regions, and contrast that with the more expensive stainless-steel versions that meet a global specification. Adherence to the global specification disadvantages many local manufacturers and vendors from bidding in a global, regional or national tender. In some instances, the recipient might prefer to receive cash in order to buy instead of a product that they are familiar with using. While stainless steel may be more hygienic than aluminium, this matters little if, as has been observed, the beneficiaries sell the pots. Similar such arguments might be made for common shelter and hygiene items.

Provided that the 'do no harm' principle is always applied (as is the case for many unacceptable locally-produced pharmaceuticals) there should be flexibility in sourcing policies. It is therefore important for logisticians to gain skills and experience in retail market analysis and monitoring.

Whether global or local, the ethical and environmental performance of suppliers or traders, as well as their suppliers and subcontractors, should be considered (UNOCHA, 2014). Although work on this important issue is ongoing, the humanitarian logistics network does not yet have a common approach. For example, compliance with the UK's Modern Slavery Act is now critical as global supply chains are at high risk for modern slavery exploitation. Also, the increase in third and fourth parties means that supply chain verification is more and more challenging as the vendor is far removed from the whole chain. What is needed is a minimum standard applicable to all suppliers and manufacturers, wherever they fit in the supply chain. In particular, registers of local suppliers that can reach, or exceed, the agreed minimum standards would improve supply chain performance and resilience, particularly at the last mile.

It is vital to understand that the definition of 'last mile' is not from port or airport to a convenient warehouse. It is quite literally the last mile, and that makes humanitarian logistics quite different from its commercial counterpart. This can mean having to use any means of transport available including bicycles, donkeys, camels and elephants. Even in the best of times, a country's infrastructure can be unreliable and, when disaster strikes, can be badly damaged or destroyed.

While some emergency supplies are managed independently, at an international level aid mechanisms are collaborative, such as with the UNHRD (United Nations Humanitarian Response Depot), which is a network of warehouse facilities around the world. From these locations, agencies can respond within 72 hours to a major disaster. This UN common service has now been in operation for almost 20 years and although its services have worked well, the level of inventory utilization has generally been limited. The strategy for the Global Logistics Cluster, hosted by the UN World Food Programme (WFP), further underpins emergency supplies and logistics preparedness and crisis response. Furthermore, as noted above in relation to the Covid-19 pandemic response, the humanitarian landscape is changing, and emergency response is increasingly tackled at a national rather than international level. Therefore, a thorough analysis of supply chain preparedness and pre-positioning impact and sustainability within countries most prone to crisis is needed to determine the potential to coordinate more effectively the sourcing, storage and delivery of emergency relief items over the next decade. For example, it would be helpful to review regional and national markets, local relief item manufacturing capacity, warehouse locations, supply lead times, effectiveness of overall emergency needs assessments and applicability to recovery programmes.

Commercial companies of all sizes and at all stages of the supply chain will continue to enter the humanitarian aid market motivated not only by the corporate social responsibility (CSR) benefits, but also potentially lucrative profit margins for customized services and the commercial leverage of their own businesses. However, all of these inputs are unlikely to address the essential, specialized, capability for local 'last mile' logistics to deliver goods and services directly to those who are in need of assistance. (Fenton, 2013).

It is important to note that while standards for products are important, so are those for key supply chain functions and cross-cutting capabilities. Launched in April 2021, the Universal Logistics Standards (ULS) draw together best practices in humanitarian supply chain management and logistics. The standards include gender and diversity as a specific cross-cutting theme, something that was previously poorly addressed in the aid logistics sector, plus a review of the elements linked with technical PARCEL (Partner Capacity Enhancement for Logistics project) standards and references to the

CHS (Core Humanitarian Standards). The ULS are intended for use by organizations, staff and volunteers as guidance to improve the quality of disaster preparedness and humanitarian response and draw on the experience and knowledge of key actors from across the humanitarian logistics sector. The intent is that they can be aspired to in order to support the integration of operations into the wider aid system. They also serve as a reference to support training, capacity strengthening and other initiatives, and have been designed for those working at a local and national level as well as donors, international humanitarian organizations, UN agencies, academics, civil society, private sector companies and training service providers.

Key cross-cutting issues form part of the standards, such as 'protection', which the Sphere Standards Handbook states should be prioritized by all humanitarian actors and built into all activities. It has been challenging for the logistics sector to identify measurable protection risks and corresponding mitigating measures. Criteria need to be established to assess the protection element of each intervention and logisticians should keep protection in mind when planning and executing operations – the aim being to prevent, mitigate or respond to identified risks. Working locally with those who need goods and services (and ensuring a gender balance) results in more positive protection outcomes.

The issue of 'safeguarding' is also included in the ULS. This relates to ensuring due consideration is made within supply chain interventions to protection from sexual exploitation and abuse (PSEA), sexual harassment and child exploitation and abuse, whether in physical or digital spaces. Safeguarding issues have become a major concern within humanitarian supply chains and logistics operations, and cases of abuse can affect both personnel and the people that they aim to assist. Humanitarian actors and their partners often rely on external providers and contractors to undertake specific functions and services including the construction of facilities, transportation and the delivery or distribution of goods, cash and vouchers. Many outsourced functions involve regular, sometimes unsupervised, contact with humanitarian staff, partners and communities (including children, young people, and other vulnerable adults). This carries with it a risk that such contact could result in the exploitation and abuse of vulnerable people by those in positions of power. To minimize the risks of exploitation and abuse it is important that organizations and external providers are aware of and adopt safeguarding best practice and ensure that staff are trained on safeguarding policies and codes of conduct.

Much of the environmental impact of the humanitarian sector can be attributed to logistics, such as through carbon (and other) emissions linked to the transport of goods and personnel, the manufacturing of relief items and waste generated through their packaging. Impact can occur long before a response and long after it (the impacts caused by suppliers, the impact of crop production, and so on). Logistics and the supply chain (procurement, transport, storage and delivery of humanitarian supplies) therefore present multiple crucial possibilities for making humanitarian action more environmentally friendly. Although a growing concern within the humanitarian community is the need for more environmentally friendly logistics, little guidance exists on how to reduce the environmental impact of supply chains and local humanitarian logistics activities, such as by recycling waste.

The right cost

Funding for humanitarian aid programmes provided through institutional donors saw a dramatic reduction in 2020 and 2021 despite increasing need. Whether it is the Gates Foundation, corporations financing aid via their CSR programmes, or individuals donating to disaster responses, it is increasingly clear that private giving is a vital element of the aid landscape. Furthermore, initial experiments with outsourcing of key tasks, such as procurement, has been found to be beneficial but does not address the long-term resource gap and capacity development challenge facing the humanitarian logistics community.

Unfortunately, the funding gap grew larger with the Covid-19 pandemic. Focus shifted worldwide to health supplies and services. Donor funding to humanitarian organizations dropped, and there were huge job losses in the public and private sector. Humanitarian response became even more complex and there were shortages of food commodities, medical and basic relief supplies. With these shortages, prices also sky rocketed. Job losses meant no income and limited money in circulation. To provide the necessary support to very needy populations required a very high level of private, public, humanitarian and UN collaboration. For instance, the private sector in Kenya played a big role during the Covid-19 emergency response. They also provided PPE, food, sanitation materials and influenced the cost of goods and services to remain low.

There continues to be a need for greater understanding between the private sector (international, national and local entities), the UN, NGOs, national public sector and local civil society. While communication and dialogue is beginning, and barriers that have existed have started to fall away as humanitarian and commercial logisticians are realizing that they have a lot in common, there remains a significant task to establish better connections. The aid community has yet to fully recognize logistics as a core strategic competence, so training and professionalism have continued to suffer as a result. However, the picture has started to change over the past five years as we've gone from the perception that a humanitarian logistician might be a former commercial truck driver wanting to work in the NGO world, to today where many INGOs (international non-governmental organizations) require humanitarian logisticians to have an accredited professional qualification. However, some aid organizations are still behind the commercial world in recognizing the strategic importance of logistics and supply chain management at all levels of society.

It is now incumbent upon actors within the aid 'system' to improve the cost effectiveness through enhanced procurement processes and establish operational efficiency that is geared towards better service delivery. Besides operational costs, how cost effectively a projects is delivered is dependent of the price at which project supplies and commodities are acquired. Price is normally the base criteria, subject to delivery of the right quality and at the right time. However, in a time sensitive humanitarian response, availability may sometimes override the price, though price remains a key consideration. Note that logistics represents 60 to 80 per cent of expenditure, particularly during the early stages of an emergency response operation. In 2016, the international community came together at the World Humanitarian Summit (WHS) to consider solutions to emergent funding gaps. Three key elements were identified: reducing needs, broadening funding sources and improving the efficiency of humanitarian aid. At the WHS, humanitarian donors and organizations pledged, under the 'Grand Bargain', to channel donor funds more directly to local actors to improve the effectiveness and efficiency of humanitarian action.

A report, 'Delivering in a Moving World', that was written with inputs from a wide range of humanitarian logistics practitioners, was presented at the WHS. It posed critical questions and provided recommendations for implementation based on humanitarian response case studies, including the Nepal earthquake response, the West Africa Ebola outbreak, Super Typhoon Haiyan and others. The intent was to raise awareness within the international aid system of the importance and strategic value of humanitarian logistics.

In the private sector, logistics has for the most part moved towards the outsourcing and pooling of services, even between competing companies, with the added objective to reduce operational costs and increase customer satisfaction levels. An initiative known as the Réseau Logistique Humanitaire (Humanitarian Logistics Network – RLH) brought together several international (mainly European) humanitarian organizations to address specific areas of work that have been orientated towards collaborative practices and ways to optimize operations.

The RLH had observed that neither the outsourcing of logistics services nor inter-NGO pooling of resources were sufficiently developed practices in the humanitarian sector. Furthermore, the limited logistics capacity of many countries where humanitarian interventions are carried out requires much stronger collaboration between organizations to optimize existing capacity. It is therefore noteworthy that some commercial logistics providers, for example in the Philippines, have begun offering their services in humanitarian contexts. There are also good examples of inter-NGO logistics resource pooling, via the RLH initiative, but this process has taken several years to gain traction. Partly prompted, and perhaps accelerated, by the Covid-19 pandemic, organizations have slowly started to reform their practices, with NGOs becoming more logistically efficient in an effort to reduce costs, and can begin to demonstrate their commitments to the Grand Bargain.

Nevertheless, aid interventions must strive to generate sustainable outcomes that span development and crisis response activities. Preparedness and resilience work are important, broad ranging, aspects of this objective. In the context of this discussion, there is an emerging role for the continuous monitoring of local markets in disaster prone countries. Making greater, faster and effective use of market data will help logisticians to make informed decisions about pre-positioning, surge capacity and specific commodity weaknesses in the most vulnerable countries. Appropriate economic analysis and methods to strengthen linkages to private sector actors will need to be accelerated.

The right place

Academics have been studying ways of making supply chains more resilient by sourcing disaster relief items closer to a potential event and by prepositioning these regionally rather than globally, the aim being to take cost, import restrictions and time out of the supply chain while encouraging resilience in local economies (Taylor, 2012).

According to an article published in November 2021 by McKinsey and Company, titled 'How Covid-19 is reshaping supply chains', different industries were seen to have responded to the resilience challenge in very different ways (Alicke et al., 2021). Medical supply chains were observed as being very resilient, with 60 per cent of respondents to a health sector survey reporting that they had regionalized their supply chains, and 33 per cent had moved production closer to end markets. Respondents from other sectors reported that they had struggled to find suitable suppliers to support localization plans, but almost 90 per cent advised that they expected to pursue some degree of regionalization by 2025.

The Covid-19 crisis has put supply chains into the spotlight and, in response to the challenges, adapted effectively to new ways of working, enhanced inventories and improved digital and risk-management capabilities. However, supply chains remain vulnerable to shocks and disruptions, with many sectors having to overcome supply shortages and logistics capacity constraints. Human resources have been a major barrier to accelerated digitization and the skills gap has been widening. The McKinsey supply chain survey in 2020 found that only 10 per cent of companies had sufficient in-house digital talent and in 2021 the figure had dropped to only one per cent.

Nevertheless, over the past decade, the humanitarian logistics sector has, in most instances, demonstrated that it can quickly deliver commonly used relief items at acceptable specifications to where they are needed, when they are needed. However, this has usually been achieved on a largely ad-hoc basis as, despite initiatives such as RLH (discussed above), the wider aid sector finds it challenging to enable the conditions to effectively consolidate its supply chains. There are many reasons for this, but most issues relate to agencies' differing funding sources, and also to poor donor coordination. Across multiple agencies, identical items are often sourced from the same suppliers with the same packaging destined for the same point of entry to respond to the same crisis. Herein lies the challenge: for a number of agencies, their logistics resource is a support function. This is not meant in any detrimental sense, but simply a reflection that logisticians are often overlooked as an important team to engage in the all-important needs assessment, which hopefully identifies, with the beneficiaries themselves, not only what is needed but also when and potentially from where – for example, local vendors. Thereafter, logisticians are consulted to make sense of the resulting plan of action and often inaccurately formulated budget. The programme teams identify what's needed, where and when; the logisticians figure out how to meet the need while delivering the most effective and efficient, donor compliant supply chain. It would appear that no matter how much the logisticians might lobby for collaborative supply planning and unified supply chains, it would be remarkable that the same items were identified by different programme teams within different agencies as being needed all in the same place at the same time. Unified, coordinated or collaborative supply chains may still be desirable, but their starting point could well reside less with the logisticians and more with coordinated needs assessments and more effective plans of action, in particular at the national or local level.

This challenge can only be overcome with slow and incremental changes in the culture of emergency work, away from panicked and questionably effective responsiveness and towards short-term forecasting and collaborative design. Changes in technology can also contribute to this change, but currently there is no system-wide framework for judging relative severity and aligning decisions about response accordingly. The result is typically a patchwork of macro- and micro-level analysis, which is hard to aggregate, rarely provides a comprehensive overview and serves as an inadequate basis for decisions about the prioritization of a response (Darcy et al., 2003).

Humanitarian actors have been exploring ways to use new technologies to deliver aid more effectively. This includes artificial Iintelligence (AI) and machine learning (ML). AI/ML helps companies like Amazon and Netflix develop and deliver personalized recommendations to their consumers and increasingly helps determine who gets a job, who gets benefits, who gets a loan, who gets a vaccine, etc. A decade ago, AI/ML was more likely to feature in science fiction films than conversations about economic growth, unemployment and humanitarian action. But it is now more accessible than ever, driven by advances in computing power and software, better infrastructure, improved ML algorithms – especially deep learning – and larger, more widely available datasets. Technologists and innovation enthusiasts around the world are increasingly exploring ways to use AI/ML and other frontier technologies to solve humanity's greatest challenges. The Chief Information Officer of the World Food Programme (WFP) has said the agency has a 'moral imperative' to leverage technology to achieve efficiencies (Spencer, 2021).

Advocates of and enthusiasts for AI/ML note that the effective application of these systems could improve humanitarian action by doing less with more and ultimately save more lives. AI/ML present a range of opportunities and benefits to early warning and humanitarian preparedness, assessments and monitoring, service delivery and logistical efficiency. A range of AI/ML tools are currently being used across the humanitarian aid industry to predict and analyse trends, improve organizational functioning, personalize service delivery and allocate resources more efficiently. Tools can find latent patterns in large datasets, including images, videos and freeform text as well as numbers, to make predictions about hazards, population movements and food insecurity. Keeping up with the pace of change in such technology poses considerable challenges. NGOs and donors cite a lack of awareness about what new technologies are available and lack the time to procure and adapt these to on the ground needs. New competencies within the humanitarian logistics community are emerging to bridge the gap between programme implementation and technology providers in order to ensure that solutions respond to the needs and reality of humanitarian response (G Smith et al., 2011).

Digitalization provides the opportunity to analyse quickly, standardize and communicate assessment findings. Interactive dashboards tailored to context allow different users to focus on the information that they find most relevant, supporting critical analysis and the ability to prioritize resources. Sample size and consistency would improve as more agencies use new AI tools developed, thus helping to benchmark and build a more in-depth picture towards impact and response.

In a bid to support this approach and help to overcome the coordinated needs assessment challenge, World Vision International's design monitoring and evaluation team has developed a simple, smart-phone-enabled, basic rapid assessment tool (BRAT) that can be used by emergency response team members. Global Medic's RescUAV project demonstrates the value of using small unmanned aerial vehicles (UAVs), or drones, to provide assessment information. Through UAV technology, they can gather better information that is used to make aid delivery more efficient, by providing search and rescue guidance, situational awareness and emergency mapping. Now, not only can critical assessment information be gathered for programme planning purposes but also simultaneously for logistics operations. Primary data can be gathered in an accelerated time frame that provides an emergency response with decision-making information on both the disaster context and immediate needs in households and communities. Such tools have been designed to be flexible and easy to use and contextualize, and can be used in the very earliest days of a disaster response, before the cluster system may have been set up or joint needs assessments launched. Mobile technologies can also be used where the context of a disaster is dynamic, say where people are moving or a major change affects the population's recovery, for example when the national authorities allocate a new area for resettlement, etc.

Unified or not, all logisticians will be looking at their supply chains and asking whether they are genuinely efficient and effective. Some may have the information management systems, such as the BRAT, to provide data and metrics with which to effect some degree of measurement and reporting. The key performance indicators (KPIs) need not be complicated nor numerous; a starting point may simply be whether the right item was delivered to the right place at the right time and at an acceptable cost. It's important not to lose sight of the critical need to assure, and measure, effective programme outcomes. Therefore, the challenge is not so much defining the required data, collecting and analysing it to give the required operational information, but more what the logistician does with the information. Initially data could serve to identify trends within their own organization, but how do logisticians know whether they are delivering items for half the cost and twice as fast as another agency, or at twice the cost and half the speed?

For almost two decades, the Fritz Institute has been collaborating with humanitarian organizations and donors to conduct supply chain assessments. These assessments demonstrate how critically important to the improved and long-term effectiveness of humanitarian organizations it is to establish and disseminate sector-wide humanitarian supply chain management performance measures. A follow up survey concerning the status of humanitarian supply chain KPIs revealed that the majority of aid organizations have no KPIs in place, but of those that do almost 50 per cent say that metrics are not linked to their organizations' programmes or mandate. Almost all organizations say that they would like to establish KPIs. Critically, many larger organizations have been hindered by their inability to gauge their own performance against others, which therefore impedes their capacity to measure efficiency.

Value for money (VfM) has rapidly gained prominence in the documentation of humanitarian agencies and donors alike. Until recently, this was largely considered to be about procurement by ensuring that competitive tendering processes had been conducted to provide evidence of VfM. This approach fails on at least two fronts: the first is akin to that described for KPIs – measuring your own performance using only your data will not let you assess your key performance across the sector, any more than declaring that an organization can deliver VfM simply because it follows its own procedures; the second re-adjusts the perspective of VfM where the logistician could deliver class-leading performance in terms of cost whilst meeting all the required delivery dates and specifications, but if in so doing the agency could have achieved the desired impact by a more efficient means, then VfM has not been achieved. Therefore, good project design is critical to ensure that all elements of assistance are planned to maximize cost efficiency.

As an example, consider an efficient humanitarian supply chain in Haiti that was able to continue to deliver hygiene parcels six months after the January 2010 earthquake, when the local markets were re-established in February, and cash transfer programmes established in March. When regarded in terms of the whole, end-to-end programme, such a supply chain is not delivering VfM. It may have done something right, but it did not necessarily do the right thing. Therein lies the crux; VfM needs to be measured and reported as more than best practice in procurement. All the functions within an operation must determine not only the options required to assure efficient and effective impact, but also how peer organizations are delivering their interventions, and at what cost. Some donors now consider VfM as a key aspect of awarding grants. It is not too far-fetched to see a donor, in receipt of multiple proposals for responding to a crisis, applying their own comparative evaluation of VfM in their grant allocation process. It may be preferable for agencies to agree on a methodology for comparing and adjusting in order to deliver collective VfM. Otherwise, there is a risk that a donor methodology will either be enforced, or indeed donors themselves will actually direct or manage operations in order to assure *their* perception of what can often be complex and contextual 'value for money'.

The right time

Perhaps more than most other single factors, in addition to planning, it's the combination of item, place, quality and quantity that challenges timing. Depending on the relative weighting afforded to each factor, the pressure to deliver relief items in response to an emergency because they are available immediately, or deliver to areas in lesser need because they can be reached more easily, or to deliver late, or over a prolonged period, because the timetable is driven by the lead-time for the right item, all contribute to a less than optimal response. An example that can be drawn on here is supply chain feedback loops. Say the design of emergency shelters requires parts that are unavailable locally and must be imported with a two-month lead time – such valuable supply information should have an effective feedback loop to prompt a change in shelter design, and ultimately result in faster service delivery to beneficiaries.

Both the humanitarian and commercial logistics sectors could be seen to have identical objectives – improve the supply chain to deliver maximum dividends to shareholders – with just the dividends (dollars versus blankets) and shareholders (holder of shares versus. those affected by the crisis) that differentiates them. In reality, the finer differences, such as the willingness of individuals to approach a humanitarian agency knowing that their legal status is unlikely to be challenged, their details won't be forwarded to the authorities and knowledgeable individuals will be able to help beyond a straightforward distribution, or at the very least signpost to where help can be found. As such, cooperation with commercial sector last mile logistics has real potential, but wholesale delegation or sub-contracting the role less so.

Commercial supply chains are becoming more vulnerable, partly due to the implementation of 'just in time' techniques, globalization policies and 'no-stock policies'. Although humanitarian logistics has much in common with commercial logistics, good practices from the corporate world have not fully crossed over. Furthermore, given that disasters are now more frequently affecting the developed world and are having a direct and often dramatic effect on global business, there is also much that can be learnt by companies from humanitarian logisticians about how to operate creatively in chaotic environments and with markedly restricted resources. Supply chain risk management (SCRM) may be the most critical area for humanitarians to learn from the business sector, argues Paul Larson from the University of Manitoba. He notes that humanitarian action is the ultimate risky business. Whether a pandemic, volcanic ash cloud or war is causing a crisis, SCRM is about minimizing interruptions either through avoidance or effective response. Humanitarian logisticians need the latest technical knowledge and business techniques and should develop risk management skills, rather than be forced simply to take risks.

Long supply chains, which can be costly and inefficient, are prone to disruption in a crisis, as exemplified by the Covid-19 pandemic. Furthermore, the impact that imported aid can have on local markets can compound the effects of the immediate shocks of a humanitarian crisis, as local businesses compete with free items flooding the market and are unable to access procurement contracts. On the other hand, rapid advances in digital manufacturing technologies and their spread throughout low- and middle-income countries are opening up opportunities to make anything, anywhere.

Traditional manufacturing processes require economies of scale to be cost-effective, but new economies of scope are now possible with, for example, one injection moulding machine or 3D printer being used to make a wide range of products. Access to open-source design files for hardware removes intellectual property constraints and the high costs of the initial design of an item. Together, these factors can create ample opportunity for local manufacturers to meet the aid market's demand for relief items. While humanitarian sector actors at all levels are, in principle, committed to the localization of aid, to localize procurement of aid items and capitalize on improved local manufacturing capabilities, buyers (donors, governments, international and implementing organizations) must adopt policies and practices to procure locally-made supplies.

The private sector plays a crucial role in disaster preparedness, response and recovery and this factor is becoming more critical as the severity and frequency of disasters increase around the world. A joint initiative by the United Nations Development Programme (UNDP) and the United Nations Office for the Coordination of Humanitarian Affairs (OCHA), known as the Connecting Business Initiative (CBi) was set up in 2016 to support both crisis response and development efforts by engaging the private sector in disaster prone contexts.

CBi aims to become the go-to hub for business networks involved in disaster management, both strengthening their collaboration with governments, development and humanitarian actors and contributing to save the lives and livelihoods of people affected by crises. CBi works directly with business federations around the world, representing over 4,000 members and reaching more than 40,000 micro-, small- and medium-sized enterprises (MSMEs) and by the end of 2021 had responded to more than 100 crises, mobilized US $52 million and assisted around 17 million people.

Increasing urbanization poses multiple challenges to the delivery of material as well as cash-based assistance. For example, solutions to quickly minimize traffic congestion in order to increase the efficiency of emergency urban freight delivery. Urban Logistics should focus on service level and contract performance analysis. Information sharing and service contract design will help to enhance collaboration and coordination. Service level contracts have proven to be an effective tool to motivate logistics service providers (LSPs) to enhance their service quality. The key will be to figure out how to do this as part of preparedness measures so that appropriate emergency delivery services can be activated during a crisis. Therefore, to encourage collaborative urban logistics service planning, a framework to share information and contract guidelines is required to ensure effective coordination among beneficiaries and aid organizations (the customers) and suppliers and service providers to the market – in this situation 'the market' being an emergency response operation.

For the mega disaster, the aid sector can now rapidly mobilize and deliver to point of entry, globally prepared stocks of relief supplies. It's the last mile that typically presents a challenge, and the more densely populated and therefore more congested and disrupted the destination, the bigger the challenge. Compounding this scenario is the consideration that urban populations tend to have fewer opportunities for coping strategies, certainly in comparison with those rural areas. This, coupled with reduced capacity to store, for instance, a month's worth of supplies at a time, adds to the strains on the supply chain. As has been seen during the Covid-19 pandemic, supply chains need to be more agile, more flexible and better able to respond to frequent distributions.

The right quantity

After 1945, the world ushered in an era of open trade, creating easier access to goods and commodities from external sources. It opened borders, fostering preferential trade agreements, supporting deficit areas and re-distributing surpluses. Effective and efficient response to disasters is dependent on the speed of access to basic goods, commodities and services at affordable costs. From a global perspective, we rely on multi-lateral trade agreements and a connected world where access is unlimited, compared to economies where local capacities are inadequate. The drive for globalization grew tremendously. Systems and practices were built and aligned to embrace globalization right through to 2019. In 2020, the worst fears were realized: a world where countries could not produce enough to sustain their own population and were not able to access other economies. Covid-19 became a burden, prices went up, commodities became scarce and needs were not met.

The number of disasters worldwide has increased five-fold. Climate-related disasters happen daily in one country or another in the world. This frequency and continuous intervention by humanitarian logisticians has given birth to a new generation of logisticians who are experienced, innovative and tech savvy, continuously developing solutions and applying existing supply chain techniques to overcome constantly emerging challenges.

Due to the complexity and frequency of emergencies and the desire to meet basic humanitarian needs, through collaboration organizations have established minimum standards to facilitate responses. For instance, minimum quantity requirements can easily be derived for the Sphere Handbook (2018).

Determining the right quantity is often the role of programme or project departments. Information is gathered from assessments at the onset of an emergency, during programme design and agreed upon by the proposal developers, fundraisers and donors. The most ideal situation in emergencies is to have an item available as and when it is needed, in whatever quantities. This scenario has cost implications that most organizations can either not afford or would make their operations very costly. Inventory holding not only increases the cost of operations, but also increases the risks. In complex humanitarian emergencies, stock-holding sometimes attracts negative attention (Besheer, 2021). To cater for the needs of a rapid onset emergency, organizations draw on minimum stocks held. The humanitarian logistics profession is opening up to specialists from the private and public sectors (principally health) who are bringing in new skills and thinking. Research suggests that aspiring humanitarian logisticians, when compared to those in commercial roles, need to possess a broad range of skills and should consider the importance of contextual knowledge before entering the profession, as should academics before attempting to conduct research in this field. There is a strong requirement for technical and functional knowledge and educators need to place a stronger emphasis on appropriate training in the technical and programmatic aspects of the role, in logistics administration and on educating future humanitarian logisticians in how to train others (Kovács and Tatham, 2010).

Several initiatives have influenced the development of the sector. Of note are the practical logistics training courses run by the Bioforce Institute – an organization that aims to increase the impact and relevance of emergency action and development programmes – and the Fritz Institute/CILT(UK) certificate in humanitarian logistics (CHL), which has become a highly regarded qualification. The creation of the Humanitarian Logistics Association (HLA), which was registered in 2009 as the first professional association within the aid sector, has been serving as a catalyst to enhance the professionalization of humanitarian logistics and the recognition of its strategic role in the effective delivery of relief during humanitarian crises. With several thousand members based in over a hundred countries, the association acts as a neutral platform to provide technical advice to empower a new generation of humanitarian logisticians. The HLA aims to enhance and complement knowledge exchange, network interaction and coordination, supporting training initiatives, the development of good practice and broad representation for a growing worldwide community of practice.

Humanitarian logisticians need to develop competence in a range of skills and technologies including information management, market assessments and cash and voucher distribution, as well as the more typical procurement, transport, tracking and tracing, customs clearance and warehouse management functions. Not only must they demonstrate technical competence but also broader competence as humanitarian professionals. According to the Consortium of British Humanitarian Agencies (now the START Network) Core Humanitarian Competency Framework, humanitarian workers need to demonstrate competence in: understanding humanitarian contexts and how to apply humanitarian principles; achieving results; developing and maintaining collaborative relationships; operating safely and securely; self-management in a pressured and changing environment, and; leadership in humanitarian response (CHS Alliance, 2017). They could be expected to support economic recovery projects that may include activities such as infrastructure rehabilitation, loans or grants to traders, transport subsidies, etc. Market-based programmes aim to help protect, rehabilitate and strengthen the livelihoods of people affected by crises. It is important for humanitarian logisticians to develop skills to support such interventions as these increasingly will include value chain or supply chain projects. They also need skills to work with local partners to help capacity build, for example to address the risk management skills gap identified above.

The complexities and constraints of managing an effective supply operation cannot be underestimated. Indeed, the concept of a humanitarian intervention is now fundamentally challenging the more 'traditional' direct provision of food and non-food items (NFIs) to those in need. While the range of relief items has always been varied, from water tanks to goats, medicines to tools and equipment, food to fishing nets, but now the very nature of goods and services in the humanitarian sector is changing. What if disaster affected people don't need material assistance? Are logisticians still needed?

The challenge is clear: programme activities and modalities are changing. It is therefore critical for aid organizations and disaster management authorities to examine the likely impact now and prepare a fit for purpose workforce for the future. Several emerging trends and their impact on the role of humanitarian logistics includes:

- Cash as a modality
- Engaging in markets
- Emergency and development work along a continuum (often comprising the same workforce).

Where the local market serving a disaster-affected community has capacity and access to appropriate supplies, aid organizations increasingly provide financial assistance, not only because this approach is now many donors' assistance modality of choice, but more importantly because it is a method to support market resilience. Rather than undermining market conditions by bringing in commodities and services from outside, conditional or unconditional financial transfers, either electronic or in the form of vouchers or cash, is a modality which strives to re-establish normal commercial transactions as quickly as possible after a disaster.

Cash and market-based assistance is complementing or replacing in-kind assistance in many contexts. The role of the typical humanitarian transport and logistics function is therefore changing in relation to programme support, with supply chain teams challenged to become more 'market aware' to avoid destabilizing local economies and to enable crisis-affected communities to seek greater choice of assistance. Provision of assistance to affected populations can take the traditional form of in-kind goods or cash and voucher-based assistance. There can potentially be two pipelines: commodities and cash. There is now a point in the planning and modality selection where a decision must be made as to whether either or both of these modalities should be selected. Logisticians must be involved early in planning processes to support programme design choices.

The Universal Logistics Standards incorporate references to support this requirement and recognize that logisticians should be as competent and confident about delivering cash at scale as they currently are at delivering goods. Cash and market-based assistance in particular can present high risks for safeguarding and protection, and it is important that these risks are identified and mitigated through thorough and documented risk assessment processes to ensure safe programming.

It's easy to think of cash programming as being coupled to livelihoods, food assistance or access to relief items. However, what humanitarian organizations have seen in the last few years is an emergence of cash programming within new sectoral themes, including, for instance, access to water through the issuing of vouchers and private sector led water-trucking enterprises. Unconditional digital cash transfers are now more widely used for humanitarian assistance and the trend continues to grow. The argument about whether or not cash transfers should form part of humanitarian action has largely been won. What is less clear is whether or not cash is being provided as efficiently or effectively as it could be and at the right scale, and whether cash transfers have transformative implications for the future of humanitarian aid (including managing supply chains), given that they challenge the main ways that aid has been delivered over the past several decades (Cash transfer programming and the humanitarian system Background Note for the High Level Panel on Humanitarian Cash Transfers). Humanitarian assistance should be 'market aware' so that local markets are understood and accounted for in the design, implementation and monitoring of humanitarian interventions. Market analysis is therefore an essential precursor to defining response modalities and response options and to identifying opportunities to support local markets and the supply chains that underpin them.

The impact this has on logistics is complex. The modus operandi has been designed and described, in significant detail, around the direct delivery of 'goods' and now has reached a point where a revision of these new modalities is needed. It will be a challenge moving forward to design procedures and a control framework that enables effective delivery and accountability while appropriately managing risk. What should procedures and controls look like in a humanitarian programme delivery mechanism of cash? What other programme activities may be changing towards cash modalities and how can humanitarian logisticians prepare for these changes?

Market engagement has taken on several new dimensions in emergency work. Instead of being driven towards those markets where logistics can most easily and quickly meet supply demands, the market is now being considered as an integral part of an affected community and a critical driver both in the short-term recovery of beneficiaries and in the long-term sustainability of humanitarian impact. International surge capacity will always be needed for the Asian tsunamis, the Haiti and Nepal earthquakes, Pakistan floods and Philippines typhoons, but increasingly aid organizations are responding to more mid-sized, often protracted, emergencies where the volume of supplies needed does not always exceed the national markets' capacities. In these scenarios, there is now the opportunity to make more strategic decisions about market engagement and the use of donor funds to create valuable micro-economic impact. Spending locally to rebuild the market and ensure the local appropriateness of what is procured for beneficiaries is important. But what is the right international versus local procurement balance and how will this balance be further challenged as local spend is increased?

The aid sector has developed numerous market assessment tools, many of which have now had several years of testing, but there is a widely acknowledged market analysis gap that inhibits effective decision-making and subsequent market intervention monitoring.

How do logisticians and procurement managers decide what to buy and where to buy? How do they determine, rapidly and per commodity, whether the disaster-affected market really has the capacity to meet supply needs? Operating models will need to be changed to enable financial and procurement processes to be flexible and adaptable in situations where local markets can no longer provide supplies sustainably.

Scaling up cash-transfers is a critical issue for organizations as they must develop capability to use more complex cash delivery mechanisms, e.g. banks, post offices and mobile phones. Building on research into programme design and 'value for money', it will be important to explore options for creating tools that help to compare the cost-effectiveness of different electronic cash transfer mechanisms in relation to their impact on programme objectives as well as their impact on financial inclusion and longer-term beneficiary protection.

Another dimension to consider is the categorization of traders (local suppliers) as beneficiaries. That is, those small-scale local traders within a beneficiary group who are themselves seriously affected by the emergency and who have ceased or reduced trading as a result. In these cases, supply chain interventions need to be geared towards 'market support' activities, which have started to include loans or grants to boost market activity. Here the classic definition of the 'private sector' is called into question. How can logistics best place itself to adapt to trader beneficiaries and to other private sector actors in the humanitarian markets?

So, what does this mean for humanitarian logisticians of the future? They need to be market savvy, able to assess the impact of an emergency on local market conditions and to understand the connections of a 'target' market to the upward national and international supply chains. They need to be able to monitor commodity variations and market triggers to be prepared for a large range of response scenarios. They need to be agile and able to switch modalities where needed; moving with ease and with prior preparation and planning between the use of 'cash', vouchers, cash for work (CFW) and local, national and international commodity supply, reacting swiftly to changes in market circumstances. Logisticians need to be able to work across functional silos to redefine the nature of 'programme logistics' in order to dispel the perception that it is simply a programme or project service, and must become an integral and essential component in the delivery of humanitarian assistance. Most importantly, such assistance will increasingly be in the form of financial transfers (cash) to beneficiaries, and it is important to consider the need for aid organizations to find new ways to partner with private companies that can provide the technical inputs, on a sustainable for profit basis. The role then of humanitarian logisticians and supply chain managers would be to identify need, facilitate supply and monitor quality, performance and risk.

Across much of Africa today more than 75 per cent of the population own phones and 25 per cent have access to the internet. Thousands of kilometres of fibre optic cable now join Africa, Europe and the Middle East. It is interesting to note that in Somaliland, where regulation is very unrestrictive, companies have been able to exploit the space that this provides for innovation. Now Somaliland has one of the world's highest rates of digital transactions using the cash-free Zaad service. This is used to receive remittances, pay even very small amounts and has reduced risks of theft. Cash is disappearing, credit cards are unnecessary and daily shopping is speedy and digital. Almost every merchant, even hawkers on the street, accept payment by cell phone (The Globe and Mail, 2013).

This is an innovation that is transforming the continent as well as the way in which humanitarian logisticians and supply chain managers operate. Mobile money has also drastically reduced the cost of crime and security for consumers, private companies and government offices. The Coca-Cola branch in Somaliland, for example, is the only cashless Coca-Cola company in Africa. About 80 per cent of its sales to retail distributors are performed through Zaad, while the remainder are achieved via electronic bank transfers.

As a community of practice, logisticians pride themselves on being able to deliver goods in enormous quantities, with great precision and, crucially, at an unparalleled speed. What lies ahead for the humanitarian logistics community will be the need to gain confidence with a new definition of success – one where market recovery and resilience is the target and where surge capacity is a fall-back measure, not the default. Perhaps it's not a question of maintaining an identity, but rather maintaining critical surge capacity and skills in a new era with more complex economic targets becoming the biggest humanitarian logistics challenge yet.

Beneficiaries may receive a smart card (with a closed information database for protection and acceptance purposes) with which they can access in-kind items from a distribution centre (their allocation recorded onto the card). Thereafter, they may be able to access locally available materials (such as shelter items) from vendors with the card acting as a voucher, and finally draw from banks or remittance offices a fixed amount of cash credited to the card to be used for unconditional support. Depending on the duration and evolution of the response, the same card can be re-credited – perhaps even by several collaborating agencies – for more of the same services, or for new types of support as appropriate.

Since the global financial crisis, and exacerbated during the Covid-19 pandemic, funding for humanitarian assistance has been declining. Although logisticians may increasingly be asked to take on work beyond their original, more technical, mandate with the shrinking funding pool the question is: how do they undertake all aspects of this ever-increasing portfolio effectively? More innovative capabilities are certainly needed. Logisticians are central to effective, fast, disaster relief as they serve as a bridge between disaster preparedness and response, between procurement and final distribution and between headquarters and the field programmes. As logistics operations are inherently costly and since logisticians must track goods through the supply chain, the function is often the repository of data that can be analysed to provide post-event learning. Such data reflects all aspects of execution, from the effectiveness of suppliers and transportation providers, to the cost and timeliness of emergency responses, to the appropriateness of donated goods and the management of information.

The role of the humanitarian logistician, far from being reduced with the onset and introduction of cash transfer programming, is expanding to include building preparedness mechanisms with local actors and establishing contracts with local suppliers and manufacturers. While the need for large-scale supply chain preparedness and pre-positioning of standard goods will remain, consistent validation and monitoring of suppliers will enable variations in specifications. The humanitarian logistician will be central to the evolution of the response programme and integral to the relief teams in facilitating not what's required, when or where, but rather feeding in the most effective and efficient means by which to meet those requirements. The future suggests that skills needed by the next generation of humanitarian logisticians will include not only the means to manage complex supply chains and the procurement of locally-sourced goods and services, but also the capability to manage pre- and post-crisis market assessments and methodologies to generate comparative KPIs that guide programmes towards the achievement of qualitative and cost-efficient assistance. Logisticians will have an integral role in programming to ensure that the means to meet what beneficiaries actually need is delivered, rather than hoping that they need what aid pipelines happen to be able to deliver.

The definition of the logistics and supply chain 'rights' is changing, not just in terms of the right product, place, time, cost and quantity but towards a deeper synergy between 'do no harm' and how aid organizations should spend donor funds responsibly in order to achieve the greatest market impact.

References

Alicke, K, Barriball, E, and Trautwein, V (2021) How COVID-19 is reshaping supply chains, *McKinsey and Company*, www.mckinsey.com/business-functions/operations/our-insights/how-covid-19-is-reshaping-supply-chains (archived at https://perma.cc/YU3P-CUXN)

Australian Red Cross (ARC) (2020) The challenges of Unsolicited Bilateral Donations in Pacific humanitarian responses, 16 December, *Australian Red Cross*, reliefweb.int/sites/reliefweb.int/files/resources/161220%20Report%20-%20Challenges%20of%20UBD%20in%20Pacific.pdf (archived at https://perma.cc/V5T3-SVXV)

Besheer, M (2021) UN Food Stocks Looted in N. Ethiopia; Some Aid Distribution Halted, *VOA News*, 8 December, www.voanews.com/a/un-food-stocks-looted-in-n-ethiopia-some-aid-distribution-halted-/6344726.html (archived at https://perma.cc/7CEA-NXLU)

Cash transfer programming and the humanitarian system Background Note for the High Level Panel on Humanitarian Cash Transfers

Centre for Research on Epidemiology of Disasters (CRED)(2021) Year in Review, May

CHS Alliance (2017), Core humanitarian competency framework, chsalliance.org/get-support/resource/core-humanitarian-competency-framework/ (archived at https://perma.cc/6RTB-3ZRB)

Darcy et al (2003), 'According to Need. Needs assessment and decision-making in the humanitarian sector'

Fenton, G (2013), 'An Evolving Sector: Managing Humanitarian Supply Chains, Strategies, Practices and Research', *DVV Media Group GmbH.*

HELP Logistics, KLU, Action Against Hunger (2017). Supply chain expenditure and preparedness investment opportunities.

IASC (nd) 'About the Grand Bargain', https://interagencystandingcommittee.org/about-the-grand-bargain (archived at https://perma.cc/EBM9-8BSY)

Inter-Agency Standing Committee, Six Core Principles Relating to PSEA

KEPSA (2020) Private sector – government partnership in response to Covid-19 pandemic, https://cdn.nation.co.ke/pdfs/KEPSA_Advertorial.PDF (archived at https://perma.cc/9FUL-67S6)

Kovács, G, and Tatham, P H (2010) What is special about a humanitarian logistician? A survey of logistic skills and performance, *Supply Chain Forum: An International Journal*, **11**(2), pp 32–41

Larson, P (2011) *Humanitarian Logistics: Meeting the Challenge of Preparing For and Responding To Disasters*, Kogan Page, London

LogCluster, Delivering in a Moving World, Available at: www.logcluster.org/sites/default/files/whs_humanitarian_supply_chain_paper_final_24_may.pdf (archived at https://perma.cc/5NYA-365N)

Lora, A M, Ali, M, Spencer, S, Takhsh, E, Krill, C and Bleasdale, S (2021) Cost of Personal Protective Equipment During the First Wave of COVID-19, Antimicrobial Stewardship & Healthcare Epidemiology, 29th July, Cambridge University Press

Policy Department for External Relations Directorate General for External Policies of the Union (2021) 'The future of humanitarian aid in a new context full of challenges', October

Smith, G et al, (2011) 'New Technologies in Cash Transfer Programming and Humanitarian Assistance'

Spencer, (2021), 'Humanitarian AI: The hype, the hope and the future'

Sphere Standards (2018) Sphere Standards Handbook, Available at: https://spherestandards.org/wp-content/uploads/Sphere-Handbook-2018-EN.pdf (archived at https://perma.cc/N8XX-HBYA)

Sphere Standards (nd) Sphere Standards Handbook, Available at: https://spherestandards.org (archived at https://perma.cc/2VHW-8SH5)

The Globe and Mail (2013), How mobile phones are making cash obsolete in Africa, 21 June

ULS (2021) 'The ULS Handbook', Available at: https://handbook.ul-standards.org/en/humlog/#sec001 (archived at https://perma.cc/3RR8-FYE2)

UNEP Resource Centre, Mainstreaming Environment into Humanitarian Action

UNOCHA (2014) ENVIRONMENT AND HUMANITARIAN ACTION Increasing Effectiveness, Sustainability and Accountability, Available at: https://www.unocha.org/sites/unocha/files/EHA%20Study%20webfinal_1.pdf (archived at https://perma.cc/R8EP-32NJ)

WHO (2020) 'Shortage of personal protective equipment endangering health workers worldwide', Available at: https://www.who.int/news/item/03-03-2020-shortage-of-personal-protective-equipment-endangering-health-workers-worldwide (archived at https://perma.cc/4DU3-CDSR)

WHO-PCI (2020) Private sector engagement in the covid-19 response: Kenya country experience, August, Available at: https://hsgovcollab.org/system/files/2020-08/Private%20sector%20engagement%20in%20the%20COVID-19%20response-%20Kenya%20country%20experience.pdf (archived at https://perma.cc/K6AG-FLJA)

13

The way forward: Current trends in humanitarian logistics

GYÖNGYI KOVÁCS

ABSTRACT

As is the tradition, the final chapter in this book does not conclude the book as much as it evaluates the way forward of humanitarian logistics research and practice. The chapter revisits earlier such predictions and adds current developments and expected, future steps in humanitarian logistics and supply chain management. Some stem from systemic changes, either from within the discipline as the move towards cash-based interventions, or others from outside of the discipline as in the large-scale impact of the Covid-19 pandemic on not just pandemic response but how aid is delivered overall.

Introduction

Now, as always, there is a discussion of what humanitarian logistics entails versus what it does not. This is not only a question of definitions, but also mandates. Humanitarian organizations are often specialized not only in specific thematic areas (from shelter to water and sanitation to health, etc.) but also in whether they respond to 'natural' disasters or conflicts, and whether they are involved in disaster response to sudden-onset disasters, and/or in development programmes. To some extent, these mandates also impact on their resource configurations and how they are activated (Kovács and Tatham, 2009).

More recently, their mandates in specific geographical areas were more in the limelight. Many 'donor' countries do not grant any operational mandate to United Nations (UN) agencies in their own countries, for example, the refugee operation of the UN Refugee organization UNHCR in Greece required a special exemption from this from the European Union (EU). Similarly, the Covid-19 response stretched the limits of who can operate where, and what is regarded as 'humanitarian'. Pandemics do not know borders, even if the same response is called a humanitarian operation in one part of the world, and a *civil protection* or crisis management one in another. Let it suffice that regardless of national pride and choice of wording, from a logistical perspective, the same principles apply.

Access to specific operational environments is not confined by mandates only, of course. Conflict zones have put the humanitarian imperative to test. Complex emergencies have been in the limelight as a combination of where natural hazards occur in a conflict environment; adding the Covid-19 pandemic to the mix, all humanitarian efforts had to be considered under constraints from many different directions. Extreme situations include that in Afghanistan in 2021/2022 that combines elements of conflict, famine, displacement and the pandemic, while humanitarian organizations struggled to get access to people affected by any of these. In short, now more than ever, humanitarian logistics is highly complex.

At the same time, certain trends prevail, including that of localization, sustainability, the embrace of cash-based interventions, etc. This chapter will therefore revisit old predictions and trends first and turn to new developments and a glance towards the future.

Don't throw the baby out with the bathwater

The previous editions of this book have put a long list of trends forward, fifteen of them, in fact. These are as follows, with a numbering that indicates the edition of the book and the trends:

1. a shift in focus from inter-agency coordination towards *relationship management* in the supply chain;
2. a renewed emphasis on the *sustainability* of aid;
3. the development of *specialized humanitarian logistics services* that organizations offer each other;
4. an emphasis on process as well as product and packaging *standardization and modularization*;
5. the use of *new technologies* to capture data;
6. a shift in focus towards *cash transfer programmes*;
7. a stronger focus on *supply chain visibility* as well as visibility across humanitarian organizations;
8. an extension of *professionalization* endeavours to include (implementing) partners;
9. a stronger focus on *interoperability* across humanitarian organizations;
10. a heightened focus on *security*;
11. humanitarian (product and process) *innovations*;
12. *scalability*, i.e. humanitarian supply chains adding surge capacity to existing ones in a disaster area;
13. a heightened awareness of the *humanitarian imperative to be able to deliver* (back to basics);
14. the need to cater to displaced *people on the move*; and
15. new *systemic changes* such as co-creation, and the sharing economy.

In addition to these, Altay et al. (2021) highlighted the *disruptive force of the pandemic* (trend #16) on humanitarian supply chains, which could be added as a sixteenth trend to the above. Their research was based on earlier trends as well as new ones, as identified in literature as well as the Global Logistics Cluster meetings. Interestingly enough, the third edition of this book has noted pandemics as a disruptor and an area in which humanitarian logistics is of essence. Not surprisingly, this fourth edition has an entire chapter devoted to the topics (Kovács et al., chapter 12).

The latest updates to these trends have been published earlier in Altay et al. (2021), based on the latest discussions at the Global Logistics Cluster meeting in April 2021. Even newer flavours and additions can be found in the other chapters of this very book.

Kim et al. (chapter 2) begin by reconsidering the very first trend (trend 1) on this list and rightly problematize the matter of supplier relationships from a public procurement perspective. The question remains, when to initiate procurement processes in light of their extensive lead times versus the shorter lead times between early warnings versus disaster occurrence. This is also encompassed in early warning-early action programmes, and even in trend 12 on the scalability of disaster relief as to how and when to add surge capacities (Annala et al., 2014). Humanitarian organizations have come a long way with forecast-based financing and forecast-based actions, especially under the umbrella of both the Red Cross Red Crescent Climate Centre and the more recent Anticipation Hub. It's high time to take their learning also on board in procurement.

Besiou et al. (chapter 11) revisits trend 2 when looking at the ecological dimension of aid, and further relate it to trend 7 on supply chain visibility – albeit from a different, transparency perspective. As they mention, several big projects are under way to look into the ecological dimension of humanitarian aid, including projects focusing on packaging, reverse logistics and waste management. There are also new alliances in this area, including some that focus on environmental audits upstream, which gives new flavour also to trend 1. The renewed focus on ecological questions brings systemic changes such as the questioning of what to deliver in the first place (trend 15), and when humanitarian organizations need to follow their first imperative to deliver (trend 13), but add details and nuance to many others as well. Given the concerted activities in the humanitarian context in this space, this will be an interesting area to watch. And while there has been the lingering question, 'But who is going to pay for it?' it is worth noting that the ECHO's new (2022) humanitarian logistics policy includes greening as a key emphasis.

Vega and Roussat (chapter 3) gives an overview of new sorts of logistics service providers, which are blurring the view between humanitarian and commercial ones. The various new types of organizations change both which services are provided and who provides what (trend 3) but also highlight a new type of innovation in the humanitarian context, that of business model innovation (trend 11).

Banomyong et al. (chapter 6) also look at humanitarian logistics services, adding a quality perspective. Their emphasis on quality in delivery is replicated in Håpnes' (chapter 5) imperatives from UNHCR's perspective.

Larson (chapter 4) adds to trend 4 in his focus on process standardization, to trend 9 on interoperability, and trend 11 in the area of process improvements. Indeed, process modularization has been recently on the agenda, also in relation to trend 12 and the possibility to plug and play disaster relief to extant processes in host countries and thereby scale up existing activities without disrupting them (Saïah et al., 2022).

Fenton and Muyundo (chapter 12) start with a quote on how artificial intelligence changes the world of humanitarian logistics. While there have been many different technologies for trend 5 highlighted in the past (with a recent peak on blockchain-related research, e.g. Kumar, 2021), relatively little can be found on artificial intelligence in humanitarian logistics. Rodríguez-Espíndola et al. (2020) provide the most recent overview of potentially disrupting technologies in humanitarian logistics, citing blockchain, artificial intelligence and additive manufacturing. According to them, the largest potential benefit of AI lies in the area of needs assessment, whereas additive manufacturing bears most positive effects when used for on-site production.

Harpring (chapter 9) elaborates on where we stand with cash and voucher assistance (CVA, trend 6). Cash based interventions (CBIs) have been on the agenda for a long time but have now also seen a boost under the pandemic, not the least due to disruptions in other types of humanitarian deliveries. Recognizing the time sensitivity of humanitarian deliveries regardless of modality or mechanism, and learning from preparedness overall, a more recent focus is on that of cash preparedness. Similar to supply chain relationships in trend 1, the idea is to identify and screen potential supply chain partners and enter framework agreements that will facilitate financial transactions if and when needed. Ali (2022) adds nuance to financial transactions from a different perspective: the difficulties of transferring money to support medical deliveries in various countries, whether due to a shortcomings in banking systems, trade restrictions or conflicts. Such difficulties will impact on both in-kind and cash deliveries also in the future, and should be reflected on more in such choices in humanitarian logistics.

Håpnes (chapter 5) provides a historical case study of UNHCR, which also highlights turns in focus and concludes in a number of imperatives for the future. Perhaps one of the most surprising ones relates to trend 8 where the imperative of professionalization of humanitarian logistics is not met in the latest restructuring of the organization.

Even the way pandemics disrupt humanitarian logistics (trend 16) is highlighted and explained in this book, both with a dedicated chapter (Kovács et al., chapter 10) and in the focus on resource scarcity (Gonçalves, chapter 8), which is also reflected in a funding crisis (more on funding in Wakolbinger and Toyasaki, chapter 1). Two trends remain without being revisited: the ones related to security (trend 10) and to serving people on the move (trend 14). Neither of them can be disregarded, however, neither in the aftermath of the international retreat from Afghanistan nor the crisis in Ethiopia and Ukraine, not to mention the never-ending crises in Syria and Yemen, to highlight but a few. Even though 2020 marked the year with more displacement due to weather and climate than conflict (Climate Centre, 2021), sadly, both climate and conflict continue to contribute to displacement since. Both deserve ample and urgent attention.

What is new under the sun?

Revisiting old predictions has left us with surprisingly little room for eliminating or even consolidating them. There are some shifts in emphasis within those, but altogether, it can be stated that humanitarian logistics is full of activity, reform and great ideas.

Apart from cyclical disasters and disaster relief cycles, some *trends also come in cycles.* There are constant shifts between grassroot endeavours and the focus on local capacities and implementing partners, versus renewed recognitions of the need of also physical preparedness outside regions, as after Cyclone Idai. Preparedness, and preparedness stock has seen a renaissance due to the pandemic as well, since countries that had some medical preparedness stock fared better in the first wave. Fleet management has come full cycle between centralizing fleet to buying local fleet again. There are good reasons for both. Centralizing fleet has come with the benefits of economies of scale in leasing deals to dream of that have brought financial stability alongside quality repairs. Local fleet, on the other hand, can ensure that spare parts exist in the country, that fleet does not require remote assistance to be serviced and that it is appropriate for the conditions on the ground, as well as instantly available.

A somewhat new concern on the localization agenda stems from ecological considerations. Local procurement can significantly cut transportation emissions, but at the same time, buying items that long commercial supply chains have imported to a country instead of humanitarians might raise new questions of the long tail of those supply chains instead. The same concern has been voiced in the area of cash and voucher assistance. Beneficiaries buying items locally may have other, and disconcertingly unknown, economic, social and ecological repercussions.

Similarly, there are shifts between cash-based interventions (CBIs) and in-kind deliveries. These do not need to be conflicting messages, however. *Both are needed*, the question is rather which is appropriate under what circumstances, and for which products and services. Indeed, there is a growing recognition of CBIs and in-kind deliveries occurring in parallel, though for different items perhaps. Some items may, after all, be available in local markets while others are not. Similarly, preparedness is key for both, and even the professionalization of humanitarian logisticians need to embrace knowing what to deliver and how to deliver it. At the end of the day, it is the same procurement officers who need to manage both.

Blockchain, AI and 3D printing notwithstanding, digital humanitarians clearly caught off during the pandemic. At a time when humanitarian 'expats' had to be evacuated, remote working first came to replace and later to still support field work. Local agencies and implementing partners have at the same time gained even more importance, since the actual implementation of any humanitarian delivery needed them even more. There is surprisingly little focus on them in global meetings or academic research, however. This is clearly a gap. Crucially, while remote work may be feasible for many other areas of humanitarianism, from cartography to policy, logistics remains the area that needs delivery capacity on site. Back to basics it is, once again.

The last edition of this book has ended with an outlook on incident evolutions and cascades. There has been significant work in that area since, with numerous special issues of journals and numerous projects being dedicated to this area. At the same time, there is much to be said about overlaps between disasters. Also in this space, there is considerable work around complex emergencies, as to say, the overlaps between conflicts and natural disasters. Adding the pandemic to this mix introduces another layer to an already complex picture. If anything, the context of humanitarian logistics is not getting any easier.

External drivers such as resource scarcity, funding scarcity, conflicts or the pandemic all go to re-emphasize the importance of the sector needing to be strong, together. On top of the by now long-standing coordination mechanisms of cluster meetings, more and more, it is groups of organizations negotiating with one another instead of individual ones. At the Global Logistics Cluster meeting in April 2021, these were called 'ecosystems of co-ordination'. They can stand for negotiations with suppliers, logistics service providers, governments, or funding agencies. Some years ago, in the third edition of this book we have discussed the aid matching of individuals from beneficiary to beneficiary. This comes now as a cycle back to the strength of the sector and the need for combined negotiation power in both economies of scale and access to specific regions, but also for a wiser way of using already scarce resources, from cross-utilization and interoperability to consolidation.

Famous last words

Hans Rosling's *Factfulness* (Rosling et al., 2018) was a refreshing and welcome view on the actual statistics versus people's feelings about trends. We are so wrong on so many accounts if trusting just the weak signals – however important – of new trends. Facts speak for themselves, after all. Global poverty has steadily been decreasing. The world is getting better, not worse, is a mantra repeated in Rutger Bregman's (2020) history book. Both books provide hope, and yet ironically, both are also trendsetters. It may be difficult to pin one's hopes on a brighter future at times with newly inflamed wars, concurrent disease outbreaks, massive droughts on several continents, heatwaves and with those, the foresight of potentially more devastating cyclone, hurricane and typhoon seasons to come. Climate scientists have documented ample evidence for the changes happening around us and even how they will impact on the patterns of hydro-meteorological hazards.

Evidence-based medicine speaks volumes for looking at facts. Evidence-based research, and evidence-based action have also been spoken for in humanitarian logistics. We need to stay to the facts to see where the impact is largest, also when systemically rethinking what to do. And that is exactly what humanitarian logistics is facing: a big, systematic rethink. It is not only about cash versus in-kind, but also ecological footprints and handprints, and strengthening local capacity and decolonizing aid at the same time as saving the planet. At the end of the day, the imperative remains *do no harm*.

There is some hope to be offered for this: now when this book goes to print, Help Logistics and the Kuehne Foundation have come to an agreement with Emerald Publishing to flip the *Journal of Humanitarian Logistics and Supply Chain Management* back to being fully open access. The facts and the research are there – and now they will also be available to everyone, everywhere. Time to use them in our decisions.

Acknowledgements

This research could not have been achieved without the kind support of the H2020-SC1-PHE-CORONAVIRUS-2020 project No. 101003606 HERoS (Health Emergency Response in Interconnected Systems).

References

Ali, G N (2022) Medical supplies agencies and access to foreign currency in resource-limited settings: Case studies from Sudan, *Eastern Mediterranean Health Journal*, **28**(1), pp 74–77

Altay, N, Kovács, G and Spens, K (2021) The evolution of humanitarian logistics as a discipline through a crystal ball, *Journal of Humanitarian Logistics and Supply Chain Management*, **11**(4), pp 577–584

Annala, L, Tabaklar, T, Haavisto, I, Kovács, G and McDowell, S (2014) Supply chain scalability, in: B Gammelgaard, G Prockl, A Kinra, J Aastrup, P Holm Andreasen, H-J Schramm, J Hsuan, M Malouf and A Wieland (eds.) *NOFOMA 2014 Proceedings. Competitiveness through Supply Chain Management and Global Logistics*, pp 297–313, Copenhagen, Denmark,

Bregman, R (2020) *Humankind: A Hopeful History*, Bloomsbury Publishing, PLC

Climate Centre (2021) Three out of four displacements in 2020 were weather-related, Climate Centre, *The Hague*, www.climatecentre.org/4978/three-out-of-four-newdisplacements-in-2020-were-weather-related/ (archived at https://perma.cc/LK3T-SWBX)

ECHO (2022) *Humanitarian Logistics Policy*, Directorate-General for European Civil Protection and Humanitarian Aid Operations (ECHO), at https://ec.europa.eu/echo/files/policies/sectoral/humanitarian_logistics_thematic_policy_document_en.pdf (archived at https://perma.cc/CUL7-6796)

Kovács, G and Tatham, P H (2009) Responding to disruptions in the supply network – from dormant to action. *Journal of Business Logistics*, **30**(2), pp 215–229

Kumar, A (2021) Improvement of public distribution system efficiency applying blockchain technology during pandemic outbreak (COVID-19), *Journal of Humanitarian Logistics and Supply Chain Management*, **11**(1), pp 1–28

Rodríguez-Espíndola, O, Chowdhury, S, Beltagui, A and Albores, P (2020) The potential of emergent disruptive technologies for humanitarian supply chains: The integration of blockchain, Artificial Intelligence and 3D printing. *International Journal of Production Research*, **58**(15), pp 4610–4630

Rosling, H, Rosling Rönnlund A, Rosling O (2018) *Factfulness: Ten reasons we're wrong about the world and why things are better than you think*, Flatiron Books, New York

Saïah, F, Vega, D, De Vries, H and Kembro, J (2022) Process modularity, supply chain responsiveness, and moderators: The Médecins Sans Frontières response to the Covid-19 pandemic, *Production & Operations Management*, DOI: 10.1111/poms.13696 (archived at https://perma.cc/ES5G-FGJL)

ANNEX 1: Operational assessment

To view Annex 1, please go to: http://koganpage.com/humanitarian-logistics

INDEX

Note: Page numbers in *italics* refer to tables or figures

'3Cs' 50
'6W Problem' 2
'6W' question 107
Accra 163, 183, 185
Action Contre la Faim (ACF) 277
Adapting humanitarian action to the effects of climate change (ALNAP) 273
Afghanistan 87, 88–90, 318, 322
Africa 17, 77–78, 159, 312
   centre-of-gravity analysis application 181–85, *182*, *183*, *184*
   Covid-19 crisis 269–70
   UNHCR 101–03
agility 3, 14, 186, 187, 305, 311
   ability and 126
   agile strategy 163
   commerce and humanitarian logistics 11–14, *12*
   managing supply networks 6–11
   supply networks 6–11
artificial intelligence (AI) 285, 298–99, 321, 324
aid agencies 2, 4, 7, 22–23, 35–36, 87
   Alexander and Parker 84–85, *84*
   donors 30–33
   funding systems and 24–30
   fundraising 33–34
aid Worker Security Database (AWSD) 83, 91, 95
aid workers 15, 79
   academic literature 85–91, *86*, *90*
   attacks on 83–85, *84*
   research agenda 95–96
   security evaluation and management 91–95
Al Za'atari camp 120
Alexander, J 84–85, *84*
Algeria (Algiers) 183
Ali, G N 322
Altay et al 319–20
altitude sickness 86

Amazon 298
Amman 104, 105
Anglia Ruskin University 15
Angola (Luanda) 183
Annala et al 79
Anticipation Hub 320
Arikan et al 34
Asia 162, 286
assisted natural regeneration (ANR) project 274
AstraZeneca 257, 259
Atlas Logistique 278
Australia 104, 120
Austria 15

Bailey, R 57
balancing loop B1 206
Balcik, B 29, 48
Banomyong et al 16, 132, 140, 321
basic rapid assessment tool (BRAT) 299, 300
BBB Wise Giving Alliance 31–32
Beamon, B M 29, 48
Beichuan 196
Beirut 104–05
Beirut port 104–05
Belaya et al 61
Belgian court 257
beneficiaries 2, 131, 132–33, 137–39, 150, 289–90
   'battle' for change 107–09, *109*
   capacities and capabilities (case study) 235–42, *236*
   complex systems components 203–11, *207*, *208*, *210*
   emergencies, humanitarian response to 195–97, *196*
   funding systems structure 23–24
   HSCS opportunities 276–81
   HSCS, impact of 270–76
   humanitarian logistics 163–64, *164*

HUMSERPERF questionnaire
  developing 140–49, *140*, *142*,
    144–45
  network design 166–69, *167*, *168*, *169*
  security evaluation and
    management 91–95
  supply chain management and
    partnership 45–51, *48*, *51*
  supply chain planning, right
    place 296–302
  UNHCR 101–03
  see also supply chain planning, 'rights' of
Beresford, Anthony 15
Besiou et al 29, 320–21
Besou, Maria 17
Bioforce Institute 76, 306–07
Black Sea ports 161
Bosnia 102
Boston 112
Bowersox, D J 178–80
BP 9
Brangeon, S 278
Bregman, Rutger 325
British government 77
Budapest 108, 109–14, 117, 123, 127, 128
Bulawayo 183
'bullwhip' effect 11
Burkart et al 33
Burkina Faso 183
Burnham, G 89
Burns, L 89
business sector 44, 47, 53, 303

CALP Network 226, 231, 242
Cambridge 76
camp coordination and camp management
    (CCCM) 106
Canada 15, 237, 274
carbon footprint 268, 271, 274
  HSCS opportunities 276–81
Cardiff University 15
Cardoso et al 137–38
CARE 174, 202
Caribbean 289
cash and voucher assistance (CVA) 217,
    220–26, *220*, *221*, *222*, 242–43,
    321–22
  capacities and capabilities (case
    study) 235–42, *236*
  capacities, capabilities and
    competencies 228–35, *229*,
    232, *234*
  cash and voucher assistance
    programming, phases 226–28, *227*

Cash based interventions (CBIs) 321–22,
    323–24
Cash Coordination Caucus 235, 243
cash for work (CFW) 311
Cash Learning Partnership Network 228
cash transfer programmes (CTPs) 23–24
cash working group (CWG) 219, 242–43
  capacities and capabilities (case
    study) 235–42, *236*
  capacities, capabilities and
    competencies 228–35, *229*, *232*, *234*
Castillo, J G 223–24
Central Emergency Response Fund
    (CERF) 22
centralization 166–69, *167*, *168*, *169*
Centre for Research on Epidemiology of
    Disasters (CRED) 170
centre-of-gravity analysis 159, 166–67, *167*,
    178–81, *179*
  application 181–85, *182*, *183*, *184*
certificate in humanitarian logistics
    (CHL) 307
Charity Navigator 31–32
Chartered Institute of Procurement and
    Supply (CIPS) 121
China 196–97, 252, 255–56
Chiu, C H 95
cholera 248, 251
CHS (Core Humanitarian Standards) 292
Cisco 13
Climate and Environment Charter for
    Humanitarian Organizations 272–73
Closs, D J 178–80
CO2e 280
Coca-Cola 312
Coleman, Andrea 77–78
Coleman, Barry 77–78
Collaborative Cash Delivery Network
    228, 243
Colombia 177
commercial supply chains (CSCs) 47–49,
    48, 131, 133–36, *134*, *135*, 138, 150,
    160, 266, 303, 323
Committee of Contracts (CoC) 122
communal SDGs 269
Competition for scarce resources loop
    (R5) 209–11
Competitor HO's relief capacity reinforcing
    loop (R2) 211
complex systems
  behaviour in 200–03
  characteristics 197–200, *199*, *200*
  components 203–11, *207*, *208*, *210*
Connecting Business Initiative (CBi) 304

Consortium of British Humanitarian Agencies 307
Copenhagen 163
Core Humanitarian Competency Framework 307
Corey et al 87
corporate social responsibility (CSR) 70, 78, 291, 293
Country-Based Pooled Funds (CBPFs) 22
COVAX 193–94, 257–58
Covid-19 pandemic 15, 17–18, 127, 159, 248–49, 286, 318
 capacities and capabilities (case study) 235–42, *236*
 cash and voucher assistance 218–19
 Covid-19 vaccine inequity failure (2021) 193–94
 disasters, response to 305–14
 emergency response and 302–05
 funding gap 293–94
 global supply chains and 127
 humanitarian and 1–3
 humanitarian crisis 269–70
 impact of 162–63
 pandemic response insights 251–53
 personal protective equipment 253–56
 relationships between HOs and suppliers 54
 right place 296–302
 SDGs and 268–69
 vaccination programmes 256–62
 see also COVAX; supply chain planning, 'rights' of
Covid-19 vaccination programmes 249
Covid-19 vaccines 256–62
Coyle et al 178
criminal attacks 91–95
crisis (Rwanda) (1994) 95, 192
Crisis Country Cluster map 170
critical path analysis 254, 256, 258, 259
Croatia 102
Cronin et al 136
Cronin Jr, J J 134–35, 141
Crown Agents (CA) 76–77
Curling, P 87
customer integration 49
customer relationships 186
customers 2
 customer integration 49
 HSCS opportunities 276–81
 humanitarian logistics 163–64, *164*
 humanitarian organization (HO) 55–58, *58*

HUMSERPERF questionnaire developing 140–49, *140*, *142*, *144–45*
logistics services 67–72, *71*
service quality performance measurement 133–39
supply chains configuration 185–88, *187*, *188*
Cyclone Idai 323

Danish Refugee Council (DRC) 120, 182, 276
Darfur 26, 90, 274
Darwin 104
Day et al 46–47
DDPM-SAOs 142, 147–48
de Villiers, Gerard 16, *164*, 166, 174, 187
decentralization 127, 166–69, *167*, *168*, *169*
Delft University of Technology 17
'Delivering in a Moving World' 295
Democratic Republic of the Congo (DRC) 182
dengue 248
Department of Disaster Prevention and Mitigation (DDPM) 150
 HUMSERPERF questionnaire developing 140–49, *140*, *142*, *144–45*
Department of Peacekeeping Operation (DPKO) 102
DHL 13
Dili 104
Direct Relief (DR) 77
Disaster Relief Emergency Fund 27
distribution centres (DCs) 175, 178, 180
Division of Finance and Administration 103, 105, 107, 108, 109–10
Division of Human Resources Management 107, 108, 109
Division of Operational Support 105, 107–08
'do no harm' principle 290
donors
 collaborative supply chains 296–302
 disaster relief and limited resources 35–36
 disaster response 24–30, *30*
 distribute Aid 76
 funding systems structure 23–24
 humanitarian operations interface 33–34
 HUMSERVPERF applications 149
 incentives provided by 30–33
 Intended rationality 209–11, *210*

organizational factors 59–60
relief operations, customers in 138–39
scarce resources 206–09, *207*, *208*
security evaluation and management 91–95
situational factors 55–58, *58*
UNHCR budgets 114–17, *115*, *116*
UNHCR supply chain management function 117–22, *119*
see also cash and voucher assistance (CVA); supply chain planning, 'rights' of
DRC. *See* Democratic Republic of the Congo (DRC)
droughts 170, 185, 325
Dubai 104, 163
Duran et al 174

earthquakes 170, 185–86, 192–93
  Gujarat earthquake 27
  Haiti earthquake (2010) 192
  Indian Ocean earthquake (2004) 25–26
  Kobe earthquake (1995) 201
  Nepal earthquakes 295, 310
  Pakistan earthquake (2005) 212
  Sichuan earthquake (2008) 196
East Africa 17
  centre-of-gravity analysis application 181–85, *182*, *183*, *184*
Ebola 74, 86, 248, 252–53, 295
  STRIVE Ebola vaccine 261
ECHO. *See* European Civil Protection and Humanitarian Aid Operations (ECHO)
economic metrics 267
economic sustainability 268, 276
Egypt (Cairo) 120, 183
El Niño effect 286
Emerald Publishing 326
Emergency Events Database (EM-DAT) 170, 181, 188
'emergency procurement' 54–55
emergency relief 159–60, 169, 170, 185, 202–03, 220, 291
  Covid-19 context 162–63
  emergency relief items 174–75
  see also Covid-19 pandemic; pandemic
Emergency Supply Pre-positioning Strategy (ESUPS) 177
EMMA Partners 228
employees 230, 252
  employee training 28

HUMSERPERF questionnaire developing 140–49, *140*, *142*, *144–45*
past pandemic response insights 251–53
environmental degradation 271, 276, 277–78
environmental metrics 267–68
Epidemiology of Disasters 132
ERP (enterprise resource planning) 110, 112–13, 127
Ethiopia 322
Ethiopian Red Cross Society 202–03
Europe 102, 162, 286, 312
European Civil Protection and Humanitarian Aid Operations (ECHO) 321
European Union (EU) 1, 256–62, 318

Factfulness (Rosling) 325
FedEx 13
feedback 16, 232–33, 302
  complex systems 197–200, *199*, *200*
  complex systems, components 203–11, *207*, *208*, *210*
Fenton, George 17, 321
Fernandez, Thomas 16
financial service provider (FSP) 236, 239–41
Finland 15, 17
Fleet Forum 113–14
Fleet management 323
Fleet Management System (FMS) 113
Food Aid 202
Forrester, J W 201–03
fourth party logistics (4PL) provider 13, 74
France 76, 255
French Red Cross 281
Fritz Institute 110, 121, 307
  Fritz Institute report analyses 117–18
  and humanitarian organizations 300
Fuchs et al 33
funding gap 25, 278, 293–95
funding systems 21–22
  donors 30–33, 35–36
  financial flows, impacts of 24–30, *30*
  humanitarian operations interface fundraising 33–34
  structure 23–24

G3 Security Limited 89
Gates Foundation 293
Gattorna, J L 186–88, *188*
GAVI 74
General Assembly resolution (46/182) 212

Geneva 106, 110, 111, 117, 128, 274
  'battle' for change 107–09, *109*
Geneva headquarters (HQ) 107–08
'geological hazard' 181
'German Air Force C-130/160s' 105
Germany 17, 137, 177
Ghana 175, *176*, 183
Give Well 31–32
GIZ 274
GLC network 275–76
global 'war on terror' 83, 87
Global Cash Working Group 235
Global Fleet Management 114, 121, 124–25
Global Logistics Cluster 291, 320, 324–25
Global Medic 299
global SDGs 269
global warming 271, 277–78, 281
global warming potential (GWP) 281
globalization 303, 305
Goentzel, J 136–37
Gonçalves, Paulo 16, 201
Good Humanitarian Donorship Initiative 35
Grand Bargain 218, 225–26, 295–96
Grandi, Filippo 127
Gray, R 136
Great Lakes region 102, 103
Greece 122, 318
greenhouse gases (GHG) 271, 276, 280
Guidero, A 90
Gujarat earthquake 27
Guterres, António 103, 126–27

Haavisto, I 136–37
Haiti 96, 192–93, 202, 301, 310
Haiti earthquake (2010) 192–93
Hanken School of Economics 15, 17
Hanken University 17
Håpnes, Svein 16, 321, 322
Harmer, A 87
Harpring, Russell 17, 321–22
Heaslip et al 75
  cash and voucher assistance 220–26, *220*, *221*, *222*
Heigh, I 29–30, *30*
Help Logistics 325–26
Herzegovina 102
Hoelscher et al 95
Hoelscher, K 86
Honduras 177
HO-supplier relationships 50–51, *51*, 62
  organizational factors 59–60
  situational factors 55–58, *58*
  supplier relationships, key issues 52–55

Houldey, G 87
'How Covid-19 is reshaping supply chains' (article) 296
Hughes, K 137
humanitarian agencies 4, 88, 90, 95, 105, 228, 231, 301
humanitarian cluster system 235
'humanitarian common logistic operations picture' (H-CLOP) 5
humanitarian landscape 291
Humanitarian Logistics Association (HLA) 6, 17, 307
humanitarian organizations (HOs) 3, 15, 23–24, 55–58, *58*, 60–62, 192–94
  capacities and capabilities (case study) 235–42, *236*
  cash and voucher assistance 217–19, 223, 225, 231–32
  commercial providers 69–72, *71*
  complex systems components 203–11, *207*, *208*, *210*
  decentralized nature 125
  emergencies, humanitarian response to 195–97, *196*
  healthcare logistics 249–50
  HSCS's environmental impact 270–76
  humanitarian providers 72–75, *73*
  HUMSERVPERF's Applications 149
  logistics services 67–69
  pandemic response insights 251–53
  personal protective equipment 253–56
  see also supply chain planning, 'rights' of
Humanitarian Recognized Logistics Picture (HRLP) 113, 126
humanitarian reform agenda 102, 104
humanitarian sector 18, 43, 60–62, 218, 225, 265, 269
  HSCS opportunities 276–81
  HSCS's impact 270–76
  humanitarian providers 72–75, *73*
  humanitarian supply chain management and partnership 44–51, *51*
  logistics services 67–69
  power dynamics 52–55
  situational factors 55–58, *58*
  see also cash and voucher assistance (CVA); supply chain planning, 'rights' of
humanitarian service provider (HSP) 72–75, *73*
humanitarian supply chain management (HSCM) 22, 33, 44, 62, 291, 300
  and partnership 45–51, *48*, *51*

Humanitarian Supply Chain Service
    Performance (HUMSERVPERF) 131,
    133, 150
  questionnaire developing 140–49, *140,
    142, 144–45*
humanitarian supply chains (HSCs) 3, 17,
    44, 162, 222, 266, 278
  environmental impact 270–76
  financial flows, impacts of 24–30, *30*
  humanitarian operations interface
    fundraising 33–34
  humanitarian supply chain
    management 291
  and KPIs 300–01
  network design 166–69, *167,
    168, 169*
  opportunities 276–81
  relationship dynamics 61–62
  supply chains configuration 185–88,
    187, 188
  see also sustainability
Humanity & Inclusion (HI) 74, 278
HUMLOG Institute 15, 17
Hungary 108
Hunt, S D 53
Hurricane Irma 289
Hurricane Mitch 26
hurricanes 170, 289, 325

India 170, 259
Indian Ocean earthquake (2004) 25–26
individual SDGs 269
Indonesia 177, 272
INGOs (international non-governmental
    organizations) 74, 84, 102, 294
INSEAD 106, 114
'interagency emergency health kits' 250
internal integration 49
internal Policy Development and Evaluation
    Unit (UNHCR) 113
internally displaced people (IDP) 102, 248
International Committee of the Red Cross
    (ICRC) 74, 272, 273, 275, 277,
    279–81
International Federation of Red Cross and
    Red Crescent Societies (IFRC) 23,
    27–28, 34, 59, 113–14, 121,
    270–73, 288
  HSCS opportunities 276–81
  Red Cross and Red Crescent Movement
    (RCRCM) 218, 222–23, 228, 275
  supply chain organization
    building 109–14

international humanitarian organizations
    (IHOs) 45, 55, 60, 218, 292
international non-governmental
    organizations (INGOs) 74, 84,
    102, 294
Irani, Z 49
Iraq 88, 104, 120
Italy 9, 175, *176*

Jacmel 193, 202, 205
Jahre, M 29–30, *30*
Japanese government 201
Javidan, M 230
Johannesburg 183, 185
Jordan 104, 105, 120, 122
Joseph, Sarah 17
*Journal of Humanitarian Logistics and
    Supply Chain Management* 326
Julagasigorn, Puthipong 16
just-in-time (JIT) 8, 55

Kahn, K B 49
Kamal, M M 49
Kenya 87, 183, 294
key performance indicators
    (KPIs) 300–01, 314
Kim et al 320
Kim, C 169
King, D 95
Kisangani, E F 89, 95
kitting strategy 166, 254, 256,
    258, 259, 262
'know-your-customer' (KYC) 239
Kobe earthquake (1995) 201
Kovács, Gyöngyi 17, 26, 61, 121
Kuehne Foundation 325–26
Kühne Logistics University 17
Kunz et al 125

Laguna-Salvadó et al 280
Lancet (Peer-reviewed journal) 257
Larson, Paul D. 15–16, 86, 303, 321
Lassa fever 248
Lebanon 76, 104–05, 120
Léogâne 193, 202, 205
Léon, V 278
Li & Fung 12–13, *12*
life cycle analysis (LCA) 281
Life cycle impact assessment
    (LCIA) 281
Lim, Jihee 15
links *12*, 12, 160, 261
Lischer, S K 89

logisticians 285, 286–90
    collaborative supply chains 296–302
    supply chain planning, right
        quantity 305–14
    supply chain planning, right time 302–05
logistics cluster (LC) 104–05, 106, 161
logistics service providers (LSPs) 68, 69–72,
        71, 75, 78, 250, 304, 321, 324–25
logistics services 78–79, 78, 161, 185, 295
    commercial providers 69–72, 71
    humanitarian logistics services 319, 321
    humanitarian providers 72–75, 73
    non-profit organization 75–78
London 89
low- and middle-income countries
        (LMICs) 260, 270, 303
low-income countries (LIC) 15, 193–94
Lubbers, Rudd 103
Lund University 111–12

Macau 103
machine learning (ML) 298–99
Madagascar 177
Mae Sot district 147–48
Maghsoudi et al 228
Makepeace et al 46–47
malaria 193, 279
Malaysia 175, 176, 272
Mamola, Randy 77–78
Managing System for Resources and
        People (MSRP) 112–13, 118, 122,
        124–25, 127
    UNHCR budgets 114–17, 115, 116
Marburg 248
Martin, R 87, 88
Martinez et al 29
McKinsey and Company 296–97
measles 248
Médecins sans Frontières (MSF) 68, 87–88,
        90, 110, 271, 274–75, 277
Medina-Borja, A 137
Mentzer, J T 49
Mérieux, Charles 76
MERS 248
Mexico 235–42
micro-, small- and medium-sized enterprises
        (MSMEs) 304
Middle East 120, 312
Miklian, J 86
mile-centre solution 178–80
*Minimum Standard for Market Analysis
    (MISMA)* 225
Mitchell, D F 87, 89, 95
Mizushima, M 45

Modern Slavery Act (UK) 290
Moderna vaccines 257
modus operandi 310
Morgan, R M 53
Morocco (Rabat) 183
Morokuma, N 95
Mövenpick Hotel 104
'must-meet-demand' criteria 112
Muyundo, Tikhwi Jane 17, 321

Nairobi 183, 185
Nakuru 183
National Migration Institute (INM) 238
natural disasters 141, 184, 186, 248,
        270–71, 286, 318, 324
    disaster events 170–73, 171, 172, 173
    see also earthquakes; hurricanes;
        tsunami; typhoons
NEAT+ 281
Nepal 177
Nepal earthquake response 295, 310
Netflix 298
Netherlands 17
network design 159, 160–62, 166–69, 167,
        168, 169
    centre-of-gravity analysis
        application 178–81, 179, 181–85,
        182, 183, 184
    disaster locations 170–73, 171, 172, 173
    emergency relief items 174–75
    ESUPS project 177
    humanitarian logistics context
        163–64, 164
    humanitarian logistics planning
        164–66, 165
    supply chains configuration 185–88,
        187, 188
    UNHRD network 175–77, 176
Niemann, W 164, 166, 187, 187
9/11 85
nodes 12, 12, 160, 261
Nolte et al 137
non-food items (NFIs) 164, 308
Northeastern University 112
Northern Triangle of Central America 237
Norwegian Business School 112
Norwegian Refugee Council (NRC) 120,
        274
Novartis 259
Nvivo software 237
Nygård, H M 86

OIOS Audit 121
Oloruntoba, R 136

Operation Provide Promise 102
Operations Management scholars 201
organizations
    aid organizations 61, 76, 288, 294, 300, 304, 308, 310, 312
    bureaucratic organizations 117, 125–26
    cash and voucher assistance 220–26, *220*, *221*, *222*
    community-based organizations 23
    Crown Agents 76–77
    disaster response 24–30, *30*
    Distribute Aid 76
    donors 30–33
    humanitarian organizations, decentralized nature of 125
    international non-governmental organizations (INGOs) 74, 84, 102, 294
    non-governmental organizations 175–76, 218, 254, 269
    organizational factors 59–60
    plan and re-plan 279–80
    power, trust and commitment 52–55
    relief organizations 132–33, 137, 139, 141–42, 146, 150
    Riders for Health 77–78
    right cost 293–96
    right product 287–93
    situational factors 55–58, *58*
    uncertainty 6–11
    UNHCR supply chain management function 117–22, *119*
    UNHRD network 175–77, *176*
Oslo 112
Oslo Guidelines 104
Ouagadougou 183
Oxfam 4, 110
Oxfam equipment catalogue 4

Pakistan earthquake (2005) 212
Pakistan floods 310
Palestine refugees 102
Panama 163, 175, *176*
pandemic
    humanitarian healthcare logistics 249–50
    past pandemic response insights 251–53
    personal protective equipment 253–56
    vaccination programmes 256–62
    see also Covid-19 pandemic; supply chain planning, 'rights' of
Parasuraman et al 133–34, 137
PARCEL (Partner Capacity Enhancement for Logistics project) 291–92
Parker, B 84–85, *84*

payment service provider (PSP) 236, 239–41
performance measurement tool (HUMSERVPERF) 16
personal protective equipment (PPE) 1, 247, 249, 253–56, 258, 287, 294
    capacities and capabilities (case study) 235–42, *236*
    see also Rwandan refugee crisis
Pettit, Stephen 15
Pfizer BioNTech 257, 259
Philippines 177, 289, 295, 310
Philippines typhoons 310
Pilch, F T 89
polio 248
Port-au-Prince 192–93, 202, 205
Portuguese Refugee Council 103
post-traumatic stress disorder (PTSD) 87
potential regional clusters *184*
PricewaterhouseCoopers (PwC) 107–08
private sector 23–24, 47, 77, 269, 275, 292–96, 304, 311
Procurement Management and Contracts Service (PMCS) 112, 117, 118, 120–22, 123
production capacity 257, 259, 260, 262
Progresa (Oportunidades) 237–38
Prospera. *See* Progresa (Oportunidades)
protection from sexual exploitation and abuse (PSEA) 292
provincial reconstruction teams (PRTs) 87

Red Cross Red Crescent Climate Centre 320
RedR Australia 120
Refugee Convention (1951) 102
Refugee Protocol (1967) 102
refugees 89, 238, 248, 274, 318
    capacities and capabilities (case study) 235–42, *236*
    refugee crises 26, 137, 142–43
    Sahrawi refugee camps 103
    UNHCR 101–04
    see also Rwandan refugee crisis
reinforcing loop R1 206
ReliefWeb 95, 248
Renouf, J S 89
RescEU 254
RescUAV project 299
Réseau Logistique Humanitaire (RLH) 278, 295

'the resource-based view' 228–30, *229*
resources 8–9
  cash and voucher assistance 220–26, *220*, *221*, *222*
  commercial sector 69–72, *71*
  Direct Relief 77
  disaster response 24–30, *31*
  donors 30–33
  financial resources 52–56
  human resources 105–07, 111–13, 197, 297
  improve supply chain management 105–07
  natural resources 272–73
  supply chain partnership 49–51, *51*
  see also cash and voucher assistance (CVA); scarce resources
Réunion (Saint-Denis) 183
Richardson, B 198
Riders for Health 77–78, *79*
Rodríguez-Espíndola et al 321
Roh, S 169
Rondeau, D L 88–89
Rosling, Hans 325
Roussat, Christine 69–72, *71*, 321
Rowley, E 89
Rwanda 95, 253
Rwandan refugee crisis 192, 204, 206

Sadako Ogata 102–03
Sahrawi refugee camps 103–04
Sanofi 259
Sarajevo 102
SARS 248
  SARS-CoV-2 virus 258
Save the Children 276, 277
scarce resources 3, 16, 160, 169, 192–94, 325
  complex systems characteristics 197–200, *199*, *200*
  complex systems, behaviour in 200–03
  complex systems, components 203–11, *207*, *208*, *210*
  emergencies, humanitarian response to 195–97, *196*
  see also complex systems
Schmenner, RW 68
SCM efficiency 54
Seatzu, F 92–93
security evaluation and management (SEAM) 91–95
security triangle 83, 88, 90–91, *90*, 94
Serum Institute of India 259

Service Performance (SERVPERF) 133–36, *134*, *135*
  HUMSERPERF questionnaire developing 140–49, *140*, *142*, *144–45*
service quality 131
  commercial supply chains 133–36, *134*, *135*, 304
  questionnaire developing 140–49, *140*, *142*, *144–45*
  service quality performance measurement 136–39
SERVQUAL 133–36, *134*, *135*
  HUMSERPERF questionnaire developing 140–49, *140*, *142*, *144–45*
shareholders 302
  see also stakeholders
Sheik et al 95
Shell 9
Sheu, J B 137
Siam Bioscene 259
Sichuan earthquake (2008) 196–97
Sigala, Ioanna Falagara 17, 70
Simmons, K B 87
Skelly, J 88
Smith, R J 47, 48
social metrics 267
Somalia 88, 89
Somaliland 312
South Africa 183
South America 116
South Sudan 274
Southern Algeria 103–04
Spain 175, *176*
Spekman et al 50
Spens, K M 26, 72
Sphere Handbook 225–26, 306
  Sphere Standards Handbook 292
stakeholders 36, 48, 52, 94, 138–39, 267
  emergencies, humanitarian response to 195–97, *196*
  emergency relief items 174–75
  funding systems, structure of 23–24
  humanitarian relief 59–60
  vaccination programmes 256–62
'standard' relief items 288
Stephenson, J 231
Sterman, J D 201
stock keeping units (SKUs) 4
STOCKHOLM (STOCK of Humanitarian Organisations Logistics Mapping) 177

stocks 104, 203–04, 205, 206, 289
  emergency stocks 106–07
  pre-positioned stocks 163
  safety stocks 250
Stoddard et al 86, 88
Stoddard, A 84, 87
Strategic Network Optimization
  module 118
STRIVE Ebola vaccine 261
'structural flexibility' 13–14
Stumpf, Jonas 17
sub-Saharan Africa 77
Sudan 89–90
  South Sudan 274
  Western Sudan 26
Sudanese people 274
Supatn, N 139, 140
supplier integration 49
supplier relationship management
  (SRM) 50–51, 54–55, 62
Supply Chain 2015 Plan 111–12
Supply Chain Competency Framework
  document 110
supply chain feedback loops 302
supply chain integration (SCI) 49
supply chain management (SCM) 11, 13, 44,
  107–09, 109, 251, 294
  HSCS opportunities 276–81
  humanitarian operations interface
    fundraising 33–34
  and humanitarian organizations
    (HOs) 55–58, 58
  key issues 52–55
  organizational factors 59–60
  and partnership 45–51, 48, 51
  supplier relationships 60–62
  supply chain management and
    partnership 45–51, 48, 51
  supply chain organization
    building 109–14
  UNHCR 101–07, 117–22, 119
  UNHCR budgets 114–17, 115, 116,
    123–28
  UNHCR supply chain management
    function 117–22, 119
supply chain network redesign 277
supply chain planning, 'rights' of
  right cost 293–96
  right place 296–302
  right product 287–93
  right quantity 305–14
  right time 302–05
supply chain risk management (SCRM) 303

supply chain visibility 261–62, 319, 320
supply chains
  collaborative supply chains 296–302
  commercial supply chains 131, 133–36,
    134, 135, 138–39, 150, 162, 266,
    303, 323
  configuration 185–88, 187, 188
  Covid-19 pandemic 17, 127
  food supply chains 270
  global supply chains 290
  health supply chains 247–48
  humanitarian healthcare logistics 249–50
  humanitarian logistician 15–16
  integrating 50
  linear supply chains 2
  medical supply chains 296–302
  PPE supply chains 256
  private sector and humanitarian
    sector 47
  vaccine supply chains 256–62
  see also humanitarian supply chains
    (HSCs)
Supply Management Logistics Service
  (SMLS) 112, 114, 118, 120, 123, 127
Supply Management Service (SMS) 103–04
  'battle' for change 107–09
  supply chain management
    improvement 105–07
  supply chain organization
    building 109–14
  UNHCR budgets 114–17, 115, 116
survival SDGs 269
sustainability 3, 86, 127–28, 265, 281–82,
  291, 310, 318
  as a complex system 268–69
  Covid-19 crisis 269–70
  definition 267–68
  HSCS opportunities 276–81
  HSCS's environmental impact 270–76
Sustainable Supply Chain Alliance 275, 277
Swanson, R D 47, 48
Sweden 76, 111
Switzerland 16
Syria 322
Syria crises 116, 121

t'Serstevens, Sophie 17
Tak province 147–48
Tangjianshan quake lake 196
Taylor, S A 134–35, 141
technologies 2–3, 259–60, 298–300,
  307, 321
  digital manufacturing technologies 303

Thailand 16, 133, 141–42, 146, 147–48
Thammasat University 16
Thomas and Mizushima 45
Timor-Leste 103–04
Tina Comes 17
Tindouf 103–04
TLI 174–75
Tomasini, R M 195–96
Total 9
Toyasaki, Fuminori 15, 29
Triantis, K 137
Triple Bottom Line' framework 267
tsunami 25–26, 212
Turkey 120
Turrini et al 34
typhoon Haiyan 289, 295
typhoons 170, 289, 310, 325
    typhoon Haiyan 289, 295

UAV technology 299–300
UK Government 289
UK. *See* United Kingdom (UK)
Ukraine 1, 159, 161–62, 322
Ülkü et al 33
ultra-cold chains 260–61
UN Cluster Approach 212
UN Common Cash Statement 235, 243
UN Office for the Coordination of Humanitarian Affairs (OCHA) 104, 281
UN Office of Internal Oversight Service (OIOS) 107, 113, 121
UN Protection Force (UNPROFOR) 102
UN Relief and Works Agency for Palestinian refugees in the Near East 102
UN Sustainable Development Goal number (17) 242
uncertainty
    commerce and humanitarian logistics 11–14, *12*
    managing supply networks 6–11
    sustainability and 267–70
UNICEF Supply Division 162–63
United Arab Emirates 175, *176*
United Kingdom (UK) 15, 76–77, 255, 258, 307
United Nations (UN) 88, 268–70
    agencies 218, 222–23, *222*, 318
United Nations (UN) system 67–68
United Nations Children's Fund (UNICEF) 74, 102, 110, 122, 162–63, 226, 228, 249
United Nations Children's Fund (UNICEF) 74, 102, 110, 122, 162–63, 226, 228, 249

United Nations Development Programme (UNDP) 304
United Nations High Commissioner for Refugees (UNHCR) 16, 101–03, 223, 274, 275, 318, 322
    'battle' for change 107–09, *109*
    budgets 114–17, *115*, *116*
    change and improvement 126–28
    strategic plans developing 123–26
    supply chain management function 117–22, *119*
    supply chain management improvement 105–07
    supply chain management understanding 103–05
    supply chain organization building 109–14
United Nations Humanitarian Response Depot (UNHRD) 174, 175–77, *176*, 183
United Nations Office for the Coordination of Humanitarian Affairs (OCHA) 249–50, 279
United Nations peacekeeping forces 89
United Nations Population Fund (UNFPA) 277
United Nations Humanitarian Response Depot (UNHRD) 9, 159, 174, 175–77, *176*, 183, 290–91
United Nations Office for the Coordination of Humanitarian Affairs (OCHA) 304
United Nations Population Fund (UNFPA) 277
United Nations sustainable Development Goals (SDGs) 265, 266, 268–69
United States (USA) 12–13, 87, 88–89, 112, 237, 286
    vaccination programmes 256–62
Universal Logistics Standards 291, 309
Universal Logistics Standards (ULS) 288, 291–92, 309
Università della Svizzera Italiana (USI) 16
Université Catholique de Louvain 170
University of Manitoba 15, 303
UPS 13, 74
UPS transportation services 74
upstream activities 271
US Food and Drug Administration 257–58
USA. *See* United States (USA)
USAID 275

value for money (VfM) 241, 285, 302, 311
van Kempen, E. A. 281
van Wassenhove, L 195–96, 198

Varadejsatitwong, Paitoon 16
Vega, Diego 15, 70, 72, 321
vendor managed inventory (VMI) 10, 289
Vietnam 177
'virtual integration' 11

Wakolbinger, Tina 15, 29, 70
Walmart 12–13, *12*
Waste management and measuring, reverse logistics, environmentally sustainable procurement and transport, and circular economy (WREC) 275–76
water and sanitation (WATSAN) 9, 174, 269, 318
Welthungerhilfe 177
West Africa 183, 295
West Africa Ebola outbreak 295
Western Sudan 26
White Paper (TLI) 174–75
WHO Model List of Essential Medicines 249
'Workstream 3' 218
World Bank 23
World Food Program (WFP) 9, 68, 74, 223, 228, 275–76, 279–80, 291, 298
  CVA 228
  emergency relief items 174–75
  humanitarian providers 72–75, *73*
  logistics services 68

supply chain management, need to improve 105–07
supply chain organization building 109–14
UNHCR budgets 114–17, *115*, *116*
UNHCR supply chain management function 117–22, *119*
United Nations High Commissioner for Refugees 101–03
World Health Organisation (WHO) 6, 249, 287
World Humanitarian Summit (WHS) 268, 295
World Vision 170, 174, 299
World Vision International 299
World War II 102
WU Vienna 15

yellow fever 248
Yemen 322
York University 15
Yugoslavia 102, 103

Zaad 312
Zara 9
Zarei et al 277
Zeithaml et al 135–36
zika 248
Zimbabwe 183
Zoiopoulos et al 230